American Household Botany

American Household Botany

A History *of* Useful Plants 1620–1900

Judith Sumner

Foreword by John Forti

Timber Press
Portland • Cambridge

Frontispiece. Late nineteenth century cottages were often built with an attached conservatory, considered beneficial in training and educating children.

Photographs are by the author unless otherwise identified.

Published in 2004 by
Timber Press, Inc.
The Haseltine Building
133 S.W. Second Avenue, Suite 450
Portland, Oregon 97204-3527, U.S.A.

Timber Press
2 Station Road
Swavesey
Cambridge CB4 5Q J, U.K.

www.timberpress.com

Printed in China

Library of Congress Cataloging-in-Publication Data

Sumner, Judith.
American household botany : a history of useful plants, 1620–1900 / Judith Sumner ; foreword by John Forti.
p. cm.
Includes bibliographical references (p.).
ISBN 0-88192-652-3 (hardcover)
1. Botany, Economic—United States—History. 2. Plants, Cultivated—United States—History. 3. Plants, Useful—United States—History. I. Title.
SB108.U5S85 2004
581.6'0973—dc22

2003026578

A catalog record for this book is also available from the British Library.

Contents

Foreword

We have always relied upon the world around us for food and flavor, shelter, medicine, and clothing. Traditional crafts and pre-industrial technologies that make use of natural resources were passed down through generations of cooks, craftsmen, herbalists, and artisans. Yet our aspirations for a better life through science and industry gradually obscured many of the direct ties to the natural world that supported our survival across centuries and continents. Over time, the more we shaped our environment, the more it came to reflect our image of civility and order. *American Household Botany* examines the ways in which we as a nation grew out of our environment. This historical perspective can help us understand not only what we perceive as useful and valuable by today's standards, but also what has held up to the test of time. A natural history of domestic plants such as this can help us look back, not through a romantic lens, but through a microscope.

In this uncommon exploration of our historical ties to the natural world, Judith Sumner offers scientific insight into centuries of plant use. The heirloom plants discussed in the book are historically significant not just because they were meaningful to those who relied upon them, but because they are open pollinated. For generations, housewives, farmers, and craftspeople selected out the seeds of their earliest or most prolific produce, of their strongest medicinal or fragrance plant, or of the plant that adapted best to their regional climate. As a result the values of each generation became apparent in the plants selected and handed down. When we preserve open-pollinated heritage plants today, we help to foster the genetic diversity that we will most certainly need to insure the safety of our agricultural crops in the future. Today we can live in a home that shows little evidence of the natural world beyond our walls. We can "create" produce developed to withstand shipping thousands of miles. We can believe that our health requires synthetic treatment of symptoms, and we can be so far removed from the sources of our food, herbs, and spices that we may scarcely imagine why the generations that preceded us understood "you are what you eat" as an adage to live by.

Historically, great pains were taken on a household level to train generations of cooks and herbalists as botanists who could safely identify plants intended for food or medicine. Too often language barriers across time and cultures can belittle or negate the initial understanding of a particular plant's efficacy. *American Household Botany* works to create an understanding of how our cultural traditions came about in an age when we were unable to scientifically comprehend the complexity of chemical constituents and phytochemicals.

As a horticulturist, garden historian, herbalist, and cook, I especially value the deep and meaningful connections that this book makes between our natural history and the domestic use of plants. In its pages the past will come to life for generations of museum interpreters and college students as well as to those of us interested in our relationship with the natural world. Going well beyond its fascinating revelations about food and medicine, it explores the ways in which plants and plant products—wood and dyestuff, soap and tools, alcohol and spices, to name just a few—evolved throughout our nation's history. While the book focuses largely on the plants used by immigrants coming to America between 1620 and 1900, it also appreciates the influence of native plants on American household botany.

It is rare to find an author who can effectively evoke modern science to help us understand the historic roots of horticulture. Judith Sumner offers valuable and seldom-explored perspectives on our past and present understanding of the plants that have surrounded us for generations. This book will challenge all of us to think in new ways about the depth of our everyday ties to nature.

JOHN FORTI
Curator of Historic Gardens & Landscapes
Strawbery Banke Museum, Portsmouth, New Hampshire

Preface and Acknowledgments

WHAT IS "household botany"? I define it as the botany of plants used historically in homes, which combines economic botany and ethnobotany with an understanding of botanical form and function. It includes the long list of grains, fresh and storable fruits and vegetables, herbs, spices, flavorings, scents, medicinal plants, wood, fibers, textiles, dyestuffs, garden plants, symbols, design elements, and botanical miscellany that were essential for survival and improved the daily lives of our ancestors. Today we continue many of the practices of household botany that began in the past three hundred years; every time we eat corn, drink tea, cook with herbs and spices, wear cotton or linen, or use boards of native pine or oak, we experience firsthand some of the historic traditions of American household botany. In the relatively recent past, all humans had to be skilled practical botanists in order to survive; we had to know which vegetables would hold up during winter storage, which herbs to use for specific illnesses, how to prepare plant fibers for weaving, and how to select the right woods for construction or cooking fires. We now forget many of the daily interactions with plants that were taken for granted during the past three centuries. Food and herb preservation, garden cultivation, brewing, baking, retting of fibers, and dyestuff preparation were once routine tasks, all demanding practical botanical knowledge.

Historically we have used various plant parts, tissues, and phytochemicals to fulfill our basic needs for food, medicines, textiles, and building materials, but we need to reject any teleological or egocentric thinking. Plants have evolved over time through natural selection to adapt to environmental pressures; we are merely exploiting the natural range of plant form and function by using and cultivating plants from the vast range of natural diversity. Of course, need begets ingenuity, which helped our ancestors discover the best plants for specific uses by exploring and exploiting the botanical traits of many plant species. In some cases, plants were obvious in the environment, salient species such as hairy mints or various fruit trees with obvious traits that invited interaction and experimentation. Other plants are more subtle; for example, neither vanilla flavor nor the indigo pigment can be sensed in tissues of the living plants that produce them. Their discovery and the ethnobotanical origins of their complex preparation methods are open to speculation.

In *The Natural History of Medicinal Plants* (Timber Press, 2000), I considered medicinal plants, their phytochemistry, and their place in ethnobotany and in the natural world, but medicinal botany is just part of the spectrum of interactions between humans and plants. *American Household Botany* is an outgrowth of my appreciation for useful plants and curiosity about their adaptations and origins, along with a long personal interest in cookery, food preservation, and other historical domestic practices and conundrums. In this book, I have examined the botanical diversity of plants that Americans knew and used in their gardens, farms, stillrooms, storerooms, and nearby fields and forests. I have tried to trace the flow of species and plant knowledge, beginning with the Puritans, who arrived with seeds and scions for their new gardens as well as food stores of wheat and legumes from England and medicinal herbs. Native American practices were an important resource, and colonists were soon using maize and medicinal plants from North American fields and forests. American plants such as sassafras and black walnut were exported to Europe from the early American colonies, and Native Americans adopted some European plants for various uses. A colonial crock of pickles involved the preservation of fruits and vegetables from different continents, the antimicrobial properties of herbs and spices, and the microbiology of vinegar production from apple cider.

In the interest of full disclosure, I should emphasize that this book does not attempt to investigate Native American ethnobotany as a subject in itself. My main concern is the use of plants by European settlers and their descendants. In my discussion of corn cultivation, for example, I do not mention the remarkable skill of the Pueblo peoples who grow corn in desert areas; these practices, perhaps regrettably, had little influence on European-Americans. I must also confess that the research materials and publications I have consulted have of necessity resulted in something of an East Coast bias—my book would be considerably longer had I had the resources to conduct research in western archives to study the uses of *Ephedra* by Mormon settlers in Utah or the harvesting of *Opuntia* fruits by the early Spaniards in the Southwest. Another problem is that I did not have the resources to document each ethnic group and the development of its unique ethnobotany; for example, Scandinavians in the upper Midwest undoubtedly deserve greater attention than I've been able to devote to them. It is unavoidably true that the majority of early "receipt" books, gardening manuals, and seed catalogs I've consulted are products of the eastern states and reflect the dominant culture, which is effectively Anglo-American. I hope readers will forgive these shortcomings and find much that is useful in the material I did collect.

My purpose in writing this book has been to illuminate the historical uses of plants in American homes and the morphological and phytochemical adaptations that explain these uses. Nothing in this book is intended as medical advice to follow today; historical medical practices and treatments are best left as history and not adopted for individual self-medication or experimentation. As with the practice of herbal medicine generally, *caveat emptor* is the watchword when considering self-medication with traditional remedies.

My colleagues, friends, and students at Garden in the Woods and the Arnold Arboretum in the Boston area have once again been an inspiration to see this project through to completion. The librarians at the American Antiquarian Society have been helpful in finding many useful documents and resources, and Waine Morse provided generous access to his remarkable collection of medical ephemera and artifacts. Neal Maillet has provided thoughtful, insightful comments and suggestions as this project unfolded and developed; when I first suggested the topic of household botany, he immediately understood the topic and its appeal. Neal once again has been a superb long-distance editor and an inspiration when I have needed encouragement, two important contributions that this writer really appreciates. John Forti generously shared his remarkable expertise, knowledge acquired during many years as a New England horticulturist and historian; he willingly critiqued the chapters and provided many valuable suggestions, insights, and corrections along the way. Lisa Theobald diligently edited the final manuscript, and her meticulous attention to detail has improved both the text and the documentation. My daughter, Catherine Sumner, willingly helped me to sort out several questions of applied chemistry as it relates to household botany. My husband, Stephen Sumner, has helped at every turn, from discovering useful resources to critiquing the manuscript as it evolved; he has been a cheerful spouse through another two years of research and writing, piles of papers and books, and deadlines, for which I am grateful beyond words.

CHAPTER ONE

The New World

SINCE prehistoric times, we have lived botanical lives, dependent on plants as sources of essential foods, medicines, timber, and fibers. Reliance on green resources is part of the human condition; we still use many plants in practical ways to provide nutrition, shelter, clothing, and some of the ingredients in pharmaceutical drugs. The basis of plant productivity is photosynthesis, in which green plants use carbon dioxide and water to make simple sugars, and these can be linked together to make complex carbohydrates such as starch and cellulose. Flowering plants provide food in the forms of edible seeds, fruits, and vegetables, and without the unique flowering plant life cycle, human cultures based on agriculture would be an impossibility. Grains such as wheat, corn, and rice contain the seeds of grasses, flowering plants that historically have provided the staple, storable core of human sustenance; we exploit grains and many other plant foods to supply our own larders and to provide forage for domesticated animals. In addition, plant conducting and support tissues provide wood and fibers for textiles and cordage, all based on the external cellulose walls of plant cells, and phytochemicals are used to cure disease, treat symptoms, and preserve food.

The studies of ethnobotany and economic botany weave together our past and present needs, revealing the diverse ways in which plants have been used since ancient times. Humans have grown plants around their homes since prehistoric times, a tradition that began with the culture of edible and medicinal species. Ethnobotany combined with waves of exploration and immigration, and vernacular gardens evolved to reflect both traditional plant uses and new introductions. During four hundred years of European settlement in North America, Native American knowledge and skill combined with Old World agricultural and herbal traditions to yield a unique New World household botany; plant knowledge and uses from both sides of the Atlantic are reflected both in carefully tended gardens and in the weedy roadside flora. American gardeners still grow many of the selfsame vegetables and kitchen herbs that grew in English kitchen gardens. Familiar weeds such as tansy and dandelions are also botanical immigrants from Europe that have naturalized into the American landscape; like rosemary and thyme, these are plants that once figured strongly in the European tradition of medicinal herbalism that was practiced in American homes.

The European herbal tradition is reflected in vernacular names such as lungwort, liverwort, maidenhair, and boneset. According to the Doctrine of Signatures, plants bear signs or "signatures" that indicate their medicinal uses—daisies with bright centers (eyebrights) were used for ocular afflictions, scaly stems or cones were likely treatments for skin diseases, and yellow juices were believed to cure liver disorders. The Doctrine of Signatures, although fallacious, lent order and purpose to the search for useful medicinal plants, as early settlers sought New World counterparts to European medicinal plants. The branching taproot of the European mandrake (*Mandragora officinarum*) vaguely resembles a human torso and suggested its use as a panacea; the so-called American mandrake is May apple (*Podophyllum peltatum*), an unrelated species that was grown in early herb gardens because of its similar branching taproot. Coincidentally, it is also a phytochemically potent plant, and Native Americans knew that although high doses of May apple roots might cause death, the cooked roots provided an effective cathartic. By the nineteenth century, Shakers cultivated and marketed May apple or American mandrake to physicians as a cathartic and to treat jaundice, liver complaints, fevers, and syphilis.

Native American Agriculture

In *The Herball, or Generall Historie of Plantes* (1633), the English herbalist John Gerard described and illustrated yellow, gold, red, "blew," and white cultivated varieties of the plant he called "Frumentum Indicum" or "Turkie Wheate." Familiar only with European grains, Gerard maligned the staple American grain crop that we know as corn or maize (*Zea mays* ssp. *mays*), an annual grass cultivated in North America since pre-Columbian times. As a food source, he considered it inferior to familiar European staple crops, claiming that it

> doth nourish far lesse than either wheat, rice, barley, or otes. . . . Wee have as yet no certaine proofe or experience concerning the vertues of this kind of Corne; although the barbarous Indians, which know

> no better, are strained to make a vertue of necessitie, and think it is a good food; whereas we may easily judge, that it nourisheth but little, and is of hard and evill digestion, a more convenient food for swine than for men.

Maize had nourished Native Americans for thousands of years and soon became a staple in the diets of Europeans settling in North America. Early maize cookery was learned from Native Americans, and in *American Cookery* (1796), Amelia Simmons specified the use of Indian meal (corn meal) in "A Nice Indian Pudding," "Johnny Cake, or Hoe Cake," and "Indian Slapjack." Her emphasis on corn as an ingredient was particularly fitting, since *American Cookery* was the first cookbook written and published in the American colonies. Colonists could also consult *The Art of Cookery Made Plain and Simple* by Englishwoman Hannah Glasse; the 1805 edition included a New World supplement, "Several New Receipts Adapted to the American Mode of Cooking," which provided instructions for cornmeal mush as well as baked and boiled versions of Indian pudding.

The cornmeal used for colonial puddings and breads had an ancient New World origin. The earliest archeological specimens of maize are from Puebla, Mexico, and date from 5500 years ago, but the physical resemblance to its wild ancestors is slight. Maize likely originated from wild populations of the Mexican grass teosinte (*Zea mays* ssp. *parviglumis*), which over time was selected for specific desirable traits: grains that remain on the ears, large grain size, high sugar and starch content, various colors, and the ability of the starchy food storage tissue (endosperm) to expand ("pop") when heated. Teosinte typically has only five to twelve grains on each ear, but this wild species has the same chromosome number as maize (ten pairs of chromosomes in the

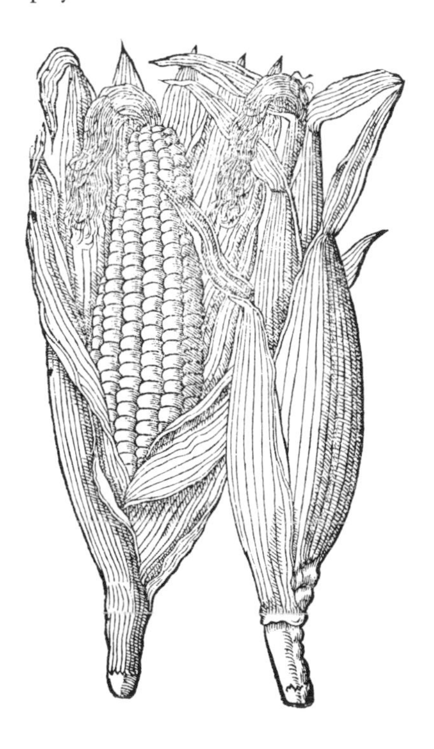

Herbalist John Gerard considered "Turkie Wheate" or maize (*Zea mays* ssp. *mays*) inferior to European crops such as oats, barley, and wheat, but the 1633 edition of *The Herball* provided this woodcut illustration.

nucleus of each cell). Cross-pollination can produce fertile teosinte-maize hybrids, suggesting a close genetic link between the wild and cultivated plants. Long before the arrival of European explorers and settlers, maize had been perfected in Mexico and had been carried across North America, a Native American cultivar so highly selected as a successful food crop that its survival as a wild plant was unlikely. Maize does not self-seed and naturalize because the cultivated ears are wrapped by long leaves, or husks, and the grains are not easily dispersed. Its survival depends on the people who plant the grains and also save grain for seed for the next crop.

Maize is a high-yield crop that can grow in fields that are too wet for wheat and too dry for rice, and the stalks are suited to polyculture with beans and squash. Its early cultivation often relied on fire as a technique for clearing and enriching land; early settlers noted the parklike landscape of New England, the result of the Native American practice of periodic burning to clear away underbrush and fertilize the soil with ashes. Amidst clearings, maize was planted in hillocks, fertilized with small fish, and interplanted with beans and squash. Squash plants covered the bare soil, and bean vines crept up the corn stalks. According to Native American legend, maize rejected squash as a potential "wife" because of the wandering growth habit of squash vines; bean plants proved a more reliable partner, since its vines would twine with fidelity up the tall stalks. Squash vines were left to meander promiscuously over the adjacent arable ground. By the time European explorers and settlers arrived in the New World, the method of planting corn and beans together was a common agricultural practice throughout the Americas, and corn hillocks blanketed the landscape. With an observer's eye, Thomas Hariot succinctly described maize culture in *A Brief and True Report of the New Found Land of Virginia* (1590): "beginning in one corner of the plot, with a pecker [dibble] they make a hole wherein they put four grains with that care they touch not one another. . . . And cover them with the mold [soil] again. . . . But with this regard that they be made in ranks [rows]."

Crops were rotated, periodic burning cleared land and enriched the soil, and in some cases entire villages relocated to new sites, depending on the season. Cultivated lands did not surround established homesteads, and fields were communal rather than fenced; in both form and method, these first New England farms resembled nothing known in Europe, and Puritan settlers perhaps took this as license to establish their own boundaries and ownership rights. Puritans commonly viewed Native American farming practices as exploitation of natural abundance, rather than real toil, and clergyman John Cotton of Massachusetts commented, "We did not conceive that it is just title to so vast a continent, to make no other improvement of millions of acres in it, but onely to burne it up for pastime." Governor John Winthrop of the Massachusetts Bay Colony saw New England as unimproved and ripe for the injunctions of Genesis 1:28: "Be fruitful and multiply, and replenish the earth, and subdue it." The cleared Native American cornfields were gradually appropriated by colonists, who used such place names as Deerfield, Brookfield, Marshfield, and Springfield to denote the cleared lands that they claimed.

Native Americans prudently buried grain stores for winter months. French explorer Samuel de Champlain described the trenches that were dug several feet deep, filled with sacks of grain, and covered with sand, comparing them to European granaries. Such was the case at Plymouth, where the Wampanoags inhabited the region extending into Cape Cod that is now southeastern Massachusetts. By the time that the Pilgrims arrived in the late autumn of 1620, Wampanoag populations had been depleted by epidemics that resulted from contact with British fishermen. Waves of disease, probably bubonic plague and smallpox, had killed most Native Americans in the area, who easily contracted viral and bacterial illnesses because they lacked the antibodies produced by prior exposure. The Wampanoag village of Patuxet had been renamed New Plymouth by English explorer John Smith, and later Puritan leaders interpreted the plagues as providential. The Pilgrims had survived their first winter there after they stumbled across and stole about ten bushels of Wampanoag corn buried in sand, described by Plymouth Governor William Bradford as "a great mercy to this poor people, that here they got seed to plant them corn the next year, or else they might have starved."

Samoset, an Abenaki leader from the north, brought Tisquantum (often called Squanto) to the Pilgrim settlement. As a boy, Squanto was captured and enslaved in England and was able to communicate with Englishmen; he was seized and enslaved a second time in Spain, escaping and returning by ship in 1619 to find his village abandoned as a result of the epidemics. School children learn about the growing season that Squanto spent with the Pilgrims, teaching them the Native American method of planting maize seeds in hillocks with fish to fertilize the soil. Edward Winslow (1622) described the Pilgrims' first planting: "We set the last Spring some twentie Acres of indian corne, and sowed some six Acres of Barley and Pease, and according to the manner of the Indians, we manured our ground with Herrings or rather

Shadds, which we have in great abundance, and take with ease at our doores." The hillocks had to be guarded for several days to prevent scavenging animals from digging up the decomposing fish. When the corn reached hand height, bean seeds were pressed into the corn hills and squash and pumpkin seeds were sown in the surrounding areas. Clam shell hoes kept the fields clear of competing weeds until the coarse pumpkin and squash vines covered and shaded the bare soil. Soil mounded around the base of each corn stalk provided support and substrate for the prop roots that grow from the lower stem. The fields grew undisturbed as self-contained habitats until the fall harvest of the maize varieties. As Gerard noted earlier, Native American corn cultivars produced grains of many hues; colonist John Winthrop recorded "a great variety of colors including white corn, black corn, cherry red corn, yellow, blue, straw-colored, greenish and speckled."

Plymouth colonists and Wampanoags shared days of feasting after the harvest in 1621, the historical basis of the American Thanksgiving tradition that is celebrated today. The foods that were eaten no doubt included corn in various forms, probably prepared with the flint corn and flour corn that were grown commonly throughout North America. Flint corn grains were white, yellow, or multicolored; the plants grew four to six feet tall and produced two ears on each plant, each bearing eight rows, with thirty to forty kernels in each row. Flint corn has a hard layer of stored starchy endosperm in each seed and was boiled to make hominy and samp, porridge-like dishes. Flour corn has a soft endosperm layer that could be hand ground into flour. Contrary to legend, during the seventeenth century sweet corn and popcorn were probably not grown in New England and did not figure in early harvest celebrations.

The common bean (*Phaseolus vulgaris*) planted in corn hillocks was also selected for various desirable traits, including growth form and edibility of the pod. The common bean probably originated as a wild legume in Mexico, Peru, and perhaps Colombia, and early cultivars were probably developed in Mexico and Central America and in the south central Andes of South America. Familiar modern varieties include pea beans, kidney beans, pinto beans, green beans, navy beans, and shell beans. As with modern beans, some of the Native American cultivars were best eaten fresh, while others were dried and reserved for winter use; their range of seed colors included red, yellow, blue, white, and spotted varieties. Some resembled modern bean cultivars that can be grown without staking, while the varieties with indeterminate growth were grown with corn for support and strength.

A unique underground symbiosis adapts beans for culture with corn, the biological basis of the ancient Native American practice of planting corn and beans in shared plots. Typical of most legumes, beans establish mutually beneficial relationships with *Rhizobium* and a few other genera of soil bacteria that carry out nitrogen fixation. These bacteria inhabit the nodules that form on legume roots, oxygen-free microhabitats in which the bacteria convert atmospheric nitrogen into the ammonia that is needed to make the amino acids that link to form proteins. Soil nitrogen is not a limiting factor to the growth of legumes, and the plants can thrive in poorly fertilized soil. The legumes produce and store copious amounts of protein in their seeds, and the bacteria release excess ammonia into the soil where other plants absorb it. Beans benefit corn by indirectly providing the plants with ammonia, which improves soil because it is a usable form of nitrogen. Beans also provide a protein-rich, storable food source for the humans who cultivate them, and beans were also considered medicinal. John Josselyn (1672), a English herbalist and visitor to the New World in 1638 and 1663, described "Indean beans" as "better for physick use than other beans"; the color and shape of kidney beans suggested to Josselyn and other herbalists that these beans could "strengthen the kidneys."

Squash, pumpkins, and melons are all members of the gourd family (Cucurbitaceae), a family of coarse herbaceous vines that produce the robust, leathery fruits known as pepos. The Wampanoags grew several varieties, including summer squashes and pumpkins, which were all cultivars of *Cucurbita pepo*. This species may have been domesticated indepen-

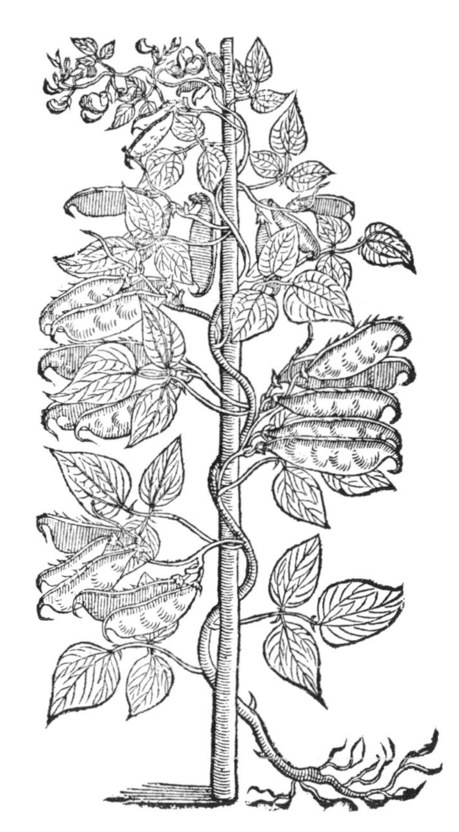

The Herball (Gerard 1633) described one variety of common bean (*Phaseolus vulgaris*) as the yellow kidney bean, which according to the Doctrine of Signatures was beneficial for kidney ailments.

dently in Mexico and eastern North America, and taxonomic confusion arises from the wide range of cultivated squashes and pumpkins that were developed through selection from wild vines. Pilgrim and Puritan settlers immediately recognized New World squashes and pumpkins as close relatives of the European "pompions" (melons), the nearest fruit to a pumpkin that they knew. Squash and pumpkins could be dried for winter consumption, and they were adopted into cookery methods as a staple or "standing dish." In *New-Englands Rarities* (1672) Josselyn noted, "The ancient New England standing Dish is made from Pompions, there be several kinds, some proper to the Countrey, they are drier than our English pompions, and better tasted; you may eat them green," but he added that these vegetables could cause worms and "provoketh Urin extremely." His method for their preparation involved hours of stewing and the addition of butter, vinegar, and spice. In *American Cookery* (1796), Simmons included recipes for puddings using winter squashes and "Pompkins," which could be variously flavored with rose water, nutmeg, ginger, mace, and allspice. In times of short supplies of fodder, squash and pumpkins were also fed to hungry farm animals.

The first edition of *The Herball* (Gerard 1633) included this illustration of "The Great Round Pompion," the pumpkins (*Cucurbita pepo*) that colonists adopted as a useful food plant and flavored with Old World spices.

Not all American food plants required cultivation; foraging and hunting supplied a substantial portion of the Native American diet. The landscape provided a rich natural garden of wild edible weeds, onions, mustards, fruits, nuts, roots and tubers, and mushrooms. In his description of the 1621 harvest festival included in *Mourt's Relation: A Relation or Journal of the English Plantation settled at Plymouth* (Heath 1622) Winslow recorded "sallet" herbs, red and white grapes, strawberries, and plums among the wild foods available for the effort of gathering them. Seventeenth century sallets were both hot and cold dishes of vegetables or fruits, and sallet herbs were various edible leaves, the origin of the modern word *salad.* Strawberries grew in remarkable profusion, and Captain John Smith named one coastal site along the Piscataqua River "Strabery Banke." In his *Key into the Language of America* (1643), Roger Williams described the Native American practice of combining strawberries with cornmeal to make bread; colonists soon made wine from wild strawberries and cherries and substituted native blueberries for currants in English pudding recipes. According to Josselyn (1672), Native Americans gathered and dried blueberries to sell to colonists and "they are very good to allay the burning heat of Feavers and hot Agues, either in Syrup or Conserve."

Wild foods also provided insurance against starvation after the harvest season, and several were rich in carbohydrates and served as suitable corn substitutes. Nuts were stored and ground to make flour for breads, and the tubers of Jerusalem artichokes (*Helianthus tuberosus*) were gathered even in winter. Groundnut or potato bean (*Apios americana*) is a wild legume with potato-like enlargements of the roots; Native Americans gathered and stored the starchy roots and may even have cultivated some of the plants. Soon colonists learned of its use, and Governor Bradford mentioned groundnut as a valuable food in his "Descriptive and Historical Account of New England in Verse" Groundnuts, along with corn, pumpkins, chestnuts, and Jerusalem artichokes, were among the earliest American species introduced to European gardens, suggesting the interest abroad in New World food plants.

Some foraged foods needed careful preparation to remove potential toxins. Depending upon the season and species, acorns required prolonged soaking or boiling in water, or their bitter tannins rendered them impalatable. Tannins interfere with digestion by binding to both enzymes and food proteins, so their removal was essential. The tubers of the green arrow arum, or tuckahoe (*Peltandra virginica*), require baking to break down the needlelike crystals that are typical of many aroids (family Araceae). Native Americans in the southern colonies gathered the tubers in later summer and ground the baked tubers into a flour used with cornmeal to lighten the texture of breads. *Tuckahoe* later became a slang term for settlers too poor to afford wheat flour, but tuckahoe starch was also used by wealthy Virginians to prepare delicate blancmange puddings and to starch their ruffs.

New World Explorations

Years before the English settlement of Jamestown, considerable interest in New World flora, fauna, and minerals already existed. Travelers and explorers described North American natural history with varying accuracy, and eager Europeans anticipated imports ranging from precious metals to medicinal herbs. Elizabethans plagued with gout and scrofulous sores needed cures, and their optimism was encouraged by Spanish physician Nicholas Monardes's 1577 treatise: "Joyful Newes out of the Newe Founde World, Wherein Is Declared the Rare and Singular Virtues of Divers and Sundry Herbs, Trees, Oils, Plants, and Stones, with Their Applications, as well for Physic as Chirurgery [surgery], the Said Being Well Applied Bringeth Such Present Remedy for All Diseases, as May Seem Altogether Incredible." Monardes described tobacco (*Nicotiana tabacum*) as a virtual panacea against ailments from intestinal worms to toothache, and he advocated the use of the Mexican jalap (*Ipomoea purga*) as an effective new purgative. Jalap is closely related to sweet potatoes (*I. batatas*), and the dried resin from its roots was later used for treating dysentery, sluggishness, and sores.

Monardes also described tea prepared from sassafras (*Sassafras albidum*) as a potent elixir, and sassafras roots soon became a valuable commodity for export to Europe. For instance, in 1630 the *Fortune* carried Puritans destined for the Massachusetts Bay colony and took on American lumber, animal skins, and "Saxefras" for the return voyage. Near the English Channel, pirates plundered the *Fortune* and brought the laden ship to the Île d'Yeu near the coast of France, where they removed the sassafras exports and everything else of value. The natural range of sassafras trees extends from Maine to Florida, and early Spanish explorers in Florida mistook it for cinnamon, also a member of the bay laurel family (Lauraceae). The Lauraceae is characterized by the production of aromatic oils, and like cinnamon,

Sassafras (*Sassafras albidum*) was an early export to Europe, where the root bark was used to prepare an aromatic medicinal tea.

sassafras was valued for its aromatic bark; sassafras root bark was brewed into a potent tea that was believed to cure venereal diseases and prevent aging, and later it was used as a flavoring for root beer and tea (see chapter 5).

Sassafras leaves are also aromatic, with a scent resembling citrus, and the trees exhibit heterophylly, the development of several leaf shapes on a single tree. One tree bears four distinct leaf types: leaves with no lobes, two lobes, or single lobes on either the left or right side of the leaf. Sassafras leaves were powdered and used to thicken soups, a practice that evolved into the use of sassafras in the filé powder used in Creole cookery (see chapter 6). Sassafras was later one of the few American plants included in the pharmacopoeia of medicines used by Washington's troops while they were bivouacked at Valley Forge. Army doctors compiled a "Pharmacopoeia of Simples and Efficaceous Remedies for the use of the Military Hospital . . . Especially Adapted to our present Poverty and Straitened Circumstances." They also used the native Virginia snakeroot (*Aristolochia serpentaria*) for bitter tonics but depended on European suppliers for most other medicines. By the time of the Revolutionary War, Virginia snakeroot had diverse medicinal uses on both sides of the Atlantic and had already been stocked by European apothecaries for more than one hundred fifty years. Europeans used Virginia snakeroot to induce sweating, Virginians boiled the plant in rum as a folk cancer cure, and the Cherokee used the roots to treat snakebites.

Medicinal plants were not the only attraction for New World explorers. Edible plants were introduced to European gardens as explorers discovered them and imported seeds to Europe. By 1600, anecdotal reports suggest that Native American foods such as corn, squash, groundnut, Jerusalem artichoke, chestnuts, and walnuts were growing in Europe, but identity of these early introductions is not always certain. In *The Herball*, Gerard described "the Virginian potato" as having hollow, trailing branches and roots that are "the shape and proportion of the Potato's, but also the pleasant taste and vertues of the same." Gerard grew the plant in his London garden, but its identity is unclear, and he may have been confused as to its origin. Food historians are uncertain whether Gerard cultivated and described the familiar South American white potato (*Solanum tuberosum*) that had been introduced to England before 1590, the sweet potato (*Ipomoea batatas*), or perhaps the groundnut or potato bean (*Apios americana*). So-called Irish or white potatoes had probably not yet reached North America during the sixteenth century, so the conundrum of the "Virginian potato" remains unresolved.

Many of the new plants introduced to Old World gardens were curiosities, plants that attracted attention because of their unique flowers, fruits, leaves, growth form, or some other notable characteristic. Several such American plants were cultivated in Europe prior to 1600. The aromatic flowers of the pearly everlasting (*Anaphalis margaritacea*) suggested its use as a household herb similar to other aromatic composites (family Compositae or Asteraceae) such as wormwood (*Artemisia* spp.). American persimmons (*Diospyros virginiana*) were unique in Europe for almost three hundred years, since Japanese persimmons (*D. kaki*, which originated in China despite its common name) did not appear in European markets until 1873. Persimmons are the attractive but often impalatable fruit of a tree with extremely dense wood; colonists may have learned from Native Americans that the first frost would convert the astringent, tannin-rich persimmons into sweet fruits that could be eaten with impunity. Tulip trees (*Liriodendron tulipifera*) were highly prized for their showy flowers and stately growth, and carnivorous pitcher plants (*Sarracenia flava* and *S. purpurea*) were cultivated as sixteenth century garden specimens. Species of *Agave* and *Yucca* were imported for their peculiar growth forms or perhaps for their fibers. *Yucca gloriosa* was collected from islands off the Carolina coast and passed from gardener to gardener as a popular ornamental; Gerard called it "Iucca" and grew it in his London garden.

Opuntia humifusa exhibited the unique flowers, spines, and succulence that characterize cacti (family Cactaceae). It was first described by Hariot, whose expedition to Virginia was financed by Sir Walter Raleigh. He returned in 1586 with a suite of impressive, potentially useful plants: Jerusalem artichokes, squash, and groundnuts, as well as the native American mulberry (*Morus rubra*), which had economic potential in the silk industry as a food source for the silk-producing caterpillars. Hariot called New World persimmons "medlars," apparently mistaking them for true medlars (*Mespilus germanica*), rose relatives native to Asia Minor that also require prolonged ripening before they are palatable. His collections also included staghorn sumac (*Rhus typhina*), which was used by Native Americans as a dyestuff, and "Black walnutt" (*Juglans nigra*). John Tradescant traveled to North America in 1637, 1642, and 1654, and he transported roots or seeds of almost one hundred Virginia plants back to the Tradescant nursery established by his father at Lambeth on the River Thames. These included plants with garden appeal such as red maple (*Acer rubrum*), Virginia creeper (*Parthenocissus quinquefolia*), columbine (*Aquilegia canadensis*), and spiderwort (*Tradescantia virgini-*

ana), a member of the genus named by famed Swedish naturalist Carolus Linnaeus for the Tradescants. The Tradescants included their American plants in the Museum Tradescantianum, an assemblage of natural and historical rarities known familiarly as "the Ark," collections that eventually constituted the core of the Ashmolean museum in Oxford.

English herbalist Josselyn, who visited his brother in the region that is now Maine in 1638 and again in 1663, made careful observations of potentially useful New England plants that he chronicled in *New-Englands Rarities* (1672) and *An Account of Two Voyages to New England* (1675). His first book described the "Remedies wherewith the Natives constantly use to Cure their Distempers, Wounds and Sores," while the *Account* provided plant descriptions, recipes, additional remedies, and "the prices of all necessaries for furnishing a Planter and his Family at his first coming." His lists include the necessary supplies for establishing a home in the New World: food, clothing, utensils, tools, and arms. Recommended foods included several tropical spices (ginger, nutmeg, mace, cinnamon, and pepper), sugar, and fruits such as prunes, "Currence," and "Raisons of the sun." Josselyn advised his readers to carry with them European medicinal remedies such as wormwood for seasickness and lemons for scurvy. Despite his loyalty to familiar foods and cures, Josselyn's commentary on New World botany is optimistic:

> The plants in New England for the variety, number, beauty, style, and vertues, may stand in Competition with the plants of any Country in Europe. . . . They are generally of (somewhat) a more masculine vertue than any of the same species in England, but not in so terrible a degree as to be mischievous or ineffectual to our English bodies.

Equipped with the 1633 edition of Gerard's *Herball*, Josselyn studied the New England flora and prepared the five plant categories that appeared in *New-Englands Rarities*. These included English plants, native North American plants, unnamed North American plants, English plants that had naturalized in America, and "Garden-Herbs (amongst us) as do thrive there and of such as do not." He supplied the first crude illustration of the purple pitcher plant that grows in New England bogs, which he named the "hollow-leaved lavender" with "one fantastical Flower" and recommended for "all manner of Fluxes," probably a reference to its Native American uses for expelling the placenta after birth and treating tuberculosis and pulmonary bleeding.

Josselyn recommended blueberries (*Vaccinium* spp.) for fevers and as an ingredient in a "Summer Dish" with milk, sugar, and spice. He also recommended a root tea of sassafras or "Ague Tree" for treating high fevers, but he incorrectly suggested that sassafras might be the same as the South American feverbark trees (*Cinchona* spp.), which are the source of quinine. As an interesting record of the changing landscape, he listed the many European species that he discovered growing wild in the New World, plants that were becoming naturalized in the New England landscape as garden escapees. These included medicinal plants such as shepherd's purse (*Capsella bursa-pastoris*), black nightshade (*Solanum nigrum*), "cheek weed" or chickweed (*Stellaria media*), comfrey (*Symphytum officinale*), white mullein (*Verbascum lychnitis*), and plantain (*Plantago major*). Plantain was used for skin ailments, and according to Josselyn, Native Americans called it "Englishman's foot," suggesting that it was spread by the treading of settlers. Native Americans soon also found several uses for plantain leaves, as poultices for burns and swellings, as a wound disinfectant, and as greens for cooking. In the *Account* (1675), Josselyn expanded his list of naturalized European plants to include flax (*Linum usitatissimum*), hemp (*Cannabis sativa*), red clover (*Trifolium pratense*), and the eglantine rose (*Rosa eglanteria*), which he recommended as a shrub that could be cultivated with juniper to form effective hedges.

In addition to transporting species between continents, European explorers also planted some New World gardens. Samuel de Champlain and his men established a settlement and garden on an island in the St. Croix River, in what is now northern Maine. An engraving of a Champlain sketch appeared in *Les Voyages* (1619) and showed the positions of the houses and their geometric garden plots. Lettuce, sorrel, and cabbages continued to self-sow and thrive even after the settlement was abandoned. A few years

The purple pitcher plant (*Sarracenia purpurea*) has tubular leaves that are adapted as insect traps, a garden oddity prized by European plantsmen and recommended as medicinal by Josselyn.

later, Captain John Smith planted another garden on an island near the coast of Maine "that grew so well, as it served us for sallets in June and July." While his garden prospered, Smith sailed in the *Pied Cow* down the Maine coast, naming "Strabery Banke" (now the site of Portsmouth, New Hampshire) for its abundant wild strawberries and observing Native American "corne fields and delightful groves" on the hills and slopes of what is now Ipswich, Massachusetts. However, his limited botanical knowledge was revealed when he described the beach plums (*Prunus maritima*) on Plum Island as "many faire high groves of mulberrie trees."

Puritan Gardens

Puritan colonists were motivated gardeners whose survival depended on the productivity of their plots. The practical role of seventeenth century gardens was to provide food and herbs for household use; the earliest New World gardens planted by European immigrants were not the grounds surrounding villas or great houses, although some pleasure was probably derived from a pleasing layout and successful husbandry. In 1639 Colonel George Fenwick, second governor of the Saybrook Colony in Connecticut, wrote to colonist John Winthrop: "We both desire and delight much in that primitive imployment of dressing a garden." Winthrop later sent saplings to George and Alice Fenwick, whose correspondence reveals that they cultivated cherries, peaches, and apples as well as thriving garden plots.

Early settlers were direct descendants of the British gardening tradition and carried familiar garden plants and traditional tools and techniques to their New World gardens. Old World practices combined with botanical and cultural information gleaned from Native American farmers, and colonial American gardens reflected these melded traditions. Field crops provided Indian corn as well as European grains such as wheat, rye, and oats. The various colored cultivars of "Turkey Wheate" became a staple in the diet of Europeans arriving in the New World, although many regarded corn as a food more suitable for cattle than people. Colonists also carried seed stock from home, including some grains that are no longer commonly grown. Josselyn described *silpee*, or naked oats, cultivated in New England; this was most likely *Avena nuda*, a close relative of common oats (*A. sativa*), in which the grains fell away from their coverings with ease. Housewives probably favored naked oats because of the simplicity of preparing them without a mill, and

Josselyn mentioned them as an ingredient in one of the "standing dishes" of New England. He described a meal of oatmeal simmered in milk and flavored with sugar and spice, similar to "white-pot," a traditional Devonshire pudding compounded of cream, flour, eggs, and spices.

Vegetables and herbs were grown in rectangular beds that were enriched with all available household, human, and animal wastes. The reconstructed gardens at Plimoth Plantation reflect the strategies for growing plants in small, intensive plots, with an emphasis on reliable crops that would fill the gaps in a diet of grains, meat, and fish. Cabbages, cauliflower, and leafy coleworts (all cultivars of *Brassica oleracea*) are members of the mustard family (Cruciferae or Brassicaceae) that were grown both as food and medicine; cauliflower in particular was considered suitable for gentlemen's tables, but acceptable dishes were also prepared by boiling cabbages and coleworts in meat broth. Coleworts are leafy wild-type forms, similar to modern kale, while cabbage develops with its leaves wrapped tightly around a giant terminal bud. Cabbages are biennials, grown for their massive buds, which unfurl and produce a flowering stalk in the second year of growth. In *The Herball*, Gerard described the leafy coleworts and cabbages that were commonly grown in England during the sixteenth century, including leafy forms such as garden and curled forms of colewort and red and white cabbages with closed heads.

Cauliflower was recognized as a cluster of flower buds and known as "Brassica florida, Cole-florie." Cauliflowers function as precocious annuals, producing a large terminal cluster of immature flowers during the first

Leafy coleworts illustrated in the first edition of Gerard's *Herball* (1597) resembled the wild species (*Brassica oleracea*), which through selection during ancient times yielded several distinct cultivars.

growing season. They are the result of horticultural selection during ancient times; invading Romans brought cabbages, coleworts, and cauliflower to England, and as Mediterranean plants, they were well known to Greek physician and herbalist Pedanius Dioscorides during the first century A.D. In *De Materia Medica*, the herbal that served as a medical reference for the next fifteen centuries, Dioscorides recommended cabbages for those with poor eyesight, palsy, and spleen troubles. Ancient Europeans also used the leaves as skin poultices on "hot swellings," and once cabbage arrived in the New World, the Cherokee adopted it as a food and also used the leaves to treat skin afflictions such as boils. For all who cultivated them, cabbages and coleworts provided reliable crops that could be consumed raw, cooked, or preserved.

In her 1796 cookbook, Simmons commented on the need to grow cabbages in "new unmanured ground," believing that cabbages grown in old garden soil developed rank odors. She also knew about the practical chemistry of the red anthocyanin pigments produced by many cultivated varieties, recommending that the "red and redest small tight heads, are best for slaw," since the water-soluble anthocyanins will tint purple any foods boiled with them in their juices. Furthermore, since anthocyanins are sensitive to changes in acidity, color changes will occur if they are cooked in reactive metals pots or seasoned with vinegar. Glasse (1805) described methods for pickling cauliflower and red cabbage in salt and vinegar, flavoring the vegetables with nutmeg, mace, cloves, and allspice.

Early root crops introduced from England included radishes, turnips, beets, carrots, parsnips, and skirret (*Sium sisarum*), a forgotten carrot relative and member of the parsley family (Umbelliferae or Apiaceae). All of these plants produce large taproots that are suitable for storage; these crops were harvested, stored in root cellars (often in moist sand where they would continue to live), and then used months later. Like coleworts, radishes and turnips are members of the mustard family (Cruciferae or Brassicaceae), and the seeds of both were used medicinally against intestinal worms. Radishes (*Raphanus sativus*) are annuals that produce edible siliques, the characteristic dry fruit of the mustard family that cracks along vertical lines of weakness in the fruit wall. They contain the mustard oils typical of the family and were used to flavor sallets. In *Acetaria: A Discourse of Sallets* (1699), diarist John Evelyn recommended "young Seedling Leaves and Roots, raised on the monthly hotbed almost the all year round." He also mentioned horseradish (*Armoracia rusticana*), a different species despite sharing the name *radish*; he noted that thin shavings of horseradish roots could be eaten with "our cold Herbs or mixed with sugar and vinegar" to make a "sauce supplying mustard to the Sallet."

Josselyn (1672) mentioned radishes "as big as a man's arm" cultivated in New England gardens; these were most likely horseradish roots, which were cultivated as much for their medicinal properties as for their culinary use. Not a root crop in the strict sense, horseradish was used traditionally in plasters and poultices to treat sciatica and gout, in which its mustard oils functioned as a counter-irritant. Children with intestinal worms were given the juice of fresh roots or powdered dry roots, or a horseradish ointment was applied to their abdomens, where it sometimes caused blisters. William Coles, a seventeenth century herbalist, also recommended horseradish to treat scurvy or malaria, which he called "The Quartan Ague." Horseradish was an early escapee from colonial gardens and soon naturalized in damp waste areas and along banks and road sides. Native Americans also adopted it as a new medicinal plant; the Iroquois treated blood disorders and diabetes with horseradish, and the Cherokee used the roots for colds, sore throats, asthma, oral diseases, and as a diuretic.

Turnips (*Brassica rapa*) are biennials that were grown primarily for their roots, but their edible leaves were also used as a green vegetable. As early as 1609, colonists in Virginia cultivated turnips, as did French Jesuit missionaries along the Saint Lawrence River. Cold weather improved their flavor, so it may not be coincidental that a November 1637 letter from John Winthrop to his wife instructed her to harvest their crop while he was away. Native Americans also adopted turnips and used both their large roots and leaves as food; fields of turnips cultivated near Geneva, New York, were destroyed in the punitive raids against Native Americans in 1779. Turnip roots were baked in their skins and were regarded as far superior to the edible wild roots that Native Americans had gathered for many years.

Beets (*Beta vulgaris*) are members of the goosefoot family (Chenopodiaceae) that were also grown for both their taproots and leaves. The red pigment that appears in the edible leaves and roots of many beets is a nitrogen-containing betacyanin (betalain), which differs chemically from the more common red, blue, and violet anthocyanin pigments. Betacyanins are aromatic compounds that characterize only a small cluster of families, including the goosefoot family, cactus family (Cactaceae), and portulaca family (Portulacaceae). Nicholas Culpepper, another seventeenth century English herbalist, recommended beets for jaundice, phlegm, and "the bloody flux," which suggests that the red betacyanins may have suggested the Doctrine of Signatures; early colonists may have had both medicinal and edible properties in mind when they introduced beets to their New World plots. Wild beets are salt-tolerant, seaside biennials that were cultivated and selected by ancient Romans into the familiar red beets characterized by internal rings of cambium tissue. These are cylindrical growth zones that increase the taproot diameter and may produce some woody conducting tissue as the plant matures, particularly characteristic of the massive beets known as mangels, or mangelwurzel, that were grown for fodder. Because of their tendency to grow woody tissue, young beets have always been favored for table use, and beets seemed to grow particularly well in the New World. In her 1805 supplement on American cookery, Glasse included a method for pickling beets in vinegar flavored with pepper, allspice, ginger, and garlic.

Carrots are cultivated forms of the familiar wildflower Queen Anne's lace (*Daucus carota*), which have also been horticulturally selected for their large, edible taproots. The plants are biennials that are harvested after the first growth season. If they are not harvested, during their second year of growth carrots produce the characteristic flower clusters and small aromatic fruits known as schizocarps ("seeds") typical of the parsley family (Umbelliferae or Apiaceae). The taproot pigmentation of early cultivated varieties was variable; carrots cultivated in northwestern Europe during the sixteenth century had roots that were pale yellow or purple due to the presence of anthocyanin pigments. Yellow carrots were favored for cooking because they did not release pigments into the soup or stew, and these were the forms that were first introduced to American gardens; they were probably pigment-free genetic mutants of the original purple carrots that migrated from Afghanistan into Europe. During the seventeenth century, Dutch growers produced orange carrots by selection from the natural variation in carrot populations; these were carrots with very high concentrations of carotenes, the orange pigments that frequently occur in fruits or in leaves, where they help to absorb the sunlight that powers photosynthesis. The same species in the wild form known as Queen Anne's lace was introduced to American gardens for its medicinal properties. English herbalists such as Gerard knew it as a herb that would "helpeth conception," quite different from its ancient use for contraception and abortion, which was recommended by Hippocrates (ca. 460 to ca. 377 B.C.). In laboratory rodents, the small fruits of Queen Anne's lace block the production of progesterone, the hormone needed for completion of a successful pregnancy.

Queen Anne's lace escaped herb plots and easily naturalized as a widespread weed, and Mohegan and Delaware Native Americans later used the flowers to prepare infusions that they used to treat diabetes.

Skirret (*Sium sisarum*) was a common seventeenth century root crop that has vanished as a familiar vegetable. In 1631, seeds of "skerwort" were listed on the bill of sale from English seed merchant Robert Hill, who sold seed to John Winthrop Jr. along with onions, cabbages, parsnips, radishes, and beets. Originally from Asia, skirret is a perennial species that was grown as an annual crop, but it disappeared from American horticulture in the early nineteenth century; no evidence indicates that any of the plants naturalized into the North American flora. Bernard McMahon listed it in *The American Gardener's Calendar* (1806), and later references suggest its extensive use in French cookery. The finger-sized taproots resembled white carrots and were prized for their sweetness when cooked. Gerard described skirret both as a "medicinable herb" and as an edible root that could be "boyled, peeled, and pithed" or eaten "as a sallet, cold with vinegar, oyle etc." He considered skirret roots superior to parsnips, and he regarded parsnips as more nourishing than turnips or carrots.

Cultivated parsnips (*Pastinaca sativa*) have larger taproots than the wild plants and were prepared for the table similarly to carrots and skirrets, all members of the Umbelliferae (Apiaceae). Wild parsnips naturalized in North America and are probably the descendants of the wild type parsnips that were grown for their medicinal properties rather than as a root crop; seventeenth century herbalists recognized the wild plants as having greater medicinal properties than garden parsnips. Members of the Umbelliferae produce toxic secondary compounds known as furocoumarins that protect them from both herbivores and fungi; this is the chemical basis of the phytophotodermatitis caused by contact with parsnip foliage, which is interesting because of its common use as a poultice. In the presence of ultraviolet light from the sun, the furocoumarins in the leaves cause severe burns, discoloration, and possible blisters on exposed skin. If they were clad in protective sleeves and long skirts, colonial women probably did not encounter this problem while tending beds of cultivated parsnips. Native Americans such as the Ojibwa considered wild parsnips toxic in large amounts but medicinal when small amounts of root were consumed for female troubles.

Other early vegetables included onions (*Allium cepa*), a bulb-producing member of the lily family (Liliaceae) that could be over-planted during the growing season with successive crops such as fast-growing lettuce. Leeks (*A.*

porrum) were used as a border around spinach (*Spinacea oleracea*), and chives (*A. schoenoprasum*) and garlic (*A. sativum*) were cultivated more commonly as medicinal rather than kitchen herbs. These *Allium* species were introduced from European gardens, but colonists also discovered wild garlic (*A. canadense*) native to North America, which they also adopted for medicinal use. Garlic was recommended for ringworm infections and sore throats, of particular interest since the allicin produced by the damaged bulbs of *A. sativum* is now known to have antibiotic properties; garlic has been used during modern times as a wound dressing that prevents infection. In *American Cookery* (1796), Simmons favored white onions as "softer flavored," but she considered "Garlicks, tho' used by the French . . . better adapted to the uses of medicine than cookery."

The lemony leaves of sorrel (*Rumex acetosa*) flavored sallets, sauces, and soups with concentrations of oxalic acid that impart the characteristic acidic taste to sorrel leaves. Gerard described sorrel as having a "sharpe, sower taste" and as "a profitable sauce in many meats." He noted its cooling medicinal properties and its use as "a pleasant relish . . . in quickening up a dull stomacke," but in colonists' gardens it was probably grown for its edible leaves. Sheep sorrel (*R. acetosella*) was also cultivated and used medicinally for fevers and scurvy and perhaps also for flavoring, but care had to be taken to avoid oxalic acid poisoning.

Sorrel (*Rumex* spp.), a coarse member of the buckwheat family (Polygonaceae), shared a similar taste and high oxalic acid content with the wildflower wood sorrel (*Oxalis acetosella*). Wood sorrel is a member of the oxalis family (Oxalidaceae) and like sorrel was also used for flavoring sauces

Gerard illustrated sheep sorrel (*Rumex acetosella*), a traditional diuretic and cooling remedy for inflammations and fevers, in *The Herball* (1597).

and improving appetites. Jossleyn (1672) listed both plants, sorrel as a "Garden Herb" and wood sorrel as a plant that escaped cultivation and naturalized into the New England flora. Both sorrel and wood sorrel are now common weeds, like the European purslane (*Portulaca oleracea*), which Josselyn also noted growing wild in seventeenth century New England (see chapter 3). John Winthrop Jr. included an ounce of "pursland" seeds in his 1631 seed order. Its growth habit is low and creeping, and the plants carpet bare soil; seventeenth century plantsman and herbalist John Parkinson (1640) noted in his *Theatrum Botanicum* (London: Thomas Cotes) that it grew "in the alleyes of the Garden between the beds" and is used "as lettice in sallets, to coole hot and faint stomackes in the hot time of the yeare."

As noted earlier, pumpkins and squashes were acquired from Native Americans for early culture, and European "cowcumbers" or cucumbers (*Cucumis sativus*) were also cultivated in hills. Cucumbers were used fresh or pickled, and pumpkins were commonly stewed. As with many foods, both were also considered to have medicinal properties; Gerard recommended pumpkins for treating kidney stones and cucumbers for inflamed "red and shining fierie noses."

New World Herbalism

Garden herbs were the primary source of flavorings, household herbs, and medicines, and early colonists grew plots that supplied these herbal needs. Many of these are discussed in chapter 7, but a few examples are noteworthy for their chemistry, lore, or family associations. The herbs known to New England Puritans were familiar English medicinal plants that they carried with them and planted in North American gardens. Opium poppies (*Papaver somniferum*) are arguably the most important medicinal plant in human history, in light of the pain that has been alleviated by their complex analgesic mixture of twenty-six alkaloids, including morphine. Poppy seeds were grown for use in bread and comfits, and the plants were used for their medicinal properties; poppy leaves and capsules were boiled in water and sugar to make a sleep-inducing infusion. According to Gerard, "It mitigateth all kindes of paines, but it leaveth behinde it oftentimes a mischiefe worse than the disease." He mentioned the danger of death from opium and perhaps understood the dangers of opium addiction.

The composite family (Compositae or Asteraceae) was represented by several aromatic species, including yarrow (*Achillea millefolium*), feverfew (*Chrysanthemum parthenium*), southernwood (*Artemisia abrotanum*), wormwood (*Artemisia absinthium*), mugwort (*A. vulgaris*), lavender cotton (*Santolina chamaecyparissus*), tansy (*Tanacetum vulgare*), and garden chamomile (*Anthemis nobilis*). All contain essential oils that are composed of terpenes, and they had diverse medicinal and household uses in the earliest colonial homes. Yarrow had been used since ancient times to stop the flow of blood from wounds, and feverfew was used for inflammations and as an antidote to opium. The various *Artemisia* species and lavender cotton were used to treat intestinal worms, soothe insect stings, and also to kill household insects; they were also favored as border plants for garden plots, perhaps for their insecticidal properties. Tansy and chamomile were both considered "hot" by early herbalists and had several uses, from treating stomach complaints, colds, and jaundice, to flavoring foods.

The mint family (Labiatae or Lamiaceae) in New World gardens included sage (*Salvia officinalis*), rosemary (*Rosmarinus officinalis*), thyme (*Thymus serpyllum*), ale hoof (*Glechoma hederacea*), calamint (*Satureja calamintha*), catnip (*Nepeta cataria*), horehound (*Marrubium vulgare*), hyssop (*Hyssopus officinalis*), lavender (*Lavandula vera* and *L. spica*), marjoram (*Origanum onites*, *O. majorana*, and *O. vulgaris*), pennyroyal (*Mentha pulegium*), and selfheal (*Prunella vulgaris*). All have epidermal hairs (trichomes) that contain pungent terpenes and discourage attack from herbivores; these selfsame protective molecules impart the remarkable tastes and scents associated with various mints. Many of their early medicinal uses have been forgotten, but Gerard recommended sage to "cleanseth the bloud"; rosemary for the "head and braine"; thyme for "Lethargie, frensie, and madnesse"; and ale hoof for burns, wounds, and ringing in the ears. Ale hoof has naturalized as a common lawn weed, and sage, rosemary, and thyme now have primarily culinary uses. Simmons (1796) recommended pennyroyal (presumably European pennyroyal, *M. pulegium*) as a kitchen herb along with other European herbs such as parsley, thyme, and marjoram, apparently overlooking the use of pennyroyal as an abortifacient. Gerard recommended pennyroyal for stillbirths and delivery of the placenta, but we now know that an effective dose is likely to be toxic or even lethal. Remarkable healing properties were ascribed to selfheal, now merely another naturalized lawn weed; according to Gerard, a decoction of the plant in wine or water will "joine together and make whole and sound all wounds both inward and outward." Some early uses of household mints are still familiar, such as traditional use of horehound for treating coughs and the household use of dried lavender flowers with stored linen, clothing, and leather.

Purveyors and Plantsmen

The first settlers carried seeds with them, dried and saved from their own gardens or perhaps purchased from seedsmen such as Robert Hill, the English grocer who supplied seeds to John Winthrop Jr. in 1631. His "bill of garden seeds" provides a useful list of seventeenth century household plants, including basic foods such as cabbage, carrots, and parsnips and medicinal herbs such as rosemary, thyme, and tansy. Hill even supplied Winthrop with seeds of New World pumpkins, which apparently were already being grown for seed in England. Once their American gardens were established, colonists exchanged garden seeds freely among themselves, and soon generations of American gardeners were a continent removed from their ancestors' geographical and botanical heritage. Nevertheless, food preferences and household gardens remained essentially English in nature through the next century, evolving slowly with the inevitable commingling of traditional English plants with New World species and explorers' exotic botanical booty.

During the eighteenth century, several American seed houses and nurseries were established. All houses had gardens, whether a workman's dwelling with modest plot used for kitchen vegetables or a sea captain's mansion enhanced with parterres and terraces. Elaborate landscaping included ornamentals, English berries and fruits, and glass-enclosed beds for tender plants. Homeowners, whether of moderate means or wealthy, needed suppliers of garden plants, and cities also functioned as the centers of horticultural activity. In the Boston area, nurserymen and truck farmers grew plants for seed, propagated plant stock and root stock, cultivated plants, and imported plants and seeds for the growing horticultural market. One such Boston entrepreneur was Richard Francis, who established a nursery under the sign of the Black and White Horse in 1731; he imported and sold seeds and garden plants, first in Long Lane and later on Newbury Street (the present-day Washington Street). John Little moved his nursery from Milk Street to Francis's first site in Long Lane, where he had various enclosed hot beds that he probably heated with the warmth generated by decomposing manure. More growers thrived near the lower end of the Boston Common. As early as 1719, Evan Davies sold propagated plants, berry bushes, and various roots near the landmark powder house, in proximity to the "Great Garden," a market where gardeners sold produce and propagated garden stock. His offerings included "English Sparrow-grass Roots" (asparagus) and currant and gooseberry bushes. In 1734, William Grigg sold fresh peas there for five shillings a peck, and two years later he installed the gardens at the mansion built by Thomas Hancock overlooking the Common. His assignment was to "layout the upper garden alleys, Trim the Beds and fill up all the allies . . . and prepare and Sodd the Terras adjoining." By 1770, this garden functioned as a commercial nursery. Following Hancock's death in 1764, the site passed to his nephew, Declaration of Independence signer John Hancock; his gardener, George Spriggs, advertised grafted English fruit trees, perhaps the propagated offspring of the original dwarf fruit trees and espaliered plants obtained from England by Thomas Hancock. Spriggs's stock included cherry, pear, plum, apricot, nectarine, quince, lime, and apple trees. English walnut trees were also sold by Elizabeth Decoster, a seed importer who operated during the mid–eighteenth century under the "Sign of the Walnut-Tree."

American gardening moved beyond mere subsistence during the eighteenth century, at least for prosperous urban dwellers. Yet even gardeners with small, utilitarian plots had vast options in selecting plants to cultivate and consume. The diversity represented by eighteenth century vegetable varieties far exceeded what is currently available through commercial sources; peas alone were selected into multiple distinct types, including several subvarieties of the hotspur, rounceval, and marrowfat cultivars. Cultivars differed in maturity time, growth form, leaf form, coloration, and virtually any other genetically controlled trait. Eighteenth century gardeners could also select from among hundreds of varieties of herbs, lettuce, cabbage, beans, and other vegetables to supply their kitchen plots.

A familiar irony of horticulture is that gardeners often desire plants that are the most difficult to obtain. By the late eighteenth century, English gardeners were cultivating American species in their gardens and landscapes and relied on explorers such as John Fraser, who made four collecting forays to America and introduced magnolias and other prized North American plants to his London nursery in Sloane Square. Many American colonists often ignored the native North American flora and planted exotics such as the Persian irises, Peruvian nasturtiums, and South African geraniums offered by Virginia plant purveyor Minton Collins in 1793.

Purveyors of American plants sought customers abroad and at the same time gladly marketed exotics to Americans; the commercial trade in cultivated plants distributed American plants worldwide and brought the world's flora to the United States. Some horticulturists boasted that their stock came from the finest European sources, while others such as early nineteenth century seedsman John Russell advertised that most of his seed stock was produced in the United States. The eighteenth century was an era of unparalleled geographical and botanical exploration, and the results of this scientific curiosity influenced household botany, both with respect to the plants that were available for American gardens and how these plants were named and classified.

One such botanical explorer was Swedish botanist Pehr Kalm, who traveled to America aboard the *Mary Gally*, arriving in Philadelphia in September 1748. Kalm was a trusted student of Linnaeus, to whom he sent at least ninety American plants, and Linnaeus rewarded him by commemorating his name with the genus *Kalmia*, which includes such handsome native North American plants as mountain laurel (*K. latifolia*). The American species collected by Kalm were incorporated into *Species Plantarum* (1753), a two-volume work in which Linnaeus for the first time consistently used binomial nomenclature for each species that he described. Each binomial consisted of the generic name and specific epithet, the system of nomenclature still in use today. Binomials slowly revolutionized botanical communication, since the names are universal and utilize Latin roots and endings. Nurserymen and botanists on both sides of the Atlantic eventually embraced the new naming scheme; the old Latin phrases (polynomials) that appeared in herbals disappeared from use, and they are now ignored when botanists determine a plant's correct scientific name.

Linnaeus also developed a classification system based on the number of reproductive parts in a flower; his *systema sexuale* organized the flowering plants into convenient classes and orders based on the number of stamens and pistils in each flower. First published in his *Systema Naturae* (1735), the sexual system is no longer in use scientifically. Modern botanists regard it as an artificial system (one that groups unrelated plants together often for convenience), and it has been supplanted by evolutionary classification, but the Linnaean scheme holds historical interest because of its widespread adoption during the eighteenth century (see chapter 10).

Mannasah Cutler published the first scholarly study on American plants, "An Account of some of the Vegetable productions, commonly growing in this part of America, botanically arranged" in the first volume of the *Memoirs of the American Academy of Arts and Sciences* (1785). Cutler used vernacular names rather than binomials, but he organized his American plants using the Linnaean sexual system. In his writings, he recorded a visit to the home of Ezra Stiles in New Haven, where he demonstrated to Stiles, his wife, and "the young ladies" how a flower could be classified "by separating and demonstrating the parts" and that the company was "highly amused." In fact, during the eighteenth and nineteenth centuries, botanical studies became a well accepted activity for young people, young women in particular (see chapter 10).

Nurserymen also used Linnaean classification; the Prince family nursery of Flushing, Long Island, established by Robert Prince in 1737, was eventually renamed The Linnaean Botanic Garden by son William Prince Sr. For more than one hundred years, the Prince Nursery was a leading exporter of American plants to Europe and importer of exotic flora into America. The name change may have hinted at the botanical diversity and the rare plants offered through its broadside catalog, a listing that included European roses and lilacs, English passionflowers, and Italian poplars, as well as North American woody plants such as tulip tree, sassafras, and Carolina allspice.

American household botany began with the first curious European explorers to encounter the virgin wilderness and Native American agricultural practices of the New World. Survival required a knowledge of practical botany, and colonial American gardens and agriculture fields reflected the needs and tastes of their cultivators. In New World homesteads, plants provided not only foods and medicines, but also flavorful herbs, textiles, dyestuffs, timber, and fuel, as well as the creative inspiration for garden design, home decoration, and literature. In short, American gardens began with survival and evolved into a cultural expression of our interaction with new botanical discoveries worldwide. Humans are inherently curious about new plants and their possibilities; gardeners often desire what is uncommon, and unfamiliar, even if our survival does not depend upon its successful cultivation. This may have been the vision of William Penn, when he surveyed the Native American crops and potentially useful plants that grew in the land granted to him in 1681 by King Charles II. Penn urged that colonists create gardens and orchards with the notion that each settlement "might be a green Country Towne, [that] will never be burnt and will always be wholesome." He envisioned the ideal North American colony with cultivated plants in well-tended domestic gardens, a New World landscape defined by labor and a sense of rural stewardship.

CHAPTER TWO

Grains

THE DOMESTICATION of food plants has been a practical pursuit driven by hunger and necessity. Ancient hunters and gatherers developed agricultural practices in response to environmental pressures; the periodic scarcity of edible plants and the convenience of having a food supply close at hand probably motivated the first gardeners to begin growing plots of food plants. Perhaps they observed discarded seeds germinating in the domestic trash piles known as "middens," and from that essential clue they learned to save and sow the seeds of familiar wild foods.

These early cultivated food plants were the botanical ancestors of many of the plants that make up our modern diet. Now we have the luxury of selecting food plants based on taste, texture, and color, but the earliest edible vegetables, fruits, and seeds were domesticated based strictly on practical criteria: the absence of toxic secondary compounds, the ease of propagation and preparation, and the potential for long-term storage during winter months. Grains, defined botanically as the fruits of plants in the grass family (Gramineae or Poaceae), are particularly useful as a storable crop. Grains can be harvested and put by as a dry, inert food supply that can be used during the season of need. Presumably, the domestication of food plants in the Old and New Worlds followed similar historical pathways, as hunters and gatherers converted to agriculture and a more geographically stable life; in cultures on both sides of the Atlantic, grains were the dietary staples. Early Europeans perfected such essential food plants as wheat, rye, broad beans, and coleworts, while Native Americans sowed commensal fields of maize, squash, and beans; each culture had a fundamental grain that provided flour for bread, the most basic human food.

European settlers arriving in America from England soon learned to cultivate and consume the food plants domesticated on the American side of the Atlantic, while retaining the English traditions of kitchen gardens and cookery methods. Perhaps nothing illustrates the melding of Old and New World traditions better than the celebration that marked the Pilgrims' first successful harvest in their adopted homeland; Edward Winslow's 1621 account (see chapter 1) documents their crops of Indian corn and barley and the subsequent three-day feast with Wampanoag Chief Massasoit and ninety Native American men. The Pilgrims grew both English and American food plants, but it appears that the New World plants were more productive. Winslow mentioned their "indifferent good" barley crop and the pea plants that withered while still in flower, which fortunately were balanced by "a good increase of Indian corn."

This celebration of Thanksgiving was held sometime during the fall of 1621, perhaps close to Michaelmas (29 September), the date for traditional English harvest home celebrations. Four Pilgrim women survived the first winter in America, and they probably oversaw much of the cookery for the feast, which according to Winslow centered on wild fowl and venison. Food historians believe that they prepared New World corn, fruits, and vegetables following traditional English methods. For instance, the original hasty pudding, which in England was made with oatmeal or wheat flour, evolved into a New World version prepared with cornmeal. Corn could also substitute for rice, and puddings made with hominy (dried corn grains with the outer layers removed), dried blueberries, milk, and sweetening mimicked Elizabethan rice puddings. Perhaps the most significant outcome of the Pilgrims' first year at Plimoth was the cultivation and incorporation of corn into their New World diet, which set the culinary and agricultural stage for the diet that would be adopted by waves of new settlers arriving from Europe.

The Botany of Grains

No foods are more important to humans than grains, and economic botanists have acknowledged that human cultures depend on the grains that they cultivate for their most basic sustenance. While some may think of a grain as a seed, botanically speaking each grain is a fruit type known as a *caryopsis*. Botanists define a fruit as the mature ovary of a flower, and the fruit wall is known as the pericarp, which in many animal-dispersed fruits becomes fleshy and brightly pigmented. In contrast to the bright berries and succulent cherries that are dispersed by hungry birds, the grains have a pericarp that dries, hardens, and fuses to the seed coat that surrounds the single seed within. In cultivated grains, this fused layer of the pericarp and the seed coat is known as the bran. In terms of human diet, however, the most important part of a grain is its internal mass of stored food, the endosperm. The unique botanical phenomenon of double fertilization explains the origin of the endosperm that can be ground into the flours and meals that are used for bread baking.

Although grasses are classified as flowering plants, their flowers or florets are small and wind-pollinated. This explains the modest appearance of grass flowers; they are not "advertising" nectar or pollen to pollinators and have not evolved floral lures such as showy petals, scent, and nectar. Petals and sepals are small or absent entirely, and grass flowers are borne on inflorescences known as spikes, racemes, or panicles, depending on the extent of branching. Each flower typically has three stamens and one pistil, with a lone ovule in its ovary, although some grasses have unisexual flowers. If pollinated, the ovule will mature into the single seed that is found in each grain. The stamens release large quantities of wind-borne grass pollen, and the grains land randomly on the branched, feathery stigmas of grass flower pistils. Pollen germinates on the stigma and grows an elongated tube down to the ovule in the ovary. A cell inside the pollen grain divides to produce two sperm, and one of the sperm fertilizes the egg cell that is located deep inside the ovule; the resulting zygote develops into the embryo, the young plant that sprouts if the seed later germinates.

The other sperm produced by the pollen is involved in a second fertilization, a phenomenon in flowering plants that, along with flowers and fruit, is unique among all evolutionary groups on earth. Once the second sperm reaches the ovule, it penetrates a large central cell, where it fuses with the two central cell nuclei to produce a single nucleus. This second fertilization results in a cell with a triploid nucleus, meaning that it has three basic sets of chromosomes (one from the sperm and two from the two nuclei of the original central cell). The central cell and its triploid nucleus divide repeatedly to form the endosperm, a large mass of nutritive tissue that develops only in flowering plants. Endosperm evolved to support the growth of flowering plant embryos. It provides a repository of stored food in the form of carbohydrates, fats, and proteins that are ready to support the growth of the young seedling until it can begin photosynthesis on its own; coincidentally, we can exploit it as a reliable source of human food. The outermost protein-rich cells of the endosperm is the aleurone layer, which provides the enzymes necessary to break down starch when the embryo (also known as the germ) begins to germinate.

The humans who domesticated grasses learned to grow and use these stored energy reserves to their best advantage; the flours and meals milled from corn, wheat, rice, and other grains are the nutritious endosperm of grasses that have been cultivated since ancient times. Buckwheat (*Fagopyrum* spp.) is an exception because the plants are members of the Polygonaceae (buckwheat family) rather than the grass family. Their fruits are triangular achenes that contain edible seeds, and they are named for the similarity of these achenes to small beechnuts (the beech tree genus is *Fagus*). Common buckwheat (*F. esculentum*) tolerated poor soils and was cultivated as fodder, often on land where true grains would not thrive. Buckwheat seeds were also ground into meal and used as a human food, particularly in pancakes, because buckwheat flour did not produce satisfactory bread. Some regional cooks used buckwheat where they could. For instance, Quakers in the Delaware valley used buckwheat to prepare scrapple, a dense mush cooked from meat scraps and meal that was then sliced and fried. The method for preparing scrapple was introduced by German immigrants during the eighteenth century and soon adopted as part of regional Quaker cookery.

Corn

Eighteenth century Swedish botanist Carolus Linnaeus named the important New World grain crop *Zea mays* var. *mays*, noting its "Habitat in America" and assigning it to his Class Monoecia, along with sedges (*Carex*), cattails (*Typha*), and other plants with separate pollen and ovule producing flowers on one plant. Linnaeus classified corn apart from wheat, rye, and other cultivated grains (these were assigned to his Class Triandria) because he realized that maize plants differed from most other grasses in having an inflorescence of staminate flowers (the so-called tassel) atop each plant. The pistillate inflorescences are the thickened lateral "ears" that bear a single ovule inside each grain. The grains mature while still covered by layers of large bracts, or husks, with only a tuft of "silks" visible; each pistillate flower consists of a single pistil, and the silk is the pollen-receptive stigma and style growing from the base of one ovary. As noted in chapter 1, corn husks fully enclose the mature grains, suggesting that domesticated corn could not survive in nature without deliberate planting and cultivation. Linnaeus also mentioned the many varieties of corn, which had already been described and illustrated by Gerard in his *Herball* (1633) as the various colors of New World "Turkie Wheate."

Depending on the genes present, the aleurone (outermost) layer of the endosperm will develop different pigments, and only relatively recently has the genetics of corn color been more fully understood. Each grain represents the results of a genetic cross between the plant producing the ovule and the plant producing the pollen grain, and the complex colors of Indian corn varieties were a perplexing question in plant genetics. Geneticist Barbara McClintock received the Nobel Prize in Physiology or Medicine (1983) for postulating the existence of "jumping genes" in maize, DNA segments that could relocate randomly on chromosomes. She originally published her theory in the early 1950s, before the genetic equivalent of jumping genes had been discovered in bacteria. This genetic phenomenon, now better understood, explains the remarkable variety of the yellow, gold, red, blue, and white maize varieties that colonists adapted as a staple crop suited to a variety of recipes.

Color aside, ground corn could be used to prepare mush, spoon breads, hoecakes, pone, dodgers, scrapple, fritters, and a variety of corn breads. New World cooks also explored the potential of corn as a food plant beyond its use as a cereal or bread. Corn grains were simmered with beans to prepare succotash, such as the method for "Sackatash, or Corn and Beans" that appeared in *The New Household Receipt-Book* (1853) by Sarah Josepha Hale. She instructed her readers to boil a quantity of shelled beans or string beans with the grains cut off four dozen ears of corn and then "Pour it in to your tureen and send it to the table." In *The American Woman's Home* (1869), Catharine Beecher and Harriet Beecher Stowe also threw their support to succotash, calling it "an Indian gift to the table for which civilization need not blush." Essentially this was a Native American meal adapted to Victorian sensibilities, and Hale did know something about the history of American foods and their adaptation into colonists' diets. As the editor of *Godey's Lady's Book* during the late 1840s, she fired off editorials and petitioned President Lincoln, governors of states and territories, naval commanders, and missionaries to celebrate Thanksgiving and recognize the day as a holiday. In 1863, Lincoln declared that a national day of thanksgiving would be celebrated each November. No doubt, her vision of the Thanksgiving feast included dishes based on corn, arguably the most important and versatile of all American food plants.

Sweet corn varieties were used for boiling or roasting on the cob, which caramelizes the stored endosperm sugars, but most of the corn consumed by humans was eaten off the cob and compounded with other ingredients. Cooks knew that a good green corn pudding required slicing down the center of the kernels to release the white "milk" from the endosperm, before scraping the kernels into the pudding basin. Green corn was also used to make the small, fried "artificial oysters" described in *Confederate Receipt Book: A Compilation of Over One Hundred Receipts Adapted to the Times,* published in Richmond, Virginia (1863). Other recipes depended on the preparation of hominy, in which corn grains are scalded in hot lye to remove the hard outer layers (the seed coat fused to the pericarp) and the germ (embryo). In *Miss Leslie's New Cookery Book* (1857), Eliza Leslie recommended washing the "perfectly white" hominy twice, overnight soaking, and boiling for four or five hours for a "very good" dish suitable for serving with corned beef or pork. The next day, hominy remains could be patted into cakes for frying.

In addition to cooking in the traditional sense, some corn could also be popped. Popcorn varieties may descend from some of the oldest strains of cultivated maize, and we can imagine the reaction of the Native Americans who first tossed some grains into a fire. The softened, popped grains could be used without arduous hand grinding, which must have been an advantage. Popcorn has the unique internal feature of some starchy cells that retain moisture; when a grain is heated, this internal water boils and builds internal pressure, causing it to turn inside out with a forceful endosperm explosion. Although it is unlikely that the Pilgrims knew about popcorn, some later colonists ate sweetened popcorn and milk for breakfast, and popping grains in wire baskets became a popular pastime during the mid-nineteenth century.

Social mores dictated that bread baked with wheat flour was more acceptable than various corn breads, but many families consumed wheat bread only on the Sabbath.

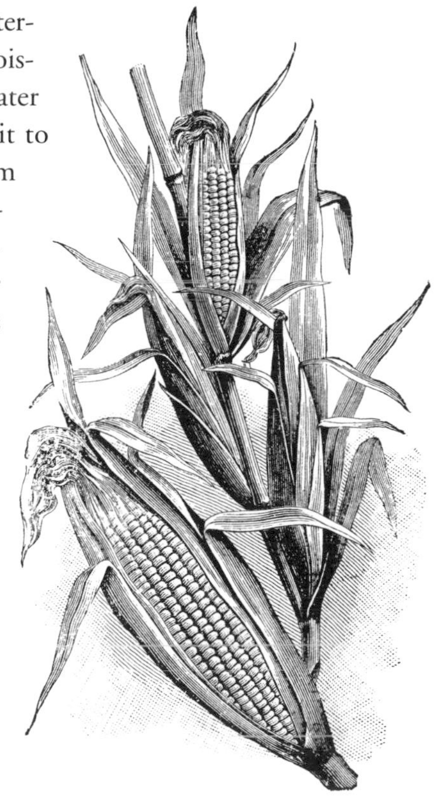

Corn (*Zea mays* var. *mays*) grains were ground as the source of Indian meal for bread baking; even in wheat-growing regions, corn breads were baked and eaten by many families six days a week.

Settlers in the American frontier often subsisted on ground corn as a staple, since the wheat that they grew could be sold as cash crop to purchase household goods, plows, and other farming necessities. Light-colored foods such as wheat breads were difficult to produce on the hearth, where ash, smoke, and iron cookware conspired to darken loaves of home-baked breads, and in many parts of the country brown breads home baked with a mixture of rye and corn flours eventually were condemned as food for the poor. In *Miss Beecher's Domestic Receipt-Book* (1858), Catharine Beecher listed a thin corn bread (she called it Indian bannock) made with "Indian meal and wheat flour" under "cheap dishes." Personal preference dictated the proportions used; in *The American Frugal Housewife* (1844), Lydia Maria Child suggested one-third or one-half "Indian" (cornmeal) and rye for the remainder of the flour, and she did not disdain dark loaves. In nineteenth century New England (the home of both Beecher and Child), brown breads were equated with Puritanical virtues and character, but in other parts of the country people desired light-colored breads; indeed, in southern kitchens white cornmeal was preferred over yellow.

Some corn was exported to Europe to meet dietary needs during the potato famine that began in 1845, and pamphlets were issued to instruct cooks how to use this grain, which was generally disliked by Europeans. Leslie, a prolific author of American cookbooks, first published *The Indian Meal Book* in London in 1847; she described a repertoire of American recipes for corn breads, mush, porridges, hominy, cakes, biscuits, and summer and winter versions of "saccatash." Leslie hoped that "this little book may be found a valuable accompaniment to the introduction of Indian Meal to Great Britain and Ireland," but later editions were published in Philadelphia for her American readers. The corn plants had myriad other domestic uses: corn cobs were used to insulate walls, in smoking meat, and as a fireplace fuel, while the husks were used to stuff mattresses and wrap foods. Corn stalks were sometimes used to thatch roofs and as an emergency building material.

Wheat

Cultivated wheat originated in the Near East perhaps nine thousand years ago, through the complex hybridization of several wild wheats (*Triticum* spp.) with goat grass (*Aegilops* spp.). Neolithic peoples carried cultivated wheat (*T. aestivum*) across Europe, and by the time farming began in Britain in about

3500 B.C., some cultivated varieties could already tolerate colder temperatures, longer winters, and greater rainfall variations than their wild ancestors. Wheat plants are typical of the grass family, with the grains growing tightly packed on the spike or inflorescence. The "bearded" appearance of each wheat inflorescence is due to the slender, stiff bristles (awns) that extend upward from the bracts (the lemma and palea) that surround each floret. These bracts make up the chaff that covers each mature grain, hence the familiar phrase "separating the wheat from the chaff," which refers to isolating the nutritious grain from its surrounding layers.

Threshing was a time-consuming task that involved flailing the mature heads to release the enclosed grains. The tendency of a grain to cling tightly to the chaff is under genetic control; an almost universal trait of domesticated grains is the tendency to be "free threshing," or easily separated from the surrounding chaff, when compared with their wild grass ancestors. When genetic mutations appeared that permitted easy threshing, ancient farmers quickly selected and bred these traits into their domesticated crops, and these traits became part of the genetic legacy of the wheat that was imported from Europe to cultivate in the New World.

At the end of the growing season, farmers had about ten days in which to harvest the mature wheat crop before the heads shattered, scattering the grains on the soil. A sickle or scythe was used to cut the stalks, which were then bound and shocked, ready for threshing. Pioneer farmers threshed their wheat with a short wooden club tethered with a piece of leather to a longer wooden handle, until horse powered threshing machines were invented in the late 1700s. The chore was best done on a dry day, since in wet or humid weather the grains did not easily separate from the heads. Winnowing then separated the grain from the chaff, which was often done by letting the wind blow through the grains as they were tossed into the air. In 1837, Hiram and John Pitts of Maine invented an improved machine that efficiently threshed the grains, removed the chaff, and disposed of the straw with a blower. Later steam threshers required the joint effort of an entire community to operate the machinery and haul the grain crops to and from the site.

Wheat was dispersed throughout Europe during the Neolithic and was a common commodity to Europeans, but the plants could not grow in many of the New World fields where corn thrived. As a domesticated crop, wheat thrived in cleared temperate grasslands, and colonial settlers accustomed to abundant European wheat soon discovered that this Old World crop did not transplant easily to many New World areas. Nevertheless, settlers had moderate success with wheat in some New England states. Seventeenth century farmers in Rhode Island and Connecticut raised wheat that they exported in trade for sugar, wine, cotton, tobacco, and indigo, but fields were quickly depleted of mineral nutrients. Wheat rust (*Puccinia graminis*) also contributed to the disappearance of wheat crops from older, settled areas, perhaps areas where cultivated barberry shrubs served as the alternate hosts for the parasitic fungus. Prior to the 1780s, wheat was planted in northern Maryland and southeastern Pennsylvania, and newly plowed land could yield forty bushels per acre. The yield then declined rapidly for lack of knowledge of crop rotation or the addition of soil nutrients. Wheat farmers migrated west into Ohio, Illinois, and the upper Great Plains to find areas for wheat farming on a larger scale; wheat was not grown successfully in the United States on a large agricultural scale until the end of the eighteenth century.

Although it was frequently more expensive, wheat became a more socially acceptable grain to most American settlers and their descendants than either corn or rye. The notable exception to this was the attitude of many nineteenth century New Englanders, who valued and esteemed brown bread as the food of their Puritan ancestors. Yet even in New England, a delicate wheat loaf could be a desirable commodity, even if not always as affordable as corn bread. In her method for "Flour Bread" Child (1844) suggested substituting other grains such as rye and "Indian" (cornmeal) for part of the flour when wheat prices were high.

Delicate baked goods such as pie pastry, gingerbread, tea cakes, short cakes, and pancakes required wheat flour to produce a refined product. Then, as now, cakes mixed up with fine wheat flour marked special occasions such as weddings and holidays. The custom of New Year's day visits began during the seventeenth century with Dutch settlers in New York. Callers were plied with little cakes, perhaps decorated with caraway seeds or imprinted with the design from a carved wooden mold, and the tradition soon

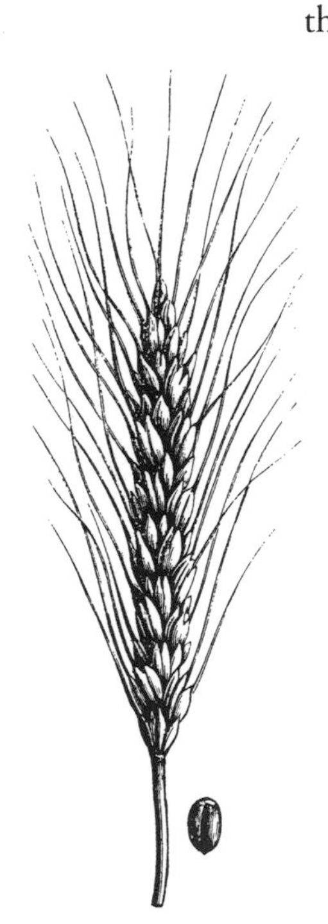

Grains of wheat (*Triticum aestivum*) required threshing and winnowing, but wheat flour was valued by cooks who desired its pale color and fine texture.

spread to other parts of the country. After the Revolutionary War, militia exercises were traditionally held in late May, on a now forgotten holiday known as Election Day; to mark the event, election cakes made of wheat flour and sometimes flavored with currants or cloves were baked and sold.

Commercially baked breads were sometimes of dubious quality and nutritive value and perhaps laced with adulterants, but city dwellers often depended on bread baked by others. Many urban households lacked ovens, which had to be built into fireplaces or in a separate building. As early as 1664, ten bakers were operating in New York City, and baking developed into an important trade in colonial America. Many bakers were reputable tradesmen, but some produced adulterated products with flour of questionable purity. Whiting (pulverized chalk), stone dust, ground bones, and plaster of Paris were all used to "stretch" flour, and household manuals described various methods for their detection. When cast-iron stoves became commonplace during the nineteenth century, Sarah Josepha Hale chastised readers of *The Good Housekeeper* (1841) to avoid unhealthy, tasteless commercial bread and bake their own loaves. She described the mixing and kneading of bread dough as a healthful exercise, which promised "to make the fairest hand fairer and softer, the exercise giving that healthy pink glow to the palm and nails which is so beautiful."

Hale also railed against the consumption of hot bread, which she claimed was indigestible, "hard and heavy in the stomach," and the dietary cause of dyspepsia and mental fatigue. The custom of eating hot leavened bread with every meal was prevalent in the South, and Hale was correct in her suspicion of its merit, but the issue was nutrition rather than digestion. The early diet of many rural families in southern slave states was appallingly poor, with corn pone and molasses providing most of the calories; seasonal vegetables and stored sweet potatoes did stave off some dietary deficiencies. Emancipation resulted in greater dietary problems because wheat assumed a more important role in many rural diets, and nineteenth century transient tenant farmers often did not bother to cultivate traditional kitchen gardens. The wheat flour that former slaves purchased from landowners was nutrient poor in comparison to the traditional southern corn-based diet augmented with garden crops; this light-colored flour was milled with the vitamin-rich embryo of the seed (the wheat germ) removed, resulting in the epidemic vitamin deficiencies that are now avoided by enriching wheat flour with essential nutrients.

Preparing a wheat flour dough and forming it into macaroni or noodles was a European concept. English colonists carried macaroni to America, and early southern cookbooks included dishes of macaroni and cheese. Mary Randolph, in *The Virginia House-Wife* (1824), also included a method for preparing vermicelli, thin noodles that were rolled and cut by hand from a dough of flour and eggs; she recommended them "as an excellent ingredient in most soups, particularly those that are thin." Some cooks also used imported macaroni, which was shaped with the sort of European "Maccaroni" machine that Thomas Jefferson sketched in his 1787 notes from Naples. He recorded the method for mixing finely ground flour with water to make a "paste," which was forced through the plates of the machine to prepare uniform macaroni shapes. The plates were interchangeable and "may be changed at will, with holes of different shapes for the different sorts of Maccaroni." Italian immigrants arriving at the end of the nineteenth century remained steadfast in their diet that relied on tomatoes, herbs, and various types of macaroni; a plain dish of macaroni and soft cheddar cheese was colonial in origin, but macaroni with spiced tomato-based sauces and hard cheeses was the food of immigrants who resisted the dietary doctrine imposed by the food-reform zealots at the end of the nineteenth century (see chapter 4).

Rye

Rye (*Secale cereale*) is a close relative of wheat, native to the same regions of Turkey, Iran, and Caucasia, where both wild wheat and barley originate. Rye often established itself in abandoned wheat fields, and it developed as a human food during the fifth century A.D. Rye arrived in America with colonists who settled both the north and south, but it was particularly suitable for cultivation in the rugged New England climate. Rye grains can germinate at temperatures close to freezing, and the plants will grow even with moderate snow cover.

During the Middle Ages, rye became an important crop in many parts of Europe, and it was an ingredient in traditional recipes that was later combined with cornmeal to bake breads in New World kitchens. Child (1844) recommended rye for brown breads and as an economic alternative for breads made only of wheat flour. Grains contain a protein mixture known as gluten, which becomes elastic when the milled flour is combined with water and yeast in bread baking; rye gluten is notoriously sticky compared to the more elastic gluten in wheat flour, which results in the different texture of rye bread. On a practical note, Child described her method for preparing household paste from rye flour, with the explanation "that grain is very glutinous" for its excellent adhesive qualities.

Many wild and agricultural plants become infected with parasitic fungi, and grains are no exception. Ergot (*Claviceps purpurea*) is a fungus that parasitizes various grains, but particularly rye, and it causes the severe, potentially lethal poisoning known as ergotism. As a grain, rye may have started as a weed in cultivated wheat fields, and wherever rye was introduced, the ergot fungus seemed to follow. Ergotism results from fungus-infected grains that make their way into the flour used in bread making; the fungus may thrive after a cold winter followed by a wet spring, but the link between the fungus in rye and ergotism was not understood until 1670, when a French physician linked the gruesome symptoms that periodically enveloped peasant villages with the presence in local fields of the characteristic purple resting stages of the fungus. European peasants consumed rye bread, and almost invariably they were affected by the disease.

Ergotism is caused by the purple fungal sclerotia, structures that overwinter in the grain head and that contain high concentrations of potent alkaloids such as ergotamine and ergonovine and the psychotropic compound LSD. The chemistry of ergot dictates the symptoms of ergotism, which include nervous dysfunction, convulsions, muscle spasms, and hallucinations. Poisoning may extend to the loss of extremities, caused by the gangrene that results from the constriction of peripheral blood vessels; gangrenous ergotism caused burning sensations. During the Middle Ages, ergotism became known as "Saint Anthony's fire" (so named because praying to St. Anthony was believed to effect a cure), and in the absence of a known cause, some Europeans perhaps suspected a link between the dreadful symptoms and the

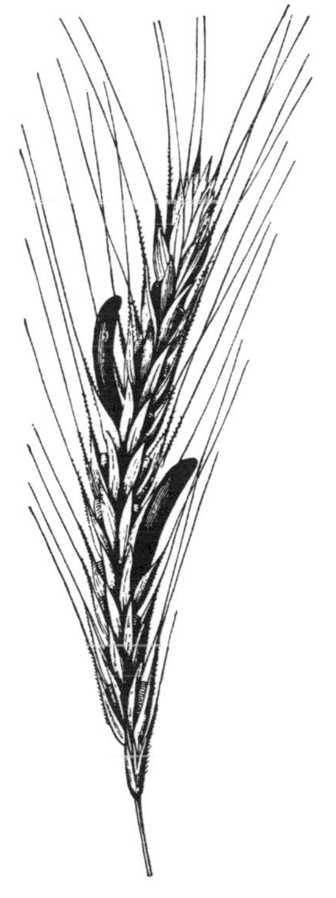

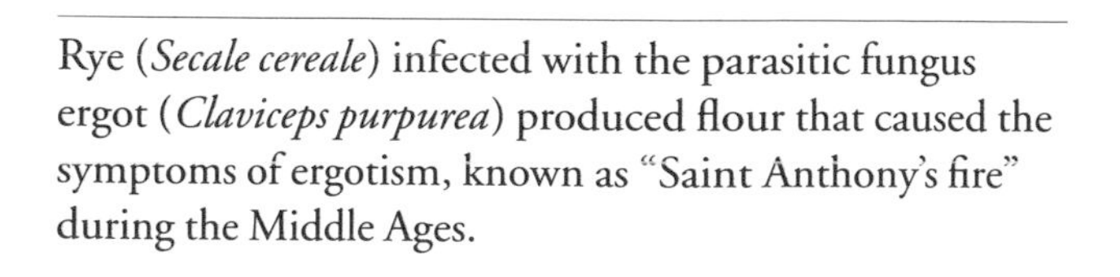

Rye (*Secale cereale*) infected with the parasitic fungus ergot (*Claviceps purpurea*) produced flour that caused the symptoms of ergotism, known as "Saint Anthony's fire" during the Middle Ages.

practice of witchcraft. In *Poisons of the Past: Molds, Epidemics, and History* (1991), historian Mary Kilbourne Matossian has traced climate, crops, diet, rye prices, and demographics and has correlated them with the incidence of witchcraft accusations and trials. She has argued convincingly that the behavior of some young girls in Salem, Massachusetts, that was recorded during 1692 is consistent with the convulsive form of ergotism. The Salem witchcraft trials resulted in twenty executions, perhaps caused by the misinterpretation of the symptoms caused by fungal chemicals from infected rye grains. Cases of ergotism were also documented during the Great Awakening, the religious fervor that reached a peak in New England during the fall of 1741, when hundreds or possibly thousands of typical citizens experienced hallucinations, terrors, trances, and fits. Matossian has argued that the rye crop, which ripened in July, may have caused these dramatic effects. We now know that the symptoms of ergotism appear when the ergot content of rye flour rises to 3 to 5 percent; perhaps the pink tinge from the milled sclerotia in rye flour was ignored, or it may have been masked by the deep color of rye flour.

As with many natural toxins, ergot compounds in minute, controlled doses are medicinal. In *The Complete Herbalist: or the People Their Own Physicians* (1897), Dr. O. Phelps Brown described administering the "degenerated seeds" of rye for inducing uterine contractions during childbirth and for treating bladder paralysis and fevers. He recommended ergot as a "valuable remedy to the obstetrician and midwife," but he noted that its use should not be extended because of potentially dangerous side effects. Ergot is still a parasite in fields of rye, but its toxicity is held in check because the maximum legal limit for ergot contamination in American rye is now a trace 0.03 percent.

Oats

Oats (*Avena sativa*) may also have evolved as a weedy grass that colonized early cultivated wheat, barley, or rye fields. The species seems to be the most recently cultivated and domesticated Eurasian grain, and based on their appearance, oats are easily differentiated from wheat, barley, and rye. Unlike grains with compact flowering heads, oat flowers are arranged on a branched inflorescence known as a panicle. In fourteenth century England, the poor were fed oatmeal gruel, which may have led to contempt for oats as an everyday food. The Pilgrims cultivated oats in Massachusetts fields, but in colonial America this grain was used more commonly for animal feed than for human consumption; a typical seventeenth century feed mixture contained a mixture of rye, field peas (*Pisum arvense*), and oats. Early American grains also included *silpee* or naked oats, probably *A. nuda* (see chapter 1), a close relative of common oats. The widespread rejection of oats was unfortunate because oats have a higher protein content (with almost 15 percent) than other grains, since oats are consumed without removing the embryo (germ) from the seed within the grain. The plants are relatively cold hardy, which may have explained the regional success of oats in Scotland. Nutritious oatcakes and oatmeal porridge were Scottish staples dating from the Middle Ages, and even though Scottish settlers in the United States continued the oat eating tradition, nineteenth century American cookbooks made scant reference to this grain.

Oats were also fed to the infirm in Europe, a tradition that carried over to American sickrooms; imported oats were marketed to invalids and convalescents. In *The Family Nurse* (1837), Child prescribed oatmeal gruel as best because it is "more cooling, and less apt to sour upon the stomach . . . bland, nutritious, and slightly laxative." Oats finally enjoyed popularity as a breakfast food by the last quarter of the nineteenth century. In *Breakfast, Lunch, and Tea* (1875), Marion Harland asserted, "Nothing is more wholesome, and nothing more relished after a little use." A sketch by Charles Dana Gibson (1898) is titled "Oatmeal and the Morning Paper," a domestic scene of a child and his father at the breakfast table, and the advertising cards distributed by Arbuckles' Rolled White Oats promised good health, a clear brain,

Oat grains (*Avena sativa*) grow in a panicle, a branching inflorescence that differs from the compact flowering heads of wheat and rye.

and a cheery disposition to those who ate oatmeal porridge. Rolled oats were prepared by using millstones to crush and remove the outer hulls from oat grains to produce groats, which were then steamed and rolled to make the familiar, fast-cooking porridge. Some may also have grown oats for home use or as an ornamental grass; in 1888, *Burpee's Farm Annual* advertised "Welcome Oats," which promised "the heaviest, handsomest and most productive variety of oats ever introduced."

Barley

Barley (*Hordeum vulgare*) is an ancient grain that was domesticated perhaps ten thousand years ago in Neolithic villages in the Near East; by the time Europeans settled in the north, the use of barley as a major grain was already on the decline. Barley arrived in America with Columbus, and until the fifteenth century, it was milled to make the most common bread-making flours in Europe; it was gradually replaced by wheat flour, which is more suitable for leavened bread because of its higher gluten content. The Pilgrims attempted barley cultivation with poor results, even though it is a crop that tolerates a wide range of environmental conditions, including low temperatures and saline soils.

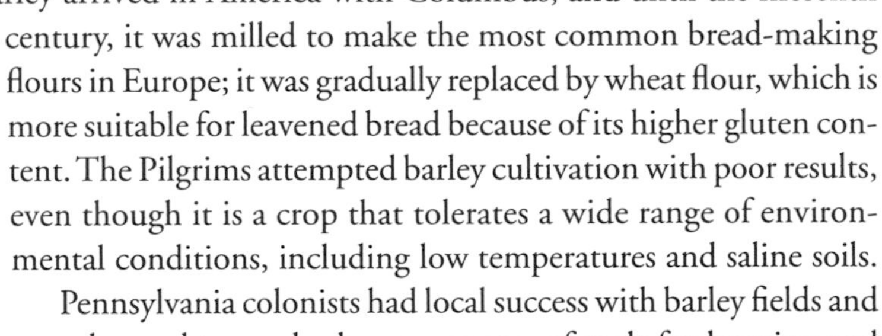

Pennsylvania colonists had local success with barley fields and soon learned to use barley as a source of malt for brewing and distilling, rather than baking. Malt consists of sprouted barley grains that are used as a carbohydrate source for fermenting yeast. Cooks have used pearl barley in soups, which was made by grinding the husks from whole barley grains that were then polished to make the "pearls." Barley was never a dominant American food crop, but barley flour did reappear in the American diet during the "wheatless" days of World War I. As part of the food conservation efforts initiated with the Lever Act, President Herbert Hoover encouraged cooks on the home front to substitute barley flour for wheat flour in preparing family meals.

Barley grains (*Hordeum vulgare*) were sprouted to make malt for brewing or ground into flour for baking.

Rice

Rice (*Oryza sativa*) is the grain that nourishes more people worldwide than any other food crop. Evidence from phytoliths, silica particles from plant cells, reveals that rice was cultivated in the area of the Yangtze River as early as 10,000–7000 B.C. The ancestors of domesticated rice probably evolved in the lowland Asian tropics, in regions with seasonal floods, as rice is most productive when it is cultivated in the flooded fields known as paddies. As an adaptation to a semi-aquatic habitat, the plants have evolved air chambers in their stems, which allow oxygen to diffuse down into the submerged roots. Under moderate temperatures, rice can be cultivated and harvested almost year-round. The grains are borne on panicles, and they require threshing and winnowing to separate the grains from their outer bracts and chaff; the term *paddy* has also been applied to an unhusked rice grain. The endosperm is composed of cells packed with starch grains, and when it is cooked, the endosperm becomes soft and pliable as the cell walls break down and the starch modifies its texture. White rice has a lower protein content than brown rice because the embryo (germ) and bran are removed by milling and polishing. Wild rice (*Zizania aquatica*) is an American aquatic grass with grains that were gathered and used by Native Americans; it is not closely related to the Asian rice that eventually was introduced and cultivated in the New World.

When the British began to colonize America, rice was still primarily an Asian food plant; the British ate relatively little rice, despite years of colonial control of India and considerable knowledge of Indian foodstuffs. Sir William Berkeley sowed a half bushel of seed rice in Virginia in 1647 and reaped fifteen bushels in return, but South Carolina became the site of New World rice culture on a large scale; once local growers realized that seed rice needed to be sown and grown in flooded fields, the coastal swamps of South Carolina proved ideal for its culture. By 1690, South Carolina growers were exporting 390,000 tons of rice annually, and slaves were doing the difficult work of planting and maintaining the flooded fields. The slaves, mostly from Ghana and Sierra Leone, probably knew something about rice culture from their African homes; some of them were also immune to malaria, a parasitic disease caused by the single-celled blood parasite *Plasmodium*, which was a serious health problem in the mosquito-ridden coastal swamps. Africans with immunity to malaria inherited a single gene for the genetically controlled blood disorder known as sickle cell anemia. With just a single gene, their red blood cells functioned normally, but they had the added benefit of malarial immunity. (They did not have sickle cell anemia, which occurs in individuals with two sickle cell genes.) Many African slaves labored in areas infested with the mosquitoes, which are vectors for *Plasmodium*, while the Charleston growers stayed far from the wet, malarial rice fields. As a result of climate, geography, and the availability of cheap slave labor, rice became an essential part of the plantation economy of colonial America and was the leading staple grain crop in South Carolina. The peak of production occurred in 1770, with the export of eighty-four million pounds mostly to England and southern Europe. When the British occupied Charleston during the Revolutionary War, they harvested and exported the entire rice crop to England, without reserving seeds for the next growing season. Thomas Jefferson later helped to revive rice growing in South Carolina, with seed that he smuggled out of Italy in 1787.

Not all American rice was exported. In *American Cookery* (1796), Amelia Simmons offered six recipes for rice pudding, generally with milk, eggs, butter, spices (such as cinnamon or nutmeg), and sometimes currants or raisins. *The Improved Housewife, or Book of Receipts* (Webster 1853) suggested that griddle cakes could be prepared for a frugal breakfast from leftover cold boiled rice, mixed with flour, butter, and eggs. Southern cooks used rice flour or boiled rice to prepare yeast breads, sometimes combining the rice with cornmeal or wheat flour, and rice was also served with colonial curry dishes (see chapter 6). Home recipes for baked rice puddings evolved into more spectacular Victorian desserts, towering molds and pyramids such as "Iced Rice-pudding" described in *Practical Cooking, and Dinner Giving* (Henderson 1878) and "Apples and Rice Ornamented" offered by *The Modern Cook* (Francatelli 1877). How often home cooks attempted such complex rice delicacies, we do not know.

Yeast, Bread, and Leavening

Yeast are microscopic, single-celled fungi that live almost ubiquitously in nature. They hail from various unrelated evolutionary groups of fungi, but most yeast are Ascomycetes, relatives of fungi as diverse as Dutch elm disease, chestnut blight, and the edible truffles and morels. Some yeast can use plant sugars and starches as foods to fuel their own energy needs, and in nature they often grow on the outermost waxy layer of fruit skins. This explains traditional wine-making, which involved crushing grapes and combining their juice with the natural yeast that colonize grape skins; the yeast then decomposed the sugars, producing alcohol as a waste. Brewing of beer and ale involves similar reactions, using grain rather than fruit as a source of carbohydrates to support yeast growth. Despite their simple growth form, yeast are metabolically remarkably versatile; they contain enzymes such as glucoamylase, which allows them to break down complex starches to liberate simple sugars for their metabolism.

Yeast spores float in the atmosphere, and they were the source of the wild yeast first used to provide leavening for breads made from ground grains. Cultivated bread yeast are now *Saccharomyces cerevisiae*, but the first yeast used for leavening were mixed cultures of wild species. In the absence of oxygen, yeast carry on fermentation, whether they are kneaded into bread dough or immersed deep in a wine barrel. Fermentation is the set of metabolic reactions in which the sugars that compose starch are broken down by yeast enzymes to release energy and produce the waste products of ethyl alcohol (ethanol) and carbon dioxide. Without some form of leavening, breads would be flat and dense; in contrast, when flour is mixed with yeast, the mixture bubbles and rises from the release of carbon dioxide as the yeast consume the starches in the flour. The reactions are the same as those involved in fermenting grapes or grain; both alcohol and carbon dioxide are the waste products. In bread making, the bubbles that make the dough rise contain the carbon dioxide from the fermented sugars; alcohol has a lower boiling point than water (approximately 173 degrees Fahrenheit, compared to 212 degrees Fahrenheit for water) and evaporates from the bread dough during baking. In wine-making and brewing, the alcohol is preserved for consumption.

We can imagine how the leavening action of yeast might have been discovered. Air-borne yeast probably landed in a flour-water mixture and had a remarkable, serendipitous effect on the final product; once the bread was baked, it was lighter because of the bubbles that were produced by the yeast and captured by the elasticity of the flour, the result of gluten proteins in the endosperm. Early bread bakers observed the phenomenon of rising dough and learned to save and culture the wild yeast to combine with their flour mixtures.

Colonial housewives who baked their own bread kept their yeast with care, and their methods were part of local lore. During the seventeenth century, a unique method of yeast culture evolved on Long Island, where pumpkins or apples were used to make a "ferment." The yeast were mixed up with flour to make a sourdough loaf known as Long Island bread, a method that was shared orally and eventually published in almanacs and local newspapers. In *The American Frugal Housewife* (1844), Child described several more common methods for yeast culture. She knew that yeast require warmth and food to multiply and suggested various concoctions of malt (sprouted grains), bran, molasses, flour, potatoes, and milk to maintain the household yeast needed for bread making. She recommended in particular a leavening mixture that could be compounded from rye flour, molasses, hops, malt, and "a cupful of good lively yeast." The yeast cells thrived and divided in the high sugar and starch content of the leavening mixture, which could be stored and mixed into bread dough, as needed. She noted that yeast could be bottled, but not with a tight lid. Fermentation occurs only in the absence of oxygen, but the bubbles of carbon dioxide produced as a waste product must have an escape route to avoid kitchen or pantry explosions; probably some of these domestic lessons were learned by trial and error. Once incorporated into bread dough, the yeast continued the fermentation reactions that made yeast dough rise. Child also advocated using rye flour to make the precursor of modern dried yeast. A mixture of hops, cornmeal, and rye combined with water and yeast would rise and thicken, and then the mixture could be rolled out, dried, and sliced into two-inch cakes. One cake was sufficient for leavening "a large loaf of bread."

Yeast was not the sole leavening agent used in American kitchens. By the late eighteenth century, some New York bakers were using baking powders made of potash, or "pearl ash," to prepare their traditional New Year cakes. This chemical leavening agent had a botanical origin, made by sifting wood ashes, pouring water over them, and collecting the water that leached through the ashes. This water, which contained lye, was boiled and reduced to salts, which were then heated until they melted. The resulting pearl ash was white or pearly in color and could be used to make breads and cakes rise because the alkaline (basic) pearl ash reacted with weak acids in the batter to release carbon dioxide; the carbon dioxide generated by an acid-base reaction replaced the carbon dioxide released by fermenting yeast. Saleratus (derived from *sal aeratus*, Latin for "aerated salt") was a related commercial product that by the 1840s was patented and marketed in several areas of the United States. It was essentially potassium carbonate, with various additives, packed in paper envelopes with printed recipes that promised reliable, fancy cakes. Using saleratus, cooks could prepare breads without yeast, and cakes could be made frugally with fewer eggs, since traditional methods depended on six or more eggs for leavening. Saleratus worked particularly well in the intense heat of the iron cookstoves that became popular during the 1860s; foundries sold their stoves with recipes for baking desserts and pastries leavened with saleratus.

Not all domestic authorities were impressed with these new baking methods, and contempt for chemical leavening appeared in popular literature; in *Ruth Hall*, an 1855 autobiographical novel by Fanny Fern (the pen name of Sara Payson Willis; see chapter 10), Ruth is questioned by her mother-in-law about her bread-making skills: "Can you make *bread*? When I say bread I *mean* bread—old-fashioned, yeast riz bread; none of your sal-soda, salaeratus, sal-volatile poisonous mixtures, that must be eaten as quick as baked, lest it should dry up; *yeast* bread—do you know how to make it?" Her character Ruth admitted that as a city-dweller she had always eaten bakery bread. Willis was educated at Beecher's Female Seminary in Hartford, where she may have learned about saleratus; she had a dynamic personality, and her school nickname was "Sal-Volatile," referring to the ammonium carbonate that was also used in some early baking powders. In *The American Woman's Home* (1869), Beecher and Stowe railed against saleratus baking powders as producing "the green, acrid, clammy substance called biscuit" and urged homemakers to reject chemical leavening and "return to the good yeast bread of their sainted grandmothers." They advocated wholesome yeast breads, rolls, twists, and tea rusks over the fancy confections and ethereal cakes leavened with saleratus. As New Englanders, they promoted the use of whole wheat and rye flours and cornmeal, and they were not alone in urging home bakers to use yeast.

Beecher and Stowe viewed yeast baking as domestic art and patriotic duty, but by the 1860s, leaders of the temperance movement opposed the use of yeast because of the perceived dangers of alcohol consumption. Temperance advocates revised traditional recipes to omit the use of rum, brandy, and wine, and leaders of the anti-fermentation movement were religiously inspired to promote the use of baking powders in place of yeast to leaven bread. Even though the alcohol evaporates during baking, yeast-leavened bread was condemned for its potential danger in leading people to alcoholism. Warner's Safety Yeast was advertised as being "used and endorsed by every leading family" and "praised alike by all classes"—supposedly it did not cause fermentation and was probably a chemical baking powder that generated carbon dioxide bubbles through an acid-base reaction. It was manufactured by H. H. Warner, a patent medicine company in Rochester, New York; potassium carbonate, the ingredient in saleratus, also had medicinal uses. Another anti-fermentation product, Horsford's Self-Raising Bread Preparation, claimed that it did not produce alcohol or "decompose" flour and also prevented rickets, cholera, and tooth decay; it was probably a self-rising flour made by combining flour with baking powder in the correct proportions.

Brewing

Brewing ale and whiskey makes use of the alcohol produced in a mixture of yeast and grains. Grains such as barley were easier to grind if they were first sprouted, and flour from sprouted barley was used for baking bread. Brewing likely co-evolved with bread making, long before recorded history, and the first beers and ales were probably made from bits of yeast-containing dough mixed into water; both processes depend on yeast fermentation. Brewed ales were part of the domestic tradition carried to America from Europe, and even though the New World provided abundant supplies of clean water, cultural patterns dictated that ale was the beverage of choice. Unsanitary conditions in English towns and villages probably meant that consuming ale was safer than drinking local water, which was perhaps fouled with animal and human waste and contaminated with pathogenic microbes. The water used in ale making was thoroughly boiled, which would have killed most bacterial cells, and the alcohol content also contributed to its relative sterility. Many of the dietary calories for people of all ages came from ale; nursing mothers, children, and even infants often drank ale rather than water or milk.

The Pilgrims arrived in the New World with a practical knowledge of brewing, in which sprouted barley (malt) was combined with hops (*Humulus sativus* or *H. americanus*) to make a carbohydrate mixture ideal for yeast fermentation. The endosperm of barley supplied the carbohydrates necessary for yeast growth; the sprouted grains already had their starch supply partially degraded into sugars by the action of enzymes released from the aleurone layer of the grains during the process of seed germination. Barley imported from Europe was used until crops were established in America, and hops grew wild in wooded thickets. American hops (*H. americanus*) has sometimes been recognized as a species distinct from European hops (*H. sativus*), but the plants are virtually indistinguishable, and early settlers were probably pleased to discover a New World supply of a familiar, useful plant. European hops had been domesticated during the first century A.D.; ancient Greeks valued hops for their bitter flavor and had added the young shoots to their salads. The part of the hops plant used in brewing is the inflorescence of pale green pistillate flowers, which are densely covered with trichomes (hairs) that contain volatile oils and bitter acids such as lupulone and humulone. These acids have antimicrobial properties, which kept the malt relatively free of bacteria that might cause the fermenting ale mixture to spoil. Reliable results were obtained when a small amount of good ale was used to inoculate the malt and hops mixture, to assure that the correct wild yeast began to grow in the brew. The addition of hops to the malt mixture for ale neutralizes the sweetness of the fermenting sugars, but their use is relatively recent in the history of brewing; until the early sixteenth century, pungent plants such as ground ivy, sage, spruce, wormwood, and ginger were used instead to flavor ales.

Brewing was a common domestic activity, and Child described beer as "a good family drink." In *The American Frugal Housewife* (1844), she provided a standard rule for home brewing, which involved boiling all of the ingredients for two or three hours, adding molasses (a half-pint to a pail of brew), and then adding a pint of "lively yeast." She recommended using plants such as hops, spruce, ginger, horseradish, and sweetfern (*Comptonia peregrina*, a resinous shrub and not a true fern) as beer flavorings. She noted that raw potatoes could be added to the boiling mixture "to make beer spirited"; no doubt the addition of more carbohydrate in the form of potato starch did enhance fermentation. In *American Cookery* (1796), Simmons described spruce beer made from hops, molasses, and the essence of spruce, fermented with a yeast mixture known as "emptins" (from *emptyings*, an archaic English term for yeast). Emptins, for which she specified careful instructions, were a live culture made from a mixture of hops, starch, and water. The essence of spruce was probably

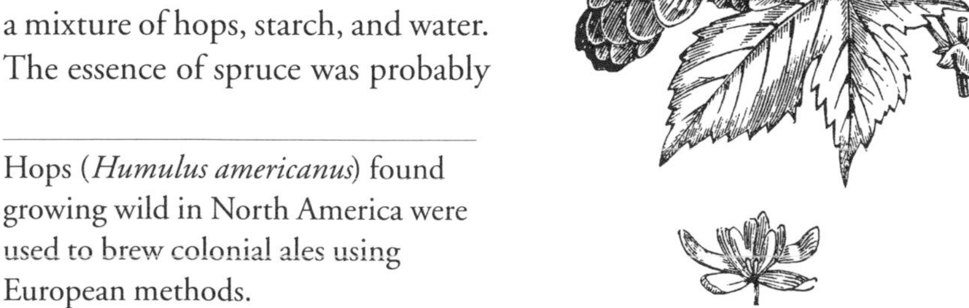

Hops (*Humulus americanus*) found growing wild in North America were used to brew colonial ales using European methods.

prepared by extracting the oil from the needles of black spruce (*Picea mariana*) in alcohol; New England spruce beer methods often also specified using the tips of black spruce branches. In *The New Household Receipt-Book* (1853), Hale described her methods for fermenting white spruce beer (made with sugar) and brown spruce beer (made with treacle or molasses). Other pungent or resinous plants such as hemlock, birch, sassafras, and sarsaparilla were used as beer or mead (metheglin) flavorings in similar traditional recipes. Both sassafras and sarsaparilla (*Aralia nudicaulis*) were among the earliest North American medicinal plants imported to Europe, and such beers were considered health drinks. They were brewed, often sweetened with the additional sugar or molasses, and then bottled in stoneware for future consumption.

Distillation

Beer and ale brewed at home probably had an alcohol content of just a few percent. Yeast eventually die from the alcohol produced in their own fermentation reactions, and early domestic yeast were probably more susceptible to their alcoholic wastes than modern cultivated strains. Child (1844) described "table beer" as "brisk and pleasant"; clearly, these home-brewed beverages bottled in stoneware jugs were relatively low in their alcohol content but high in sugar and pungency. Brewers who sought higher alcohol concentrations distilled the alcohol made by fermenting corn, rye, and other grains; this was the tradition of whiskey making that was carried to the New World by settlers from Ireland and Scotland, where by the fifteenth century brewers had learned how to distill alcohol from beer made from barley. Distillation involved heating a solution containing both alcohol and water until the alcohol (but not

Sarsaparilla (*Aralia nudicaulis*) was used to flavor colonial ales, which foreshadowed its appearance in nineteenth century patent medicines and tonics.

the water) boiled and condensed into pure droplets, which were collected. Various grains were used to make the mash, the fermenting grain mixture, and this helped to define the nature of the final distilled product; variables included aging and the type of woods used in making the storage barrels. Distinct varieties evolved, such as bourbon (named for a Kentucky county), an American whiskey developed by Irish and Scottish settlers; to make bourbon, distillation isolated the alcohol from a mash made mostly of fermented corn, and the alcohol was then aged in new, charred oak barrels. Coopers were kept busy constructing barrels to supply the growing market for the products of distilleries.

Some eighteenth century farmers realized that their corn and rye would be more valuable converted into whiskey than sold as grain, and they became small-scale whiskey producers. A 1791 federal excise tax on whiskey production met with an uprising of enraged farmers, beginning in western Pennsylvania; frontier whiskey distillers protested the efforts of government agents to collect the tax, resulting in the civil unrest known as the Whiskey Rebellion of 1794. Settlers saw the whiskey tax as detrimental to their economic welfare and personal liberty to make whiskey from their own crops, and they tarred and feathered tax collectors. Farmers from western Pennsylvania protested the arrest of rebellion organizers, exchanged gunfire, burned property, and marched on Pittsburgh. President Washington raised an army of thirteen thousand troops to quell the uprising, which stemmed from the combination of abundant grain, settlers' practical knowledge of brewing and distilling, and the tendency of governments to impose taxes. In levying the whiskey tax, Congress saw the dual opportunity to raise revenues and also discourage excessive drinking.

Temperance

In 1790, Dr. Benjamin Rush predicted the evils of intemperance in "A Moral and Physical Thermometer," published as part of *An Inquiry into the Effects of Spiritous Liquors on the Human Body and the Mind.* He charted alcoholic spirits from cider to rum and related them to vice, disease, and punishment. According to Rush, debt, jail, idleness, swearing, fraud, anarchy, murder, gout, jaundice, and melancholy all resulted from drinking liquor. Such respectable women as Lydia Child advocated home-brewed beer as suitable for the American family, but by the second half of the nineteenth century, brewing had become a major business. Commercial breweries were unrelated to the weak table beers that Child and others brewed and consumed with meals. Breweries supplied their products to saloons as a way to expand their sales, and saloons proliferated; many viewed these establishments as an affront to civil society and respectability. Salty "free lunches" enticed customers to consume more drink, and young men were drawn into saloon environs, which sometimes included cock fighting, prostitution, and gambling.

Rush's "thermometer" began the early temperance movement, which gained foothold in the mid-nineteenth century with the medical arguments against chronic drunkenness. Maine was one of the earliest pro-temperance regions, perhaps motivated by excessive alcohol consumption among large populations of single men such as sailors and loggers, both plentiful in Maine. The Society for the Suppression of Intemperance was founded in 1813 in Brunswick, and in 1851 Maine was the first state to enact prohibition legislation. Lacking direct political power, women voiced their anti-liquor sentiments with prayer, hymn singing, and temperance demonstrations and fought for the "Maine law" in other states. The Woman's Crusade of 1873–74 persuaded saloon keepers to shut down their establishments and turn to other lines of work; Midwestern housewives marched and demanded that druggists, saloon keepers, and hotel owners stop selling liquor. The women who stormed into saloons to pray and sing were not afraid to open casks and pour the contents down the gutters.

The temperance movement eventually removed alcohol from nine hundred towns and villages in thirty-one states, led by reform workers who were mostly women and included many suffragettes. Their concerns for the integrity of the American family were well founded; in some working-class areas, there was a saloon for every fifty males over the age of fifteen years. Local political meetings were often held in saloons that excluded women, and married women could neither own property nor force their husbands to provide support for their families. Popular culture reflected the temperance message, with musical ballads such as "Father, Come Home" and "Touch Not the Cup," and literature emphasized the unhappiness in homes marred by drink. The 1854 novel *Ten Nights in a Bar-Room* was eventually dramatized, and it played continually on stages across the United States and often incorporated popular temperance songs into performances. The Prohibition Party and Women's Christian Temperance Union grew from concern that alcohol was ruining American families. Temperance leaders such as Frances Willard, president of the Women's Christian Temperance Union from 1879 to 1898, saw their mission as one of both social reform and preservation of the American home; during Willard's tenure as president, the WCTU slogan was "Home Protection." In 1919, the Eighteenth Amendment to the United States Constitution brought about the nationwide prohibition of liquor, which forced brewing and distilling back into small-scale (albeit illicit) operations that were sometimes home-based.

Pledges of temperance were created for men, women, and children, including a special pledge for homemakers who vowed not to cook with spirits. Some pledges permitted the use of alcohol medicinally or on special holidays, while others such as the Philadelphia Female Total Abstinence Society Pledge demanded total rejection of "intoxicating liquors" in both beverages and "domestic cookery." Beverages suitable for temperance advocates included fruit juices and carbonated sodas, and some products were marketed specifically to those who rejected alcohol. On trade cards distributed in the 1880s, the C. E. Hires Company of Philadelphia advertised its root beer as a "delicious, sparkling and wholesome temperance drink." Charles E. Hires originally developed his product as a herb tea, sold in packets whose contents were mixed with water to make five gallons of the beverage. Although it was made with yeast, to appease temperance zealots Hires claimed that his beer was alcohol-free, but the flavor probably resembled the home-brewed yeast beers with low alcohol concentrations and sassafras or sarsaparilla ingredients. All of these were advocated and used as health drinks; the Hires advertising cards included testimonials about the ability of its root beer to "cleanse and purify the system."

Food Reform

Health issues were certainly part of the temperance message, and many viewed temperance as part of a larger movement that worked to reform the American diet. A typical nineteenth century meal was greasy and caloric, frequently lacking in fresh fruits, vegetables, and whole grains. Atherosclerosis, dyspepsia, and constipation were curses on the national health caused and exacerbated by diet. Advocates of food reform such as Sylvester Graham, a Presbyterian minister in Massachusetts, argued for vegetarian diets and against stimulants of all types. He saw grains and unrefined flours as part of the natural relationship between man and earth, and raw fruits and vegetables required the chewing that he thought was important to health. He counseled his followers to avoid bread that was more than one day old and deplored such foods as mashed potatoes that required no mastication. In 1837, disgruntled Boston butchers and bakers assaulted Graham, their hostile response to his notions of rejecting meat and eating breads baked from whole-grain flours. Flour milled from whole wheat grains became known as Graham flour, which included the bran and the germ along with the endosperm. Commercial products such as the first prepared breakfast cereals were based on whole-grain flour, and Dr. James Caleb Jackson invented the Graham crackers that were baked from the flour favored by Graham and his followers. So-called Grahamites followed their leader's injunctions to live on the "vegetable kingdom" and pure water, limiting dairy foods and avoiding coffee and tea. Graham also argued against condiments and spices, which he considered "highly exciting and exhausting" and conducive to sexual excesses.

Similar dietary methods were practiced at the Seventh-Day Adventist sanitorium at Battle Creek, Michigan, which was opened in 1866 as the Western Health Reform Institute. Dr. John Harvey Kellogg became the medical superintendent in 1876, and he advocated diets of vegetables and whole grains. Some sanitorium practices were potentially hazardous quackery that involved hydropathy, electropathy, mechanotherapy, and radium treatments; Kellogg was particularly concerned with bowel activity, and a chair that vibrated violently was used to help patients overcome intestinal paralysis. His brother, William Keith Kellogg, invented corn flakes as a vegetarian health food and founded the Toasted Oat Corn Flake Company of Battle Creek in 1906. He also developed All Bran cereal specifically as a constipation remedy. Charles Post, a former sanitorium patient, manufactured and marketed a wheat, bran, and molasses beverage that he named Postum Food Coffee and a grain-based "brain food" that he named Grape-Nuts and first sold in 1898. The cereal consisted of wheat and barley baked into sticks, which were then ground into hard particles the size of grape seeds. Yet another technique for grain preparation was developed by Henry Perkey, a Denver lawyer who suffered from digestive problems; he and William Ford invented and patented machinery for rolling and cutting cooked wheat into fine filaments, and in 1893 they first marketed the product they named Shredded Wheat. Six years later Perkey founded a school of domestic science at the site of the former Oread Collegiate Institute in Worcester, Massachusetts; among the first American women's colleges, it operated from 1849 to 1883. Perkey's students were known as the

"Shredded Wheat girls." a group of young women whom he trained in nutrition, cookery, chemistry, and public speaking.

For early colonists and later settlers, grains were the staples of a subsistence diet. By the end of the nineteenth century, the manufacture of prepared cereals converted ancient grains into convenient alternatives to the fatty, cured meats that had become typical breakfast fare. Packaged cereals spawned a dietary revolution in which nineteenth century entrepreneurship and technology launched new, mass-marketed products composed of the cooked, processed endosperm of various grains. The food reform movement of the nineteenth century promoted the same grains that were staples in the human diet since prehistoric times, and at the turn of the century they marketed these grains to an American public newly concerned with health and disease. The grains used in colonial food and ale saw a rebirth in products promoted by health zealots and entrepreneurs as antidotes to nineteenth century dietary excesses. After the Industrial Revolution, Americans had both the luxury to consider the relationship between health and diet and the income to purchase prepared foods. Grains evolved from the staples of domestic baking and brewing to the basic foodstuffs of the nineteenth century diet reform movement and the earliest "health foods."

CHAPTER THREE

Gardens, Seeds, and Vegetable Staples

After 1620, colonists arrived in America with shipboard food supplies for their New World larders that would last until gardens could be grown and the stock replenished. Typical provision lists included barrels of flour and oatmeal, sugar, dried fruits, dried peas, root vegetables, and spices. Shipments also included seeds of familiar European food plants, as well as scions from European fruit trees ready to graft on to New World root stocks. Gardens were cultivated to supply the household table, although sturdy, cold-resistant crops such as turnips and cabbages did double duty as fodder. Plowed fields were reserved for large crops or grains that required broadcast sowing, while kitchen plots were planned as needed, hand-tilled, and planted to supply daily needs.

This was indeed subsistence gardening, and colonists soon learned that the English vegetables most adaptable to the New England climate were carrots, parsnips, turnips, radishes, and cabbages; these crops could all be grown to enormous size and were tough enough to survive months of storage in root cellars. This basic vegetable diet was augmented with peas, beans, and spinach, but necessity demanded that gardeners allocate most space to the most reliable crops that produced storable foodstuffs. Survival depended on foods that could be preserved through the winter months, until the next growing season brought spring and summer crops; the lean weeks of late winter and early spring tried the patience and palates of early rural Americans. By May, foods stores were often largely depleted, and early peas, lettuce, and fruits were eagerly anticipated.

Cookbooks, whether published editions or manuscript notes kept for household use, reveal the plants that were introduced, grown, harvested, preserved, purchased, and imported to supply the table. Methods in use in early American kitchens reflect the diversity and relative abundance of various foods in both kitchen gardens and larders, and the texts often provide hints about the culture and storage of essential crops. Winter cookery relied heavily on root crops that might be lowly, both in their growth habit and as the essential vegetables in a subsistence diet, but the capacity of sizable mature taproots to overwinter in root cellars recommended them for selection and domestication. A carrot or beet stored properly during the fall harvest remained relatively dormant but alive and fresh through the winter months; the living taproot contained stored sugars and continued to respire at low rates. Come spring, it could as easily have been planted back into the ground as eaten. Cabbages kept cool but not frozen also had a similar capacity for long-term freshness during winter months; Mary Randolph, author of *The Virginia House-Wife* (1824), likened cabbage stems to asparagus and recommended white cabbage for winter use and green borecole (kale) as a late garden crop that was made more tender by frost. Summer and fall fruits provided a welcome diversion from a diet heavy in root crops and dried legumes, and the popularity of apples in particular stems from their storage potential as well as their use in cider-making. With care, apples could be kept fresh for several months, while such delicate fruits as strawberries had to be dried or converted to preserves using sugar to protect them from decomposition by ubiquitous fungi. The dooryards adjacent to kitchen gardens were often punctuated with one or more fruit trees, and dozens of apple cultivars were available to American gardeners.

Eighteenth century kitchen garden plants were generally similar down the eastern American coast, although hot weather crops such as melons and eggplants were more popular in the South, where they thrived in the heat. New Englanders favored root crops such as turnips, in which the flavor improved with frost or time spent in a cold root cellar, as stored root starches were converted to sugars. Curious gardeners could also experiment with fruits and vegetables that promised to perform particularly well in their regional climates; many species were the same food plants grown by the Pilgrims, now selected and sorted into genetically unique varieties that frequently surpassed the diversity of modern cultivars. Then, as now, many gardeners sought variety and improvement in the food plants that they cultivated, while others grew a conservative selection of fruits and vegetables in the same rows or beds for horticultural generations. In *The Natural History of Virginia, or the Newly Discovered Eden* (1737), William Byrd described the fruits and vegetables grown in Virginia gardens, the "very large and long asparagus of splendid flavor . . . beautiful cauliflower, chives, artichokes," along with varieties of potatoes, cabbages, peas, beans, greens, pumpkins, squashes, and twenty-four types of apples. We can imagine that many European fruits and vegetables seemed rejuvenated by fertile Virginia soils, and the bounty of native persimmons and pawpaws enhanced the sense of Eden. Historic records also reveal some knowledge of crop rotation, with progressive gardeners studying the benefits of periodically relocating their crops and replacing them with others.

Slave women did the cooking for southern gentry, and their domestic influence eventually led to the cultivation of African plants in Virginia kitchen gardens and elsewhere. Eggplant, okra, and sesame (often known as *benne*, its African name), as well as field peas, yams, sorghum, and watermelons all had African roots. In addition to working in the large plantation gardens, slaves at Monticello, Mount Vernon, and other plantations grew their own small plots of familiar southern vegetables, documented in the accounts of eighteenth century Virginia travelers. The gardens maintained by African American slaves also provided small incomes or food plants for bartering.

In contrast, New England women did their own domestic work, had help from their daughters or girls from neighboring families, or kept servants from England or later Ireland; in this way, northern kitchen gardens remained closer to their Pilgrim origins. The traditional New England diet relied on vegetables that were boiled or baked; plain cooking was part of a pious life, and high seasonings were generally scorned as dangerous, unhealthy, or even sensuous. Southern cookery reflected greater cultural diversity and incorporated a greater botanical diversity of both food plants and culinary herbs and spices.

Among colonial gardeners, no one wrote more about experimentation, successes, and failures than Thomas Jefferson, who adopted many of the horticultural practices at Monticello from Bernard McMahon, the Philadelphia plantsman who authored *The American Gardener's Calendar* (1806). McMahon organized his gardening advice by month and paid particular attention to the differences among various American soils and climate. He included a catalog of more than thirty-seven hundred species and cultivars of "the most valuable and curious plants hitherto discovered." In addition to ornamental plantings and "fanciful rural designs," he encouraged his readers to broaden their horticultural horizons with unfamiliar vegetables such as sea kale (*Crambe maritima*), a perennial member of the mustard family (Cruciferae or Brassicaceae) with fleshy leaf stalks and very reduced leaf blades. Following McMahon's advice, Jefferson introduced sea kale to the kitchen garden at Monticello; its culture required blanching under clay pots to prevent normal chlorophyll synthesis in the presence of sunlight, and then the pale stalks were cooked like asparagus. McMahon also offered practical instructions for manuring the garden, interplanting radishes with lettuce, and cultivating tomatoes, South American fruits that were still unfamiliar to most Americans at the beginning of the nineteenth century.

Jefferson was atypical in cultivating two hundred fifty varieties of more than seventy edible species, and his horticultural record-keeping reflects his fascination with edible biodiversity. The Monticello trials, experiments, successes, and failures were meticulously recorded in *Thomas Jefferson's Garden Book* (Betts 1944), a journal that he started in 1766. His scientific interest in botany and horticulture was apparent, but the practical side of growing food plants was also part of the gardening enterprise at Monticello; Jefferson followed a nearly vegetarian regimen and considered meat a condiment rather than a dietary staple. His garden book, letters, and papers suggest that at one time or another he grew almost every available kitchen garden plant of his time, including tomatoes, which at the time were horticulturally rare. These notes reveal fifty pea varieties, forty bean varieties, thirty varieties of cabbage, and fifteen different lettuces, including 'Dutch brown' and 'tennis ball' varieties. McMahon supplied Jefferson with cultivars including red celery, dwarf peas, Egyptian onions, and red globe artichokes, and Jefferson also obtained seed stock from American and European friends and gardeners, from seedsmen, and by saving seeds from the gardens at Monticello.

While the kitchen gardens at Monticello were on a comparatively grand scale, Jefferson's records nevertheless reveal the diversity of food plants and the expert information available to progressive growers who were keen on experimentation and improving the yield and quality of their garden produce. His gardens were also large enough to have distinct microclimates; for instance, he cultivated fig trees (*Ficus carica*) from France, which thrived in the radiated warmth of beds cultivated along the stone walls. The minute fig flowers are contained in an enclosed structure (syconium)

Sea kale (*Crambe maritima*) was a perennial mustard grown for its edible leaf stalks.

that eventually becomes the edible fig; it is interpreted botanically as an accessory fruit, a structure adapted for seed dispersal that does not originate from the ovary of a flower. Female wasps enter through a small apical pore, and they pollinate the small flowers as they lay their eggs within the syconium. The true fruits (achenes) are the small "seeds" embedded in the fig flesh, each the mature ovary of a single pistillate flower. Jefferson's figs were possibly a cultivated variety in which each syconium differentiated into an edible fig without wasp pollination, and he shared this variety with others. Other southern gardeners also cultivated figs in the warm areas of their gardens, where the shrubs sometimes died back during cold periods but soon regenerated from surviving roots.

In many kitchen gardens, small plots were allocated to delicate salad plants such as lettuces, asparagus, parsley, and mustard greens, all plants with temporal value at best. Some tender vegetables such as the European purslane (*Portulaca oleracea*) escaped garden boundaries and joined the weedy naturalized flora of North America. Purslane was prized locally as an easily cultivated leafy substitute in cinnamon-flavored spinach tarts, a traditional Elizabethan recipe that was adapted in early kitchens to New World cookery; English colonists consumed it regularly both cooked and raw, sometimes arranged in sallets with cucumbers and edible flowers. Spring asparagus (*Asparagus officinalis*) also provided welcome relief from the routine winter diet, and the species is now a naturalized roadside weed spread by birds attracted to its edible red berries. The young shoots of asparagus (also known as sparagras or sperage) were also served as a vegetable or sallet, quickly boiled and seasoned with oil, vinegar, salt, and pepper. Asparagus was prized for its tenderness and supposed aphrodisiac qualities, and like purslane, its appearance in the New World began with traditional English cookery.

Eventually, Native Americans adopted both purslane and asparagus as potherbs and for soothing poultices, but their naturalized status in North American habitats originated with Elizabethan cookery. Of course, not all edible plants required cultivation, and some settlers preferred to forage for wild plants during the spring rather than devoting kitchen garden space to leafy greens. Green vegetables gathered and cooked by the potful could vary with the season and source, but all were essential in preventing scurvy and other cases of vitamin deficiency; leafy shoots provide a necessary source of vitamin C (ascorbic acid) as well as carotenes, the plant pigments that serve as the dietary precursor to vitamin A. Nevertheless, it was generally considered more genteel to rely on kitchen gardens and cultivated crop plants to supply the family table. Although sometimes viewed as suggestive of extreme poverty, settlers in the South did rely on cattails (the fleshy rhizomes of *Typha* spp.) and wild onions such as ramps (*Allium tricoccum*) as emergency foods to forage when supplies ran low.

Of necessity, kitchen gardens were cultivated close to houses, where fruits, vegetables, and herbs could be easily gathered for home use. Design was practical, with the southern exposure reserved for the earliest spring vegetable crops and fruits. Summer crops such as corn, beans, squash, and pumpkins, and by the nineteenth century tomatoes and peppers, were planted in the cooler, moister sites that warmed by midsummer. Raised beds encouraged good drainage. To accomplish this, soil was mounded, sometimes with cut saplings tapped around the edges to hold the bed in place. With the advent of water-powered sawmills, cut boards were also used to contain beds. In design, New England gardeners favored rows that ran between the side paths, perhaps with a border composed of small fruit shrubs, while southern gardens retained traditional squares and borders devoted to one type or general category of fruits or vegetables. A few fruit trees often shaded the dooryard and also were planted along the walls that defined the kitchen garden area; grape vines grew on simple trellises and covered nearby outbuildings. Records suggest that decorative elements such as elaborate arbors and knot gardens were rare in early kitchen gardens, which were practical plots cultivated to supply a hungry family. Nevertheless, enthusiastic gardeners often planted their gardens to contrast particular growth forms and edify and delight their visitors. Jefferson recorded in his garden journal his practice of edging tomato beds with

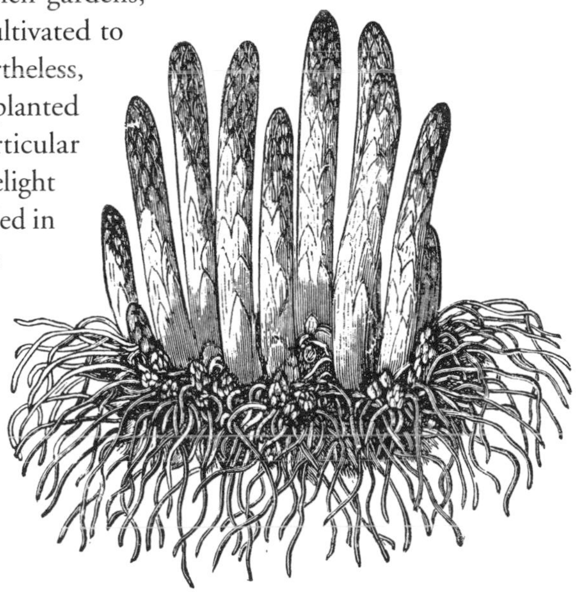

European asparagus (*Asparagus officinalis*) was a perennial spring crop cultivated for its tender immature shoots, which were harvested before they elongated.

okra or sesame and planting purple pigmented broccoli along rows of white and green varieties. Housewives tending their kitchen gardens likely planned similar horticultural arrangements to please themselves and others.

Plantsmen have practical knowledge of the structural and functional differences between fruits and vegetables, a botanical conundrum that has long perplexed gardeners and cooks alike. The common culinary wisdom is that vegetables are served unsweetened and often accompany meat dishes, while fruits are naturally sweet and often prepared in desserts and sweet side dishes. Botanists draw a different distinction and recognize a fruit as the mature ovary of a flower, regardless of its form or uses in cookery. While some fruits are fleshy and edible, many mature fruits are dry and thoroughly inedible, such as various dry capsules and nuts; the common thread is that fruits contain seeds. Gardeners who saved seeds have long understood the functional relationship among flowers, fruits, and seeds in a plant life cycle; even root crops such as turnips and radishes produce flowers followed by seed-filled fruits. Botanically speaking, the word *vegetable* is a broad term that can describe any plant part: roots, stems, leaves, flowers, or fruits. Early botanists referred to plants as "vegetables"; for instance, the early botany text *Vegetable Staticks* (1727) by Stephen Hales described physiological phenomena such as the flow of water and sap. All fruits are vegetables, but not all vegetables are fruits; edible roots, stems, and leaves fit the broad botanical definition of *vegetable* because they are plant parts.

Because of their taste and use in main dish cookery, fruits such as tomatoes, pea pods, and bell peppers are often described as vegetables, although they are mature, seed-containing fruits. The fruit and vegetable conundrum reached the United States Supreme Court in 1893, the result of litigation stemming from the Tariff Act of 1883 that levied a 10 percent tax on imported vegetables. A tomato importer challenged the tax on the grounds that tomatoes are fruits, but Justice Gray wrote

> Botanically speaking, tomatoes are the fruits of a vine, just as are cucumbers, squashes, beans, and peas. But in the common language of the people . . . all these are vegetables, which are grown in kitchen gardens, and which, whether eaten cooked or raw, are, like potatoes, carrots, parsnips, turnips, beets, cauliflower, cabbage, celery, and lettuce, usually served at dinner in, with or after the soup, fish or meats which constitute the principal part of the repast, and not, like fruits generally, for dessert.

The Court thus ruled that tomatoes are vegetables in the vernacular sense, clearing the way for the tax on imported tomatoes.

Seed Saving

Kitchen gardens were often perpetuated by the tradition of seed saving, and success of the next season's crops depended on the careful selection, drying, and storage of seed stock. Seeds were saved from particularly handsome fruits and productive plants, and gardeners knew that some biennial vegetable crops such as carrots and radishes often had to grow for a second season to complete their sexual life cycles and produce flowers, fruits, and seeds. Seed saving was possible because these were open-pollinated varieties, meaning that plants cross-pollinated with other genetically similar individuals to produce offspring that typically "bred true" for their various traits. Hybrid fruit and vegetable cultivars are the result of genetic crosses between genetically dissimilar parents, but even if they were cultivated, saving their seed for future gardens would not have been possible. Although hybrids may have the advantages of heterosis (hybrid vigor), including traits such as early maturation and large size, their seeds are frequently sterile or do not yield offspring that are reliably similar to the parent plants. Saved seeds were passed around towns and in families, and some landowners provided seeds to settlers moving out to the American wilderness. An entrepreneurial widow could set herself up in a small business, saving and selling seeds descended from early seed stock, but real excitement arose with the arrival of new seed lots from Europe. Seedsmen were few and often distant from rural towns, so ships arriving in port might often sell newly imported seeds directly from the dock, after advertising in local newspapers. The embryonic plants contained in these Old World seeds would grow and mature in kitchen gardens, providing nourishment for New World colonists and new genetic stock for the garden cycle of planting, cultivating, selecting, saving, and sharing.

Knowledge of plant life cycles and cross-fertilization was vague, but by the early eighteenth century horticultural warnings to gardeners and farmers urged them not to plant similar crops too close together lest the "farina" of similar plants combine. Even if the microscopic details of botanical sex were not yet completely understood, preserving the integrity of seed lines was important in planning and planting gardens. Hybridization of genetically similar cultivars is frequent because insects visit and cross-pollinate the similar flowers of closely related varieties, but the hybrids appear only if the resulting seeds are saved and planted. Neither gardeners nor seedsmen wanted their seed lines to deteriorate as a result of planting closely related cultivars in proximity, which often resulted in unappealing hybrid fruits and vegetables in future generations. J. H. Walden (1858) warned against planting cabbages near "anything else of the cabbage or turnip kind; they will mix in the blossoms, and the worse will prevail." He noted that, in the case of cabbage and related coleworts, "nothing hybridizes worse" and that only a single variety should be raised in a garden. Similar care also had to be taken to avoid cross-pollination between squashes and pumpkins and watermelon and related citron melons. Walden maintained that when plants hybridize, the "poorer" types usually prevail, a reversion back to their wild traits, but occasionally the results of cross-pollination were salutary—for instance, a nutmeg muskmelon planted near "the common roughskinned variety" produced an intermediate type "of great excellence." The difficulty in saving seeds produced by such accidental cross-pollinations illustrates the enormous diversity produced by horticultural selection from a single species. Crops as different as cabbage and broccoli (both *Brassica oleracea*) retain the ability to cross-pollinate despite their very different physical forms.

Growers saved seeds of their best crop plants and were advised to do this early in the season, rather than relying on the latest crops as a source of seeds for the next season. Seeds had to be dried in a warm, shaded spot and stored with care, in a dry location away from hungry mice. Growers were also well advised to test seeds for reliable germination; moist cotton wool or moss provided a suitable habitat in which to "try" saved seeds before they were sown. Another early method involved sprinkling a few seeds on a hot stove surface. Those that retained some moisture (and presumably were viable) would crack and jump as the internal moisture boiled away; nonviable seeds lacked moisture and simply burned. By the nineteenth century, commercial seedsmen were common and seed-saving was no longer a necessity; some of the most reliable garden seeds were sold by Shaker religious communities in a profitable enterprise. The Shaker seed business began in 1790 with the sale of parsnip, turnip, and onion seeds, but it soon expanded to include more of the vegetables and flowers cultivated on Shaker farms. During the nineteenth century, Shaker seed stock was shipped to stores and displayed in colorful boxes that bore labels such as "Shakers' Genuine Vegetable & Flower Seeds from Mount Lebanon, Columbia County, N. Y." Their seeds were packaged in small bags that were printed with instructions for planting and sometimes cookery.

Beans and Peas

On 4 November 1492, Christopher Columbus recorded in his diary "fabas very different from ours," referring to the varieties of common beans (*Phaseolus vulgaris*) that Native Americans cultivated in their corn fields (see chapter 1). These were New World members of the pea family (Leguminosae or Fabaceae), essential food plants already well known to Europeans who had relied on pulses such as peas and broad beans since ancient times. Their characteristic fruits are legumes, elongated pods that are each derived from a simple pistil with a single row of ovules. Following pollination and fertilization, the ovules mature into seeds, while the pistil elongates and develops into the legume. The embryo inside each seed grows and absorbs the endosperm, reprocessing it into food that is stored in the two cotyledons (the first pair of leaves) of the embryo. The two large, protein-rich cotyledons of legume embryos can be dried and stored indefinitely, and cultivated legumes were originally prized for their nutritious seeds rather than edible pods.

Wild bean and pea ancestors were twining, warm weather annuals, with small seeds that dispersed easily from pods that eventually dried, twisted, and cracked along their length. Wild beans have been unearthed in New World archeological sites, but cultivation brought about some remarkable changes in both legume fruits and seeds. As a result of selection, seed size increased, dormancy vanished, and seed dispersal out of the pods was curbed, all traits desirable to early agriculturists. Archeological excavations along the Peruvian coast reveal the cultivation of common beans and the related species *Phaseolus lunatus* (butter bean or lima bean, named for the rounded or lunar shape of its seeds) since pre-Columbian times, resulting in American bean cultivars that differed in seed size, shape, color, seed coat markings, as well as in the characteristics of their pods. By the time of the arrival of the first European explorers, beans were cultivated in Native American fields from Chile and Argentina north to the valleys of the Saint Lawrence and upper Missouri Rivers. An important factor in legume culture is the plants' ability to tolerate a wide range of soil conditions; many species have symbiotic nitrogen-fixing *Rhizobium* bacteria that colonize nodules on legume roots, allowing the plants to flourish in infertile soils and marginal sites.

Linnaeus as well as earlier sixteenth century herbalists were confused about the New World origins of *Phaseolus*, which by the eighteenth century was commonly cultivated and eaten in Europe. In *Species Plantarum* (1753), Linnaeus noted the common bean (*P. vulgaris*) as native to India and the lima bean (*P. lunatus*) as "Habitat in Benghala." Scarlet runner beans (*P. coccineus*) from the uplands of Central America had also arrived in Europe during the sixteenth century; they were first prized as garden flowers and only later used for their edible legumes. Nevertheless, despite the early introduction of American beans to European gardens and markets, colonists included familiar European peas and broad beans on their provision and seed lists. By the nineteenth century, French and English cultivars of *Phaseolus* were introduced back to New World gardens, and common beans surpassed European broad beans (*Vicia faba*) in their popularity. Broad beans cause favism, a genetic disorder in humans that was not well understood. Susceptible individuals lack a critical metabolic enzyme (glucose-6-phosphate dehyrogenase), a condition that in turn is aggravated by exposure to the alkaloids in both broad bean seeds and their pollen. The result is hemolytic anemia (a breakdown of red blood cells), and it is most common among people of Mediterranean descent. Symptoms include fatigue, breathlessness, pain, and dark urine (from blood cell breakdown), and it can be fatal in children. Favism was probably not a significant problem among early American colonists, but no doubt many European populations gladly substituted American beans for potentially toxic Old World broad beans.

Colonists made no distinction as to their origin and grew the most productive and desirable legumes in their gardens; during the eighteenth century, editions of the *New Hampshire Gazette* listed thirty-two bean varieties, including vining and bush growth forms, scarlet runner beans, broad beans, and perhaps small-seeded lima beans. Cooks favored beans for a variety of uses, with few recipes specifying particular bean varieties. In *American Cookery* (1796), Amelia Simmons recommended beans to accompany roasted lamb or mutton, but beans also constituted simple main dishes. New England cooks prepared beans in soups, stewed, or baked, various methods that began with the common method of soaking the dried beans overnight in a kettle hung aside the fire. Baked bean recipes frequently stipulated "white beans" and called for pork and molasses to thicken and sweeten the liquid used in baking. City dwellers without brick ovens depended on commercial bakers who baked beans as well as bread. Most important, dried bean cookery took time, and in addition to softening beans overnight in lukewarm water, *The Improved*

Housewife (Webster 1853) recommended overnight baking as well. A quicker dish was succotash (or sackatash), which specified that shelled beans be boiled with corn kernels sliced from the cob and the mixture then thickened with flour and butter. *Mrs. Rorer's Philadelphia Cookbook* (Rorer 1886) advised soaking, boiling, and sieving black beans for soup, and then embellishing the tureen with slices of lemon and hard-boiled egg. Large Windsor beans, a type of broad bean, were used for puree, and young shelled Windsor beans or lima beans were boiled until tender and served with butter.

Early bean cultivars were selected for seed qualities, and some of their pods evidently had very tough walls; in *Miss Leslie's Complete Cookery* (1851), Eliza Leslie encouraged her readers to boil the pods of green beans and scarlet runners for an hour or more, to insure complete tenderness. Southern cooks prepared young green bean pods, with their strings removed, as "French beans." By the end of the nineteenth century, seedsmen advertised "stringless" beans with less fibrous pods, such as the common bean cultivar known as the 'Golden Wax Flageolet' pole bean, introduced by Burpee in 1886 and described as "entirely free from strings and of superb quality." Diversity was vast; kitchen gardeners could plant beans ranging from white zulu pole beans to red- or black-seeded wax beans or red speckled Valentine beans. Beans particularly interested Henry David Thoreau, who studied botany at Harvard; he sowed and tended more than two acres of legumes at Walden Pond, with the explanation "I was determined to know beans."

Garden peas (*Pisum sativum*) also have an ancient history, a crop that probably originated in Near East Neolithic farming villages and migrated with barley and wheat as agriculture spread across Europe. Wild pea ancestors are unknown; field peas (*P. arvense*) appear to be garden peas that escaped from cultivation, naturalized, and are now grown for fodder. Peas thrived in the English climate and became a staple in the diet of rural peasants, and during the long weeks of Lent, peas substituted for meat. Like beans, peas were dried and stored or cooked fresh, although until Tudor times dried peas were the rule; these could be kept indefinitely and used to make porridges and puddings, often with their seed coats removed and cotyledons separated (split peas). Colonial varieties included large green and yellow marrowfat peas, sugar peas with edible pods, hotspur peas, and 'Early Carlton,' described by Simmons (1796) as "produced first in the season—good." The traditional method of cultivation involved growing the vines on "pea brush," dead branches that served as trellises for the vines; the branches from orchard pruning were often saved and used for supporting pea plants. Nineteenth century gardeners could select among cultivars with variations that included early flowering, high sugar content, dwarf and indeterminate growth, edible pods, and green, white, blue, and gray seeds. These are many of the same traits that Gregor Mendel, the Czech monk and botanist, used in crossing garden peas, experiments that illuminated the genetic principles of dominance, segregation of traits, and independent assortment; his painstaking work with peas cultivated in a nineteenth century monastery garden provided the essential key to an understanding of heredity in both plants and animals.

Colonists carried traditional pea cookery methods to the New World and continued to prepare dishes with Elizabethan and Jacobean roots; peas were suited for both peasant and gentry tables, and many seventeenth century families ate them three times daily. In her "Booke of Cookery," Martha Washington recorded the traditional dish "Pease Porrage of Greene Pease" and a recipe for green pea soup, flavored with garden mint. Dried peas were a staple winter food, but fresh peas were a harbinger of spring and were Thomas Jefferson's favorite vegetable. At Monticello he experimented with fifteen cultivars of English peas and harvested crops from mid-May until mid-July; he and his neighbors competed annually to see who could bring the first peas to table. In *The Virginia House-Wife*, Randolph (a cousin of Jefferson) recommended

Cultivated garden peas were a traditional European legume easily cultivated in American kitchen gardens; the seeds were eaten fresh or dried, and some nineteenth century varieties also had edible pods.

gathering young peas early in the morning, boiling them no more than twenty or thirty minutes, and dressing them with butter and mint. She added pounded celery seed to soup prepared with dried split peas, which was improved with the addition of fresh peas if they were available. Her recipe for pease pudding called for boiling young tender peas in a cloth, mashing and seasoning them, and then forming them into a pudding to be cooked and served with pork; similar recipes for puddings prepared with dried split peas and boiled in a cloth were still in use during the Civil War.

In *The Good Housekeeper* (1841), Sarah Josepha Hale described her version of "Old Pease Soup" prepared with the addition of spinach, as well as a spring dish of young peas stewed with shredded lettuce. New Englanders cooked dried peas with their salted (corned) beef to prepare a variation of the traditional boiled dinner seasoned with herbs such as thyme and marjoram. Thrifty cooks added a bit of pearl ash, an early baking powder (see chapter 2), to peas that had become tough and yellow; its caustic properties softened the seed coats and rendered them more tender. Perhaps the least expected use for peas was one of the earliest: in *American Cookery* (1796), Simmons suggested that peas could fill fruit pies in place of apples, raspberries, blackberries, or plums.

Another legume that was grown in southern gardens was the cowpea (*Vigna unguiculata*), an annual crop that was domesticated in West Africa and carried to North America by slaves. One common variety is the black-eyed pea, which has white seeds with a small pigmented spot, but black, brown, and red-seeded varieties were also known. Jefferson planted cowpeas, and Randolph (1824) called them "field peas" and was perhaps the first to describe their cookery. She recommended that cooks boil cowpeas until tender and then mash and fry them into a cake with a golden crust. Some cowpea cultivars were the remarkable "yard long beans," referring to the extraordinary length of their pods.

Arriving in the southern colonies with the slave trade, peanuts (*Arachis hypogaea*) were a South American legume that had been carried to Africa by Spanish and Portuguese explorers and slavers. Peanuts are not true nuts, which are defined botanically as the one-seeded, indehiscent fruits associated with wind-pollinated trees such as oaks and walnuts. Peanuts were known commonly as groundnuts, because following pollination, the flower stalks elongate and push the developing legumes into the soil. Once buried, the peanuts mature into a reliable crop that supplied the table and fattened hogs. Virginia and Spanish types were known; although Virginia peanuts were larger, many gardeners preferred the Spanish cultivars for their erect rather than sprawling growth and higher oil content.

Now an American staple, peanut butter is a relatively new innovation that is often attributed to Dr. John Harvey Kellogg, nineteenth century food reformer and medical superintendent of the Western Health Reform Institute, the well-known sanitorium located at Battle Creek, Michigan. George Washington Carver, the son of a slave, recognized the value of peanuts and their symbiotic nitrogen-fixing bacteria as a strategy for improving southern soils depleted by years of cotton and tobacco farming; during his long career as an agricultural researcher at Alabama's Tuskegee Normal and Industrial Institute for Negroes, he developed three hundred twenty-six domestic and agricultural uses for peanuts.

By the second half of the eighteenth century, exotic foods began to appear in New England shops and markets, including tamarinds (*Tamarindus indica*), tropical legumes valued for their flavorful fruits. Tamarinds originated in Africa, but those imported to New England were probably cultivated in the West Indies. Their fruits have a tart, astringent mesocarp layer that flavored traditional East Indian curries and chutneys; their sweet-sour taste arises from high concentration of sugars and tartaric acid. Early records indicate that by 1773 tamarinds were sold in areas as remote as Deerfield, Massachusetts, revealing the diversity of foods shipped from Boston and available to prosperous farmers in rural areas of New England who could afford imported goods. Tamarinds were also valued for their medicinal properties as "laxative refrigerants" and were grouped medicinally with other legumes with laxative and emetic properties such as various sennas (*Cassia* spp.); many of the tamarinds sold in America were probably used in the sickroom rather than in the preparation of family meals. Leslie (1851) advised tamarind water as

The flowering stalks of peanuts (*Arachis hypogaea*) elongate and push the developing legumes into the soil.

a suitable drink for invalids and described its preparation by infusing tamarinds in cold water for fifteen minutes or more. Sturtevant (1919) noted their "sweet, acidulous taste" and that "convalescents often find the pulp a pleasant addition to their diet."

Taproots and Tubers

The staple vegetables introduced by American colonists were primarily biennial crops favored for their massive storage taproots: carrots, parsnips, radishes, turnips, and beets. Several of these were discussed in detail in chapter 1 as plants cultivated in Puritan gardens, and some early cultivars are now completely unfamiliar, such as the enormous, coarse beets (*Beta vulgaris*) known as mangels or mangel-wurzels. These were "scarcity roots," a crop that could supply food in times of severe shortage. Seedsmen boasted that long, red mangels could grow to thirty pounds, and these were used primarily to feed livestock except in times of severe domestic need. American farmers also grew yellow mangel cultivars as fodder; *Burpee's Farm Annual* (1888) noted that English farmers received a higher price for milk from cows fed 'Golden Tankard' mangels, which bore little resemblance to the small, red table beets prized for their tenderness and fine texture. Jefferson grew both table beets and mangels, but his plant lists made no mention of chard, a cultivar of the same species that was valued for its white, waxy leaf stalks that were cooked or pickled. Chard (sometimes called silver beet or Swiss chard, although it is not from Switzerland) appeared in American seed catalogs by the end of the nineteenth century and may be closest to the wild ancestor of cultivated beets.

Carrots (*Daucus carota*) in purple, red, yellow, white, and orange varieties were also commonly grown for both human and animal consumption. The enlarged outer

The massive taproots of beets (*Beta vulgaris*) have concentric cylinders of cambium tissue that divides to increase the diameter of the storage roots.

layer (cortex) of each taproot stores sugars that supply the biennial plants with energy for flowering during their second growth season, and aside from beets (also biennials), carrots contain more sugar than any other vegetable. Some early vintners concocted batches of carrot wine in which yeast cells fermented the abundant taproot sugars. Carrots have been used primarily as a food, however, and while the first cultivars may have had deep purple anthocyanin pigments in their robust roots, the varieties preferred by American cooks were the yellow and orange varieties that did not tint soups and stews purple. Simmons (1796) favored "middling siz'd" yellow carrots, no more than a foot long and two inches in diameter, and she suggested their use in soups and hash and to accompany veal. She cautioned that carrots need good plowing and rich soils, implying that she had encountered the forked taproots harvested from gardens with stony plots. Carrots were grown to enormous sizes, and cooking time had to be adjusted depending on the size and age of the roots; cookbooks commonly advised that a mature root might require three hours in the pot, probably needed to soften the woody cell walls that developed by the end of the growing season.

Orange carrots, rich in carotene pigments, had the added benefit of providing dietary vitamin A; each carotene molecule is composed of two vitamin A molecules joined at their ends, essentially identical to the compound retinal present in the rod cells in the human retina. Retinal is converted to retinol, the vision pigment essential to the rods that function in low light intensities. This is the basis of the folk wisdom that carrots will improve night vision, which is true to an extent. Nutrition aside, because of their appropriate color and texture, carrots were sometimes substituted in squash pies or sweet potato pudding, and they were a standard accompaniment to dinners of boiled beef or mutton. Leslie (1851) also suggested cutting and shaping carrots and beets into fanciful flowers to adorn roasted meat.

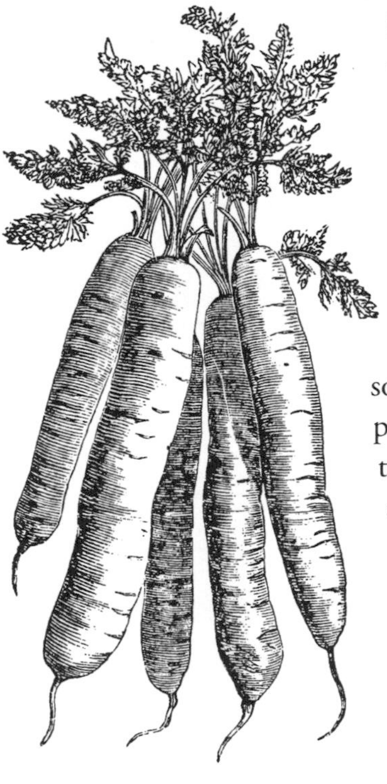

Carrots are cultivated varieties of wild Queen Anne's lace (*Daucus carota*), biennials with massive storage taproots selected for their pigmentation and edibility.

Parsnips (*Pastinaca sativa*) were grown and used like carrots; they closely resemble some yellow and white carrot cultivars, and both are members of the parsley family (Umbelliferae or Apiaceae). Their flavor and sweetness was improved with exposure to frost, and parsnips were harvested late into the season and then stored in root cellars for long-keeping, which was probably the basis of some of their popularity. In 1834, seedsmen introduced highly flavored turnip-shaped parsnips to the United States market; one such short-rooted cultivar was the round French parsnip, advertised as good to bundle along with potherbs to flavor soups and stews. Parsnips could be sent to the dinner table mashed and buttered, and *Miss Leslie's New Cookery Book* (Leslie 1857) described a breakfast dish of parsnips parboiled and fried in butter or meat drippings. Like carrots, as parsnips matured, they often became hardened as their vascular cambium produced layers of water conducting cells (xylem, or wood). Wise cooks split the taproots lengthwise during preparation, exposing the inner tissues so that the vegetables would cook and soften evenly. *Mrs. Putnam's Receipt Book* (Putnam 1860) recommended that parsnips be sliced lengthwise, rolled "in egg and crums," and fried in butter for "a very nice side dish of vegetables."

Native Americans also adopted parsnips as a reliable food, and parsnips were among the crops intentionally destroyed when General John Sullivan launched a 1779 raid on the Iroquois of western New York. Both escaped and wild types of parsnips and carrots are now familiar weeds that have been incorporated into traditional folk and Native American medicine. The furocoumarin molecules known as psoralens are found commonly in the parsley family, particularly parsnips, and cause photodermatitis (see chapter 1); gardeners and farmers must have learned to avoid the rash caused by psoralens by not handling the plants in bright sunlight.

Nineteenth century gardeners planted a variety of radish (*Raphanus sativus*) cultivars, including early, summer, and winter types; scarlet, violet, black, white, or yellow varieties; and mammoth forms as well those with finger, turnip, and olive-shaped taproots. Burpee introduced 'White Strasborg' radishes in 1884, with the claim that they would retain their crispness even when mature and overgrown. Nearly a century earlier, Simmons (1796) favored turnip-rooted radishes for winter keeping and noted that radishes "grow thriftiest sown among onions." When fresh radishes were prepared for serving, their stems were trimmed and the roots were pared, and they were served in a small dish accompanied by salt; freshly pulled radishes were quickly soaked in cold water to keep them fresh, since old radishes were considered unwholesome. Cooks also pickled radish fruits in vinegar flavored with combinations of mace, turmeric, allspice, ginger, and pepper, and these spicy "pods" were used to ornament main dishes in upper-class cookery. The elongated fruits are siliques, the characteristic fruit type of members of the mustard family (Cruciferae or Brassicaceae), and these may not be produced until the second growth season if the radish variety is biennial.

Turnips (*Brassica campestris*), also mustard relatives, were one of the most reliable garden crops grown by the first American colonists, and tons were consumed in the early colonies; many of these were the selfsame turnip varieties used to fatten livestock on family farms. Cooks who followed the advice of Simmons (1796) served mashed turnips with mutton, perhaps along with watercress, boiled onions, caper sauce, and "colliflower." At Monticello, the dish known as chartreuse was prepared using root vegetables, including white and yellow turnips, carrots, beets, and kohlrabi (the food storage stem of a *Brassica oleracea* cultivar). The vegetables were cooked separately, layered in a deep mold, and baked in a *bain-marie*, a hot water bath in which smaller pans were placed for slow cooking. Unmolded on a platter, chartreuse made an elegant dish for elaborate meals, quite different from the typical use of root vegetables in boiled dinners.

By the nineteenth century, kitchen gardens included turnip cultivars that varied in shape, color, texture, flavor, and time of maturation. 'Yellow Aberdeen' and 'Robertson's Golden Ball' were described as excellent for both table use and fodder, while 'Teltau' had small, spindle-shaped roots particularly valued for enriching soups. Young turnip leaves were stewed as greens that could be harvested from stored turnips that began to sprout by spring, and traditional southern cooks added bacon to the pot; a Civil War–era street ballad proclaimed "My woes have been solaced by good greens and bacon." Young turnips were cooked and served whole, with a few inches of their foliage left intact, and mature

In addition to their edible taproots, radishes (*Raphanus sativus*) produced small fruits that were preserved with vinegar and spices.

turnips were dressed with butter and cream and perhaps sweetened with sugar. Leslie (1851) described a dinner of roast pork served with mashed turnips, potatoes, and apples. She cautioned cooks against exposing cooked turnips to sunlight, which she believed imparted a "singularly unpleasant taste." The "turnip-rooted cabbage" described by Randolph (1824) was probably a rutabaga, a European hybrid between turnips and cabbage or kale (*Brassica oleracea*); Jefferson noted that its leaves resembled those of cabbage and not turnip, but Randolph advised cooks to prepare only the roots, by boiling thick slices until tender and dressing them with melted butter.

Salsify, or oyster plant (*Tragopogon porrifolius*), superficially resembles parsnips and is also native to Europe. Some considered its cooked taproots similar in taste and texture to oysters and prepared small fried cakes and fritters that were said to resemble oyster dishes. Like parsnips, salsify roots could be left in the ground as long as the temperature did not fall below ten degrees Fahrenheit, and they could be harvested over several months, which provided another fresh vegetable when gardens were fallow. Experienced cooks knew that the peeled salsify root darkened rapidly and so scraped the roots only after cooking; heat presumably destroys the enzymes that promote the reactions that result in discoloration, similar to the enzymatic darkening that occurs in raw potatoes. Randolph prepared salsify roots stewed, sliced, and baked "in scollop shells with grated bread." In *The American Frugal Housewife* (1844), Child suggested mixing salsify with minced salt fish, a practical way to extend the dish when supplies were short. As in other composites (family Compositae or Asteraceae), salsify "flowers" are actually heads comprising minute individual purple florets; each floret develops into a fruit (achene) with a branched pappus that enhances dispersal. Salsify easily escaped garden plots, and like goat's-beard (*T. pratensis*), it naturalized in fields and along roadsides. Cherokee Native Americans used the fluffy pappus-bearing achenes to fill pillows and mattresses and gathered wild salsify and boiled and fried the young plants, and many tribes chewed or ate the milky latex pressed from salsify leaves and stems. The leaves of salsify and goat's-beard are edible, and botanists have described them as resembling the leaves of grass or leeks (*Allium porrum*), although they are unrelated; *porrifolius*, the botanical epithet applied to salsify, translates from Latin as "leeklike leaves."

Edible composites also included Jerusalem artichokes (*Helianthus tuberosus*), which are closely related to sunflowers and were adopted from Native Americans for their substantial edible tubers. The tubers are actually edible underground stems or rhizomes, although they are usually considered along with root crops. The species originated in central North America, and Native Americans cultivated Jerusalem artichokes widely; in 1605, Samuel de Champlain observed them growing on Cape Cod. Jerusalem artichokes were soon introduced to European markets, and American colonists used them commonly until they were supplanted by potatoes. Some may have believed the legend that the rhizomes or tubers cause leprosy (now known as Hansen's disease, a bacterial disease caused by *Mycobacterium leprae*), perhaps because their bumpy appearance suggested lepers' disfigured digits. Cooks who ignored this superstition prepared Jerusalem artichokes like potatoes, boiling them until tender and then dressing the cooked tubers with butter or sauces. Like other subterranean crops, they could be harvested and stored in root cellars, where the tubers could tolerate long-term storage.

Curiously, despite their similarity to potatoes, artichoke tubers actually contain no starch; Jerusalem artichokes store their excess food in the form of inulin, a carbohydrate composed of the sugar fructose (as compared to starch, which is composed of linked glucose sugars). Jerusalem artichokes are distinct from globe artichokes (*Cynara scolymus*), edible composites that colonists introduced from England; globe artichokes were valued for the edible leafy bracts that surround their flower heads, and they were among the first crops cultivated in the southern colonies.

Potatoes

White potatoes (*Solanum tuberosum*) store excess food supplies in enlarged underground stems known as tubers, and once they were introduced to the American colonies, they supplanted many traditional root crops as dietary mainstays. Growers learned to propagate them asexually by cutting stored

The misshapen underground stems of Jerusalem artichoke (*Helianthus tuberosus*) endured long-term storage and were prepared and eaten like potatoes.

tubers into pieces, each containing an "eye," an axillary bud that can sprout aerial stems and roots, mature into a plant, and eventually yield more tubers. This method was described during the mid-eighteenth century in *A Treatise on Gardening, by a Citizen of Virginia*, probably authored by John Randolph (ca. 1770); potatoes from the fall harvest were saved and stored to plant the following spring, yielding clones of the original plants. The method was easy and efficient and produced enormous quantities of a reliable vegetable crop. Indeed it is difficult now to imagine American cookery without white potatoes as a starchy staple crop.

As a domestic plant, potatoes arrived in North America from South America by way of Europe, but the plant's origin was in the Peruvian Andes. South and Central America have more than two hundred native wild potato species, with particular diversity in the Lake Titicaca region of Peru and Bolivia, where potatoes were likely domesticated by selection and hybridization more than seven thousand years ago. The Incan empire was fueled by potatoes that were pressed, dried, and frozen to produce *chuno*, which was carried with ease and could last for years. According to legend, in 1580 Sir Francis Drake carried the first potatoes to England from Chile (where he landed in 1578), an unlikely account since the tubers probably could not survive for two years at sea. In reality, conquistadors compared potato tubers to truffles and introduced them to Spain during the early sixteenth century. European herbalists ascribed aphrodisiacal qualities to the tubers and helped to spread South American potatoes into Italy, the Low Countries, and England. As members of the nightshade family (Solanaceae), potatoes were close relatives of the toxic deadly nightshade; some suspicious Europeans believed that potatoes might cause tuberculosis and even leprosy, a theory voiced by the Swiss botanist Caspar Bauhin in his *Phytopinax* (1596). For years, potatoes were considered suitable only for poverty diets or fodder.

Potato fields may have required more initial cultivation, but the yield in dietary calories for an acre of potatoes was about four times that of grains. By the seventeenth century, potatoes became known as a safe, reliable crop that could tolerate cool temperatures and poor soils, so long as rainfall was sufficient. European populations increased in size as peasants began growing potatoes for their own use, reserving the better fields for more valuable grain crops. Ireland was particularly known for the potato crops that fed its poor, and white potatoes acquired the synonym "Irish potato." Soon potatoes were also introduced to American kitchen gardens and farms; Captain John Smith saw them in Virginia in about 1620, and William Penn mentioned potato cultivation in Pennsylvania in 1685. Scots-Irish farmers who settled in New Hampshire began potato cultivation on a large scale in 1719. In their South American habitats, potatoes are herbaceous perennials with straggling growth, but North American farmers grew them as annuals. By the late eighteenth century, the late-yielding wild type plants were replaced by varieties with early-maturing tubers.

Eighteenth century growers cultivated several varieties, including the smooth-skinned How's potato and yellow Spanish types, usually starting with seed potatoes. The vines flowered, cross-pollinated, and produced small berries containing viable seeds, which growers occasionally planted. Sexual reproduction sometimes yielded new potato varieties that could be brought into cultivation, such as 'Early Maine,' which originated from seeds of the familiar 'Early Rose' potato grown in Maine and Prince Edward Island. Since potatoes are tubers (stems adapted for food storage), by spring their axillary buds (eyes) begin to sprout into stems; prudent housekeepers kept an eye on potato stores, kept them dry, and conscientiously removed any vegetative growth. Simmons (1796) cautioned that potatoes should be harvested before the wet autumn weather and carefully dried before storing. She thought particularly well of the "genuine mealy rich" potatoes raised in Ireland and recommended that American potato stock be renewed with fresh imports, since potatoes "depreciate in America." Her observation was likely based on the effects of North American climate and depleted farm and garden soil rather than genetic deterioration of the potato clones.

Cooks boiled, steamed, or roasted potatoes and also used them to stuff fowl or thicken chowders. Cooking times had to be

Potatoes (*Solanum tuberosum*) were cultivated as genetic clones from seed potatoes, tubers bearing several buds that were cut apart and planted.

adjusted depending on size and age of the tubers, and cooks were cautioned against boiling potatoes until they became watery instead of dry and floury, a response to the early practice of boiling potatoes for hours to rid them of suspected poisons. Potatoes were then variously mashed or fried; flavored with butter, cream, onions, or gravy; or layered with leftover meats to cobble together a thrifty pie. Randolph (1824) described potato buns and potato paste, a pastry of flour and cooked potato used to encase apple dumplings. Boiled, sieved potatoes sometimes substituted for flour in bread baking, and potatoes were also used to culture the necessary bread yeast, often combined with hops that served as a preservative. In *The American Frugal Housewife* (1844), Child related the method (apparently not one that she had tried) for potato cheese, described as a European favorite that improved with age. Cooked potatoes were pounded to a pulp and combined with sour milk and salt; the mixture was kneaded, packed in small baskets, and left to dry. The milk provided natural cultures of lactic acid bacteria (*Lactobacillus* spp.), and the potato starch presumably supported bacterial growth and cheese production.

Potato starch was also prepared at home, usually a spring chore that converted softened, watery, stored potatoes to a useful household product. The tubers were sliced and soaked in water to release their starch grains, which were then dried by evaporation. The thin-walled cells that compose a potato contain thousands of amyloplasts (starch-containing cellular structures related to chloroplasts), which store the excess sugars produced by photosynthesis in the form of starch. Potato flour or starch was used to thicken puddings, and in *The Good Housekeeper* (1841), Hale described a potato pudding flavored with nutmeg, brown sugar, and currants. Thrifty housewives used potato starch to stiffen and revive old silk garments, and it also appeared as an ingredient in some early whitewash mixtures.

Overdependence on a single food crop can result in economic and social disaster if a parasitic blight or pest attacks the monocultural fields. In the case of potatoes, a serious problem was the genetic similarity of the early cultivars; they were all derived from the first potato introductions from South America and propagated primarily by seed potato clones. The Irish Potato Famine resulted from the late blight fungus (*Phytophthora infestans*), a water mold that beginning in 1845 demolished the potato crop in Ireland, at a time when laborers typically consumed twelve or more pounds of potatoes each day. Waves of Irish immigrants arrived in the United States because they faced starvation or possible dysentery, scurvy, or typhus if they remained in Ireland. The cause of the disease was not understood; most Irish assumed the blight was an act of God or the result of poor weather, rather than the result of a microbe that had infected a genetically homogeneous crop. During the 1840s, late blight also infected some potato crops in North America, but sufficient alternative staple foods were available. Eventually potatoes played a role in the Industrial Revolution, since industrialists on both sides of the Atlantic favored potatoes as a source of cheap food that allowed them to keep wages low.

The early suspicion that potatoes are poisonous is in part correct; potato cells exposed to light synthesize solanine, a glycoalkaloid that interferes with activity of cholinesterase, an enzyme associated with activity of the neurotransmitter acetylcholine. The green leaves and stems of potatoes contain solanine, as do tubers that are tinged with chlorophyll because they have been exposed to light. Household wisdom has always been to store potatoes in the dark, if not in a root cellar, and then in a covered bin, because the white tubers lack solanine and can be eaten with impunity. Solanine does not deteriorate during cooking, and symptoms of solanine poisoning include neurological and gastrointestinal disturbances. Some apparently experienced these symptoms on occasion; in their manual *The American Woman's Home* (1869), Catharine Beecher and Harriet Beecher Stowe caution readers about the nature of the potato that,

> harmless and nutritive as it appears, belongs to a family suspected of very dangerous traits. It is a family connection of the deadly-nightshade and other ill-reputed gentry and sometimes shows strange proclivities to evil—now breaking out uproariously, in the noted potato-rot, and now more covertly, in various evil affections.

They cautioned cooks about the water in which potatoes were cooked, explaining that "the evil principle is drawn off," but they praised the delights of well-prepared boiled, roasted, and fried potatoes. Cooks also knew that a dark pigment forms when raw potatoes are sliced and exposed to the air, a chemical change that occurs in the presence of oxygen when potato enzymes convert phenols into harmless melanin pigments. In whole potatoes, the enzymes and phenols are kept apart, and sliced raw potatoes that are kept cold or covered with water show only minimal browning; the process is halted entirely by cooking, which denatures the enzymes responsible for browning.

Sweet potatoes (*Ipomoea batatas*) are members of the morning glory family (Convolvulaceae) that probably originated in Central or South America. During his first voyage, Columbus carried cultivated sweet potatoes back to Spain, and Europeans considered them a delicacy and possible aphrodisiac. The plants were propagated asexually by slips, trailing stems with adventitious roots that sprouted from the storage roots. Early cultivars included roots with red, purple, and white storage tissue, in addition to the common orange-pigmented types; these were all cultivated in Spain by the mid-sixteenth century. Sweet potatoes were established in European gardens before white potatoes, but confusion arose when the Caribbean Arawak Indians used the name *batata*, which evolved into *potato*, for both plants. When white potatoes became familiar in Europe, they assumed both the name and reputed aphrodisiacal properties originally ascribed to sweet potatoes, although the species are unrelated. In his *Herball* (1633), Gerard noted that the sweet potatoes were eaten cooked with prunes, dressed with oil and vinegar, "rosted in the ashes," "sopped in wine," and included in various "sweete meates." Gerard agreed with the common wisdom that sweet potatoes had the capacity to promote "bodily lust, and that with greediness."

Sweet potatoes were probably first grown in North America only after pre-Columbian times, although a close relative with starchy roots (*Ipomoea macrorhiza*, native to Georgia and Florida) perhaps was foraged. Virginia settlers may have raised sweet potatoes as early as 1610, and William Bartram observed sweet potatoes growing in southern Native American villages. Sweet potatoes were introduced to New England gardens in 1764, and good crops were grown in the Boston area, far north of where sweet potatoes are

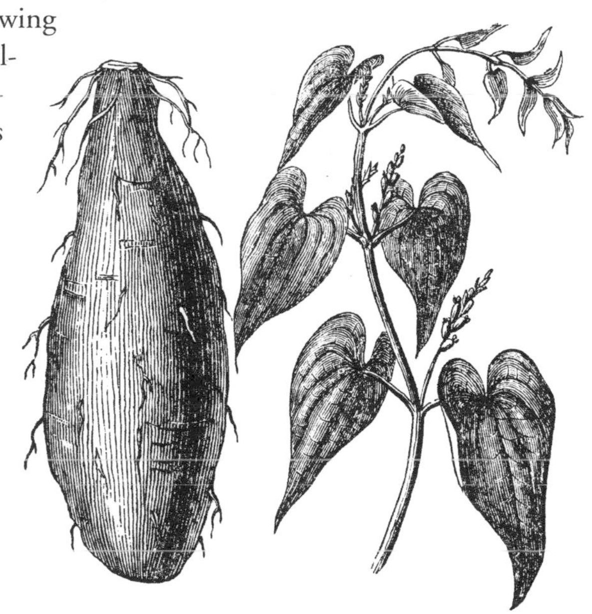

Sweet potatoes (*Ipomoea batatas*) prevented vitamin A deficiencies in the diets of many impoverished tenant farmers in the southern states; they are distinct from the true yams (*Dioscorea* spp.) that were introduced from Africa.

now commonly cultivated. In 1860, Frederick Law Olmstead described sweet potatoes in slave diets, especially during the winter months; southern cooks favored sweet potatoes in place of white potatoes, and the large storage roots supplemented the nutrient-poor diets of many subsistence farmers and their families. By the 1860s, nine different sweet potato varieties were in cultivation, and the deeply pigmented orange types with high carotene concentrations in their root cells were particularly valuable since they provided the precursor of vitamin A.

Sweet potatoes were prepared like potatoes—boiled, baked, or fried—and could be stewed with chicken and ham or made into a pudding flavored with sugar, cream, eggs, wine, nutmeg, and lemon. Southerners may have called sweet potatoes *yams* on occasion, but true yams are actually species of *Dioscorea*, a genus of tropical vines with massive underground tubers. Yams may weigh several pounds and require boiling to remove the toxic alkaloid dioscorine; they are also a natural source of steroids (sapogenic glycosides) that have been used in synthesizing both cortisone and oral contraceptives. Many yam tubers are also rendered toxic by the presence of oxalic acidic crystals in the layer just below the skin, which is easily removed by peeling or cooking. Yam tubers were a starchy staple crop in tropical Africa for more than five thousand years, and since they are easily stored, they were carried as supplies on slave ships bound for the New World. Slaves planted and cultivated various yams in New World gardens, including the water yam (*D. alata*), which was introduced to Africa from the Pacific region.

Like white potatoes and sweet potatoes, yams were boiled or roasted; along with African okra, *benne* (sesame), and eggplant, yams were established in the New World by slaves who gardened in subsistence plots. Yams were not widely adopted into traditional southern cooking, but Jefferson was curious enough to cultivate some African yams in the kitchen garden at Monticello. His 1786 letter to Anthony Giannini requested seeds for "the sweet potatoe. I mean that kind that the negroes tend so generally," most likely yams.

Edible Leaves and Stems

Cabbages and cauliflower (both cultivars of *Brassica oleracea*) were plants familiar to Puritan colonists, and eighteenth century gardeners continued to rely on varieties imported from Europe. The *New Hampshire Gazette* on 20 January 1764 advertised "Early Dutch, Yorkshire Batterica, Sugarloaf, May,

Red, Turnip, and large winter cabbage, Green, Curl'd, and Yellow Savoy, Early and Late Cauliflower." During the nineteenth century, cabbages (by then accompanied by potatoes) still provided the backbone of American diets, especially for working class families. Available were early, late, and winter varieties; delectable savoys; "surehead" types guaranteed to form heads (even in extreme southern heat); and anthocyanin-pigmented cabbages such as 'Red Dutch.' European cabbages including Dutch, Danish, French, and English varieties were still marketed to American growers, and productivity and capacity for long storage were key attributes; *Burpee's Farm Annual* (1888) boasted varieties such as 'Fottler's Brunswick' that matured into heads weighing twenty pounds or more.

Cabbages were synonymous with homely boiled dinners, but nineteenth century sensibilities were sometimes put off by the odor and taste of cabbage that had stewed for hours; cookbooks advised using plenty of water and cooking the quartered heads rapidly. An alternative to separate dishes of boiled cabbage and boiled potato was the combination known as cale-cannon, prepared by binding cooked, chopped cabbage with mashed potatoes and flavoring the mixture with pepper, salt, and butter. As an addendum to her recipe, Leslie (1851) noted, "Cabbages may be kept good all winter by burying them in a hole dug in the ground," a reminder that cabbages were fresh vegetables that could be used during the long winter.

In *The Tragedy of Pudd'nhead Wilson*, Mark Twain wrote, "Training is everything. The peach was once a bitter almond; cauliflower is nothing but cabbage with a college education." Despite the association of cabbages with working-class cookery, cauliflowers (botanically the same species, which is ironic) were regarded as socially upscale vegetables. Cauliflowers produce heads of immature flower buds

Savoy cabbage (*Brassica oleracea*) cultivars have leaves with a distinct texture and were considered particularly delectable for table use.

(the "curd") rather than the massive terminal bud that forms in mature cabbages. Like cabbages, nineteenth century varieties were selected for "reliability in heading." Some could be forced into flower early in the season, while autumn types were selected for the long, protective leaves that wrapped around the pale or white curd. Cauliflowers were boiled intact, with the large outer leaves tied in place around the curd, and they were often served with delicate sauces. Randolph (1824) suggested "cabbage-a-la-creme," a carefully prepared dish of tender creamed cabbage, in contrast to the soft cooked, unadorned cabbages and root vegetables served with northern dinners of boiled beef or mutton.

Broccoli is also *Brassica oleracea*, but this variety was poorly known and rarely cultivated until the early twentieth century. Seeds of "Baroquely" did appear in the list of imported seeds published in the *New Hampshire Gazette* on 20 January 1764, but very few gardeners were enticed to grow it. It was known to Jefferson, who planted several broccoli varieties at Monticello. Randolph (1824) described the proper cookery of two distinct varieties; broccoli that formed heads was cooked like cauliflower, boiled slowly in salted water and overcooking carefully avoided, while broccoli that matured into flowering stalks was bundled and cooked like asparagus.

Cabbage, as well as cauliflower and broccoli, lent itself to successful pickling. A typical method involved packing shredded, salted cabbage in a stone crock and then filling it with vinegar infused with black pepper, cinnamon, nutmeg, and allspice; Catharine Beecher (1858) suggested alternating layers of red and green shredded cabbage. Sauerkraut, a product favored and produced by German immigrants, depended on lactic acid bacteria to ferment shredded, salted cabbage; this mixture packed densely in a crock provided the anaerobic conditions required for fermentation by the *Leuconostoc* and *Lactobacillus* bacteria that occur naturally on cabbage leaves. These harmless bacteria ferment the cabbage carbohydrates and produce lactic acid as a metabolic waste, which lowers the pH and effectively preserves the shredded leaves. Pickled cauliflower and broccoli were steeped overnight in strong, hot brine and on the following day were drained and packed in spiced vinegar.

Other cabbage and colewort cultivars of *Brassica oleracea* in American gardens were the leafy varieties known variously as kale, borecole, or collards, which were particularly important in southern cooking. Their tough leaves were softened by frost, and seed catalogs advised gardeners to sow a late crop in the fall, cover the beds with straw, and have early greens for meals the following spring. These leafy coleworts are perhaps the closest to the wild plants from which all of the various cultivated forms have been derived by selection and hybridization. Brussels sprouts were less familiar, a long-stemmed type in which each an axillary bud forms a head that resembles a miniature cabbage. Whether they were originally from Brussels is debatable, but Sturtevant (1919) mentioned that historically they were grown in America "only in the gardens of amateurs" but were deserving of "more esteem." *Burpee's Farm Annual* (1888) described 'Perfection' Brussels sprouts as "well worthy of general cultivation," producing dozens of sprouts that also become "very tender and of rich flavor when touched by frost."

Kohlrabi, a *Brassica oleracea* cultivar with small leaves and a swollen storage stem, was also infrequently cultivated. Mature kohlrabi was used for forage, and young stems were boiled for table use. Varieties included the white or anthocyanin-pigmented cultivars such as 'Early Vienna' that were imported to American gardens from Germany and Austria, and some confusion existed as to the relationship among kohlrabi, cabbages, and turnips; kohlrabi was also known as turnip-rooted cabbage but is distinct from rutabaga, the European hybrid between turnips (*B. campestris*) and cabbage or kale.

Cultivated lettuce (*Lactuca sativa*) probably originated in Asia Minor and was introduced into England by Roman invaders, and the basic "cabbage" (head) and "cos" (leafy upright) types have been known since the sixteenth century. Elizabethans relied on lettuces in sallets or salads dressed with vinegar, oil, salt, and perhaps orange or lemon juice, and European herbalists noted their cooling, soporific qualities. Gerard (1633) described lettuce as a "cold and moist potherb," and indeed, lettuces were sometime boiled to render them more digestible and nourishing.

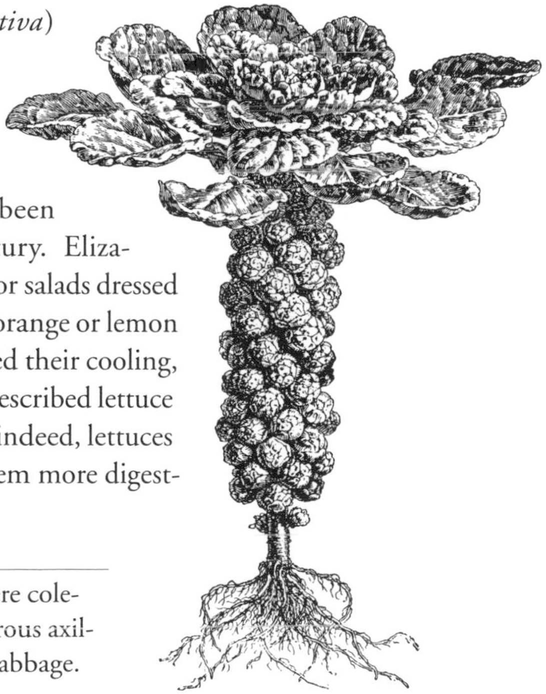

Brussels sprouts (*Brassica oleracea*) were colewort varieties selected for their numerous axillary buds that each resemble a small cabbage.

While colonists arriving from England certainly knew lettuce from their Old World kitchen gardens, such tender crops became "garden sass" to practical New Englanders; their diets relied primarily on storable root crops and coleworts that were served aside the main meat course. Tender lettuces were temporal crops and seasonal dietary diversions; they were best grown and eaten early in the season, before summer heat caused the plants to bolt and set seed. John Winthrop Jr. included three ounces of "lettice seed" in his 1631 list of garden seeds purchased from grocer Robert Hill and shipped to America, but the variety that he cultivated is unknown. Seventeenth century gardeners could select among leafy and head-producing varieties including curly, red, white, and romaine types, and some of these types were described by William Byrd in *The Natural History of Virginia, or The Newly Discovered Eden* (1737).

Watercress (*Nasturtium officinale*) had early medicinal uses, but it was probably introduced to North America for its edible leaves that contain mustard oils. In *Medicinal Plants* (1892) Charles Millspaugh listed watercress as a "relish" along with several other edible mustards and coleworts, all members of the mustard family (Cruciferae or Brassicaceae). It is distinct from the garden nasturtiums (*Tropaeolum majus*) that are native to Peru and Chile and were known as Indian cress; *Tropaeolum* is not a member of the mustard family, but coincidentally the genus also produces mustard oils (see chapter 6), and the plants were grown for their edible flowers and seeds. Watercress differs ecologically by thriving when partially submerged in running water; the plants colonize an area vegetatively and were encouraged to naturalize in many areas. The plants can be gathered during any season in regions south of Pennsylvania, which probably caused many to grow it as an available source of leafy greens. Randolph (1824) also used it like parsley, minced fine and

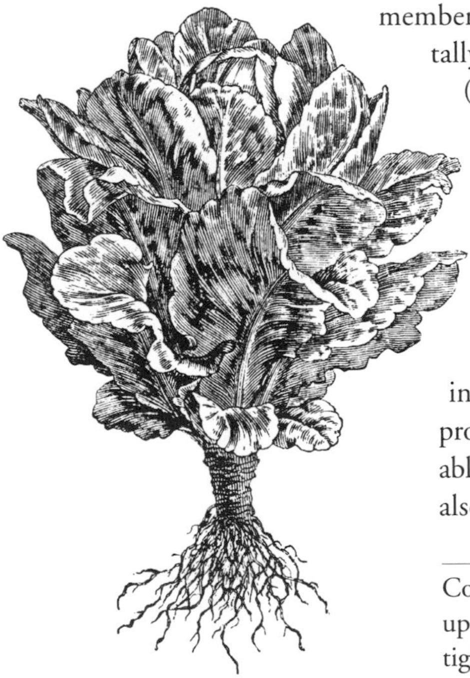

Cos varieties of lettuce (*Lactuca sativa*) had upright, leafy growth and did not form a tight head.

cooked in butter, to prepare sauces "which are excellent to eat with fish, poultry, or boiled butchers meat."

Skilled colonial cooks composed lettuce salads with watercress or chervil (*Anthriscus cerefolium*, a parsley relative) and dressed with cooked egg yolks, oil, mustard, and vinegar. Randolph recommended that her readers "lay around the edge of the bowl young scallions, they being the most delicate of the onion tribe" and dress the salad with tarragon vinegar. Her method revealed the French influence on American cookery during the early nineteenth century; such arrangements of lettuce and other greens were known as "French salads." Until the end of the nineteenth century, only well-to-do families ate salads frequently; lettuce was a perishable crop and could not survive shipping, and city dwellers lacking kitchen gardens had to depend on the marketplace for fresh vegetables. Glasshouses first appeared in the 1840s to supply prosperous Bostonians with lettuce as well as cucumbers and radishes, and they became more common by the 1870s. Of course, cooks with garden plots could cultivate lettuces ranging from tennis ball types with small, tight heads to the upright cos cultivars; thrifty gardeners heeded the advice of Child (1844) to cut back lettuce rather than uprooting the plants to encourage more growth through the summer months.

Elizabethan cookery used spinach (*Spinacia oleracea*) in sallets and tarts, and Europeans also valued its leaves as a laxative and emollient. In his *Herball*, Gerard described spinach as a potherb or "sallade" herb that yielded little or no nourishment; of course, he was quite mistaken, because spinach provides nutrients including vitamins A and C, and it probably forestalled many dietary deficiencies. Its characteristic acidic flavor is from oxalic acid, which is also present in sorrel (*Rumex* spp.); although oxalic acid is toxic, the small amounts in spinach may be eaten with impunity. In 1631, John Winthrop Jr. included an ounce of spinach seeds in his seed order, and he probably planted his spinach bed bordered with leeks in the traditional manner. Spinach often accompanied boiled meat, and in her manuscript cookbook, Martha Washington recorded her recipe for fricandeau ("fricando") of beef served with spinach or sorrel, an old English dish; poached eggs on a bed of cooked spinach was also familiar to both English and American cooks.

Cooks knew that the curled leaves required close inspection for soil and insects, and cookbook authors encouraged their readers to prepare spinach for cooking with care. Leslie (1851) advised three or four washings in water and ten minutes of boiling time, followed by quick braising in butter and seasoning with salt and pepper. Like lettuce, spinach was traditionally an early season crop that could be sowed successively to supply the table for several weeks, and hardy varieties were also available for fall sowing. By the late nineteenth century, seedsmen advertised spinach cultivars that would stay leafy longer, before reacting to hot weather by bolting and setting seed.

Puddings baked in pastry were based on spinach, sorrel, or tansy (*Tanacetum vulgare*), an aromatic composite (family Compositae or Asteraceae). Martha Washington recorded a recipe for a pudding that incorporated all three plants, combined with cream, sugar, spices, and thickened with bread crumbs. Randolph's 1824 recipe for tansy pudding called for the juice from spinach and tansy leaves mixed with eggs and cream and flavored with nutmeg, wine, and sugar. Tansy was one of the earliest cultivated plants to be introduced to American kitchen gardens as an edible plant as well as culinary, medicinal, and household herb. English herbalists prescribed tansy cakes ("tansies") as a way to eliminate the intestinal worms that they believed resulted from eating fish during Lent, and the leaves were also used to treat gout, toothache, and the tendency to miscarry in pregnancy, but later it was known as an abortifacient. Housewives used the plants as stewing herbs that would discourage insects and vermin.

The pungency of tansy leaves results from a volatile oil that contains thujone, a terpene-based toxin (see chapter 7) that is found in other composites such as yarrow and wormwood, as well as in some sages. Thujone likely evolved as a secondary compound that discouraged herbivores from leaf grazing; its toxic effects are known to include seizures and death, but tansy was grown as a medicinal herb by the Shakers and was still used in nineteenth century American homes to treat hysteria, fevers, and parasitic worms. As a result of its common early use, as both a food and a medicine, tansy escaped from gardens and naturalized over much of North America; Cherokee and other Native Americans adopted naturalized tansy plants for some of the same medicinal uses. Tansy is now avoided as a food, tea, and culinary or medicinal herb because of its potential risk, and the sale of products containing tansy oil is illegal.

Poke (*Phytolacca americana*) or pokeweed, sometimes also known as poke sallet, is another potentially toxic species that was one of the many wild leafy plants gathered for use as a potherb. The term *sallet* was used by the earliest colonists to describe edible greens (see chapter 1) and persisted in the south as a generic term for edible leaves. Poke is native to North America; Native

Americans boiled the fresh shoots or dried them for later use, and settlers in the south gathered poke in the wild. They avoided some of the toxicity by using only the young shoots, but all parts of the plants contain phytolaccin, a mitogen that increases cellular divisions (mitosis) and can cause blood cell abnormalities. The roots and berries also contain toxic saponins, glycosides that function as strong emetics. Toxicologists now recommend that the poke should be neither handled nor consumed, since phytolaccin may affect people differently. The long-term effects of handling and consuming quantities of poke are unknown but might involve deleterious cellular changes; settlers and Native Americans alike may have suffered subtle genetic effects from a traditional, repeated use of poke.

Leafy crops also included celery (*Apium graveolens*), in which the edible parts are enlarged leaf stalks (petioles) that bear a small tuft of compound leaves. Along with carrots, parsnips, and other members of the parsley family (Umbelliferae or Apiaceae), celery has internal secretory canals that contain pungent terpene-based essential oils and other secondary compounds; these oils have been prized as flavorings since ancient times, when Greeks and Romans first brought wild celery into cultivation, and they resemble the secretory canals in sarsaparilla (*Aralia* spp.) in the closely related aralia family (Araliaceae).

While wild and "cooking" varieties are more pungent than "eating" types, all celery is characterized by support tissue in its leaf stalks. The strands that run the length of each petiole are composed of overlapping collenchyma cells, a plant tissue adapted for strength and flexibility that is found frequently in herbaceous stems and leaves. The elongated cells have unequally thickened, nonwoody cell walls and develop into flexible strands that withstand mechanical harm. As in

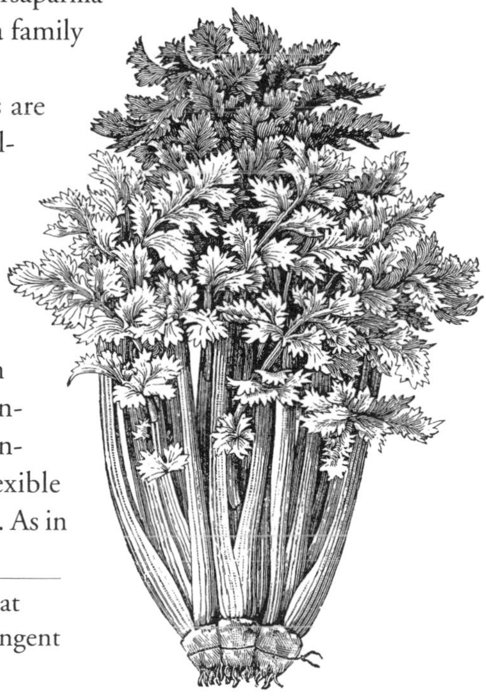

Celery (*Apium graveolens*) has leaf stalks that were selected for their large size, mildly pungent flavor, and pale pigmentation.

other leaf stalks, the chloroplasts in celery petioles develop chlorophyll when they are exposed to sunlight, but the practice of mounding soil around the stalks has long been used to produce blanched celery in which most of the chlorophyll is absent. The practice of blanching celery was developed in the kitchen gardens at Versailles, where the cooks valued the white, milder tasting stalks. Like cauliflower, celery stalks were a vegetable that denoted social status. Perhaps this was the result of the labor-intensive practice of blanching the vegetables; the prominent display of celery on a dinner table suggested prosperity.

During the 1860s and 1870s, celery was arranged for the table in a tall glass or silver celery stands made especially for this purpose, and Victorians ate improved cultivars of celery raw, after the stalks were scrubbed and soaked in cold water. A savory sauce for mutton or fowl was prepared by cooking chopped celery until tender and then thickening the mixture with flour, butter, and milk or cream. Selection during the nineteenth century resulted in self-blanching cultivars that required no banking of soil around the stalks to produce pale leaf stalks; other varieties included types with crimson or golden petioles. 'Boston Market' was valued for its white stalks and mildness, and 'White Walnut' was prized for its nutlike flavor. In the *Skilful Housewife's Book* (1852), L. G. Abell suggested that celery plants be cultivated in rich compost, either sprinkled with salt crystals or watered with a salt solution to simulate the coastal conditions of the wild European ancestors of cultivated celery; similar advice was given for asparagus, another "saline plant" with a Mediterranean ancestry. Some celery varieties kept well in root cellars if their roots were kept covered and moist.

Celery "seeds" (the small, dry fruits of the Umbelliferae known botanically as schizocarps) were also valued for their pungency; the biennial plants flowered during the second growth season, and the schizocarps were commonly used to flavor vinegar. Leslie (1851) advised that two ounces be pounded in a mortar and left to "steep for a fortnight in a quart of vinegar" before being strained and bottled. Essence of celery, prepared by steeping the bruised fruits in brandy, provided flavoring for soups when added by the teaspoon. Celeriac, the enlarged hypocotyl (lower stem) of celery, was grown from European cultivars selected for their massive "roots" (enlarged stems); Beecher (1858) described celeriac as "very good, and but little known," a vegetable that was more easily cultivated than celery. It was typically boiled until tender and served with milk and butter; seedsmen who offered celeriac, or "turnip-rooted celery," described the plants as tender and marrowlike, suitable for cooking, salads, and flavoring soups and meat dishes.

Rhubarb (*Rheum raphonticum*) is another edible leaf stalk used by nineteenth century cooks, who knew this species as "pie plant." Cultivated rhubarb probably originated in China or eastern Siberia and was grown in the physic gardens of European monasteries for its medicinal properties (see chapter 7). Rhubarb appeared in American gardens by about 1800, and McMahon mentioned rhubarb tarts in *The American Gardener's Calendar* (1806). Once established in a kitchen garden, rhubarb was a perennial crop that could be harvested annually and also dried for winter use. Although, botanically, it is not a fruit, stewed and sweetened rhubarb was used in recipes for pies and puddings. In *The American Frugal Housewife* (1844) Child referred to rhubarb as "Persian apple" and noted that it was "the earliest ingredient for pies, that the spring offers," but she described these pies as "dear" because of the amount of sugar required to make the stalks palatable. Sugar is needed to counteract the tartness resulting from the abundance of citric, malic, and oxalic acids in the plant tissues; excess oxalic acid is stored as large crystals that can cause poisoning and even death if rhubarb leaves are consumed before they are thoroughly cooked.

Old World globe artichokes (*Cynara scolymus*) deserve a parenthetical mention as a footnote to this discussion of plants with edible leaves. In this case, the edible parts were the bracts that surround the composite heads, each a cluster of small flowers, which is typical of members of the composite family (Compositae or Asteraceae). The heads were boiled intact, and the bracts were eaten with melted butter. Optimistic colonists arrived in the New World with globe artichoke seeds, but the plants proved to be somewhat tender even in Virginia; early New Englanders may have harvested a few edible heads, but the plants were not hardy perennials in New England, as they were in Europe. Globe artichokes originated in Mediterranean countries and withstood English winters, but they were not destined to be a staple crop in eastern North America. Early nineteenth century gentlemen farmers commonly cultivated globe artichokes, but they could afford the luxury of experimentation and failure without the specter of hunger. Some tried to overwinter the tender plants with straw, and Jefferson's garden notes record success with artichoke crops in thirteen out of twenty-two years at Monticello. For "Artichokes to keep all the Year," Hannah Glasse (1805) suggested cooking the artichokes and then drying the bracts in an oven until "they are as dry as a board;" these could be stored indefinitely and then plumped with warm water, whence they would "eat as fine as fresh ones."

Onions and Leeks

European *Allium* species were among the plants cultivated in early colonists' gardens, and "onyon seed" and "leekes seeds" were among the stock ordered by John Winthrop Jr. in 1631. Onions (*A. cepa*) are biennial members of the lily family (Liliaceae), bulb-producing plants that may have a Persian origin but are no longer known from the wild. Bulbs are composed of fleshy, overlapping leaf bases that store food and water and enable the plants to overwinter. Cultivated onions may be grown from either seed or small offset bulbs that develop at the base of mature bulbs; these are known as onion sets and are asexual clones of the parent plant.

Eighteenth century cooks favored onions for the flavor that they could impart to bland cabbage, potato, or meat dishes; Randolph (1824) described two methods for dressing roast duck with a savory onion sauce. During the nineteenth century, onions became widely cultivated and available to all economic classes, and providentially they were one of the fresh vegetables that helped to eliminate scurvy as a common vitamin deficiency. Their pungent bulbs provide only a moderate source of vitamin C but are suitable for long-term storage. Onions were roasted before an open fire, fried in butter or drippings, or poached in milk; Leslie (1851) recommended a French method for preparing onion soup, and she noted the French opinion that onion soup is "a fine restorative after any unusual fatigue." Spanish (yellow) onions were preferred for brown sauces to serve with roast meats, and small, hard, white onions such as 'Silverskin' were pickled in brine and preserved in vinegar with turmeric or mace, ginger, and nutmeg.

Onions are chemically complex plants; when their bulbs are cut, damaged tissues release

Yellow onions (*Allium cepa*) contain quercitin glycosides, secondary compounds that may protect the fleshy bulbs from herbivores and that were also used as dyestuffs.

volatile sulfur compounds that are both antibiotic and lacrimatory, causing the tears often associated with onion cookery. Historically they have had medicinal as well as culinary uses (see chapters 1 and 7). Red onions have glucose-containing anthocyanins (cyanidin glycosides) in their outer bulb scales, while yellow onions are pigmented by quercitin glycosides that have been used as dyes and may protect the bulbs from attack by herbivores. Pigmented onions have higher amounts of the phenolic compounds, such as catechol, that protect the bulbs against fungal diseases, but nevertheless several white Italian onion cultivars were grown successfully in American kitchen gardens. According to seedsmen's claims, bulbs of 'Mammoth Silver King' could weigh up to four pounds each, and by the 1880s W. Atlee Burpee and Company offered an annual contest for the largest specimen.

Leeks (*Allium porrum*) may have originated from wild plants of *A. ampeloprasum*, native to Mediterranean islands, but their domestication is ancient and obscure. Traditional English cookery called for leeks to add to the Lenten pea porridges eaten in place of meat, and early American colonists may have used them in the same way. Leek plants are hardy in cold weather and could be grown from seed to a large size, and in some kitchen gardens they made a robust natural edging for beds. The entire plant is edible and could be used in soups or cooked along with meats, but leek popularity remained generally low despite its early introduction to North American gardens. Perhaps this was because leeks required blanching (soil mounded around their bases) to achieve the desired whiteness or because leeks required careful preparation; garden soil sifts down between the flattened leaves, which must be carefully separated and washed before cooking.

CHAPTER FOUR

Fruits

In the *Skilful Housewife's Book* (1852), L. G. Abell described the allure of fruit:

> How the purple plum, the delicious grape, the fragrant strawberry, and all the tempting fruits and delicacies, that hang on every stem, mirror to our hearts a picture of Paradise, and make us feel how sad a curse to be driven at once from both the smiles of the Creator, and from the delights of the earthly paradise, the garden of Eden.

Like the authors of many other household manuals, she supplied instructions for cultivating apple, peach, plum, and pear trees, as well as raspberry shrubs, and she provided scores of methods for fruit cookery and preservation. Despite their short season, some fruits were dietary staples; with one hundred twenty genera and more than three thousand species, the rose family (Rosaceae) is remarkable for its edible diversity, which includes several temperate crops: strawberries, raspberries, blackberries, apples, pears, quince, cherries, peaches, plums, nectarines, and apricots.

Fruits reflect the evolution of a variety of dispersal mechanisms; while some members of the rose family have dry fruits, various edible species are specifically adapted for dispersal by animals. Cultivated fleshy fruits, from apples to cherries, have been selected from wild plants that evolved edible, attractive ovary walls as an efficient strategy for dispersal in nature. Flowers with several separate pistils may mature into clusters of individual fruits, such as raspberries and blackberries, in which each ovary develops into an individual stone fruit (drupe); these cling together weakly to form the familiar red and black "berries." Flowers with a single simple pistil can also develop into drupes, in which the inner wall of the ovary differentiates into a hardened layer of sclerenchyma cells that resemble a small stone. This is the typical structure of cherries, peaches, and other stone fruits. Flowers with a single, compound pistil in a cuplike hypanthium can develop into pomes, a fruit type found only in the rose family. The succulent flesh of apples and other pomes is derived from the hypanthium wall and botanically speaking is considered an accessory fruit because of its non-ovarian origin. The characteristic star shape in the center of an apple, revealed by slicing the apple horizontally, indicates the five original pistils in the core of the pome that fused into the compound ovary of an apple flower. Pomes are unique to apples and their close relatives, but other fruit types such as drupes can be found in other flowering plant families as well.

The Rose Family

Adaptation and evolution in the family Rosaceae are revealed by the remarkable diversity of fruits, which include aggregate and accessory fruit types as well as achenes, drupes, and pomes. Each sort represents adaptation for efficient dispersal and may coincidentally have provided a useful human food. Taxonomists divide the family into subfamilies, defined in part by their fruit types, which include the Rosoideae (strawberries, raspberries, and roses), Prunoideae (peaches, cherries, and other stone fruits), and Maloideae (apples and pears). Since the family has a worldwide distribution, with many genera present in both Europe and America, early colonists discovered plants that were close relatives of familiar European cultivars. Old and New World species were used, interchangeably in many cases, as reliable foods and to spawn more cultivars through domestication, selection, and hybridization.

Fruits of the Rosaceae were critical crops in establishing American homesteads and settlements. New England villages had houses well spaced by gardens and apple orchards, and William Penn envisioned each Pennsylvania homestead with land for an orchard. In 1787–88, the Ohio Company (a group of wealthy Virginians who bought land in southern Ohio and northeastern Kentucky in hopes of selling it at a profit to settlers) required that settlers within a few years plant an orchard (fifty apple or pear trees and twenty peach trees) and build a homestead with a cellar and brick or stone chimney to qualify for their deed to a one hundred acre "donation lot." Fruits were essential to providing for a family, and they included crops that could be dried, preserved in sugar, or stored fresh. Long-keeping fruits, so-called winter fruits, were particularly useful, unlike delicate dessert fruits such as strawberries. The earliest apples were ready in June, and later varieties could be stored successfully through the next spring; apples were practical fruits that lent themselves to storage and export, which explains the considerable amount of land and effort devoted to apple orchards. Most apples were used in the colonies, but apples from colonial orchards were also dried and exported.

Domesticated apples (*Malus pumila*) probably originated in the Caucasus Mountains of western Asia and spread through Europe during prehistory. Since apple trees typically require a period of winter cold to produce flower buds and fruit, they are particularly well adapted to survival in temperate climates and were among the first cultivated temperate fruits. Puritan colonists carried apple seeds and scions to America and soon discovered that their familiar apples grew better in North America than in the European countryside. Apples soon became the staple fruits of colonial Americans, used for cider-making, preserving, cooking, baking, and occasionally eating fresh. In 1649, Governor John Endecott of Massachusetts purchased two hundred acres to establish an orchard of five hundred apple trees, and by the mid-eighteenth century, New England farmers were exporting their apple crops to the West Indies.

Countless useful varieties were selected from seedlings that appeared in American orchards. Apple cultivars became known by regional names, such as the popular 'Rhode Island Greening' named for the tavern keeper and orchardist who identified the variety. One type might have different regional names, such as 'Newton Pippin,' which was known in the south as 'Albermarle Pippin,' named for Thomas Jefferson's home county of Albermarle in Virginia. It was cultivated by both George Washington and Benjamin Franklin, who carried this variety to London, and nineteenth century Virginians exported this variety to England. Settlers con-

Apples (*Malus pumila*) are pomes, a fruit type in which the edible flesh originates from the hypanthium surrounding the pistil.

tributed to the westward spread of apple varieties such as 'Roxbury Russet,' which originated near Boston, was cultivated in Connecticut, and then was introduced to Ohio orchards by Israel Putnam. Planting American apples was the life work of John Chapman, the Johnny Appleseed of legend who traveled by foot, horseback, and canoe as far west as Indiana establishing apple orchards. Chapman worked from the 1790s until 1845, starting the seeds and seedlings that nourished nineteenth century frontier communities; some of his apple trees still produce fruit.

Apple varieties resulted from spontaneous hybridizations among apple cultivars, and many of these young seedlings were selected for asexual propagation. Buds or cuttings grafted onto hardy woody stems or roots preserve the unique qualities of a particular variety; some orchardists grafted apple scions onto sycamore roots, but generally apple trees were used as grafting stock. Even if a farmer lacked expertise in the practice, grafting was the accepted method of propagation because one tree could then provide an assortment of fruit. New types frequently appeared as "volunteer" seedlings in local orchards, from whence they were selected, grafted, and propagated throughout a region; Baldwin apples, one such variety, were first produced by a single apple seedling in Wilmington, Massachusetts, and were spread throughout the area by 1784. Of course, care had to be taken to preserve unique trees from destruction by the over-cutting of scions, which was the eventual fate of the tree that yielded the prized 'Rhode Island Greening' cultivar.

Amelia Simmons (1796) encouraged all Americans to plant a few apple trees to supply the family table: "There is not a single family but might set a tree in some otherwise useless spot, but might serve the two-fold use of shade and fruit; on which 12 or 14 kinds of fruit trees might easily be engrafted." One curious practice involved using softwood clubs to beat the trunks of trees that did not flower and fruit abundantly; this apparently affected the flow of sap in the phloem (food conducting tissue) and encouraged flowering.

Orchards were planned to have apples maturing throughout the growing season, and scions were selected for grafting based on this important trait as well as flavor and uses. Eighteenth and nineteenth century orchardists sometimes relied on traveling pomologists to affix particular imported or nursery-selected scions to their orchard rootstocks, although some were cheated by itinerant charlatans who grafted only useless crabapple twigs to mature trees. Typical propagation methods were cleft grafts and tongue grafts, and careful preparation of the stock and scion promoted the likelihood of success. In his

Monticello orchards, Jefferson experimented with grafting codlins, yellow cooking apples described by John Lawson in *A New Voyage to Carolina* (1709), who said there was "no better and fairer fruit in the world." Codlins were just one of the seventeen thousand different apples listed in various nineteenth century farm and garden publications; even experienced pomologists differed in the names that they applied to identical varieties, and apple varieties were often poorly defined and known by numerous synonyms.

Despite the reliability of grafting, for practical reasons many pioneers carried apple seeds to new homestead sites and often grew their orchards from seed. The apples produced by seedling trees varied in quality and did not breed true for desirable traits of taste and texture, which was unimportant since many of these fruit were destined for the cider mill or animal fodder rather than table use. In fact, some orchardists propagated seedling apples by planting pomice, the discarded pulp from cider mills, which contained viable seeds. Cider, the fermented juice of crushed apples, was a staple colonial beverage that was not equated with hard liquor. The sugars in ripe apples provided an ideal culture medium for the yeast that is found naturally on the epidermal layers of fruit; the wild yeast succumbed to low alcohol concentrations before the cider became particularly alcoholic. Elegant syllabubs were made with cider combined with milk, cream, sugar, and nutmeg, and colonists of all ages consumed fermented cider daily. Cider eventually alarmed temperance zealots, and by the 1840s, some movement leaders wanted to demolish orchards as a way of eliminating even the mildest of alcoholic beverages. Of course, "new" (unfermented) cider had uses as well, as a beverage for very young children; to moisten the fillings of mince pies made of apples, beef, and currants; to mix into gingerbread and cider cakes; and to boil down for apple butter. Apple water or apple tea, a pale imitation of true cider, was made by steeping raw or cooked apples in boiling water; it was considered a good beverage for the infirm. Cider was also the basis for cider vinegar, produced when bacteria convert alcohol into acetic acid (see chapter 5).

Virginia colonists grew many apple varieties and used them in cooked compotes, fritters, custards, and baked puddings. Mary Randolph (1824) described applesauce flavored with butter and sugar as a good side dish with roasted goose, and her recipe for ice cream made with apples, quinces, or pears called for cochineal insects, a Mexican species (*Dactylopius coccus*) that contains a brilliant red pigment commonly used to tint foods or dye yarn. Cochineal pigment was used to impart a pinkish tint to apples preserved in syrup. New England cooks favored apple pies and tarts flavored with various combinations of cinnamon, mace, rose water, and wine, but Sarah Josepha Hale (1841) argued that good homemakers should avoid serving pies (she considered pastry indigestible) and substitute puddings, especially for "persons of delicate constitutions." She believed that apples were the only fruit that should be cooked and that great care must be taken to avoid eating unripe and overly ripe fresh fruits. She mused that "the Eden taste still lingers in our race" to explain our desire for fruit, and indeed, nineteenth century cookbooks featured recipes for "Eve's pudding," a dense boiled pudding compounded of chopped apples, bread crumbs, eggs, and currants. This Old Testament connection began with the Genesis story of Eve, who defiantly removed a fruit from the tree of knowledge of good and evil; early translators confused the Latin words *malam* (bad) and *malus* (apple), the starting point of the historic association of apples with the downfall of Adam and Eve.

Catharine Beecher (1858) described ten "Modes of Preparing Apples for the Table," mostly variations on the theme of apples baked or simmered with spices, citrus peel, bread crumbs, and sugar. Jelly was made from orchard fruit, wild seedling apples, or various crab apple species. The procedure required cooks to boil apples in a preserving kettle, mash the softened fruit, strain it in a flannel jelly bag, and then further cook, sweeten, strain, and skim the concentrated juice. Jars of amber jelly on the pantry shelf were the result of the high concentrations of pectin, a complex polysaccharide found in and between the cell walls of ripe apples and other fruits. Apple cookery was universally American, but eating raw apples was infrequent during the mid-nineteenth century. A forward-thinking article in *The Ladies Home Magazine* (December 1860) recommended that families keep two or more barrels of apples on hand and substitute them for pies, cakes, and confections. Constipation was a national scourge (the result of a diet heavy in meat and starch, with few fresh fruits and vegetables), and raw apples held promise for relief. In addition, apples were credited with "correcting acidities" and "cooling off frebrile conditions, more effectively than the most approved medicines."

Apples were known as a typically English fruit, while cultivated pears (*Pyrus communis*) were particularly familiar to the French; many early pear varieties originated in France, and French Jesuit missionaries introduced pears to the Iroquois in North America. Pears probably originated from the hybridization of several wild species in the same region of western Asia and the Caucasus Mountains, where apples were domesticated. They were cultivated earlier than apples and spread during ancient times through Europe and Asia. Pear seeds were sent to the Massachusetts Bay Colony in 1629, and by 1790, William Prince listed thirty-six varieties in the family nursery catalog. Pears are also pomes, and older varieties have the gritty flesh that characterizes wild pears. This is the result of scattered clumps of stone cells (sclereids) that differentiate as the hypanthium matures into the pome; each cell has a thickened, hardened cell wall and microscopically resembles a small stone. Stone cells are also found in seed coats and in the stony layer surrounding the seed in peaches and other drupes; some botanists have hypothesized that stone cells in pears hasten digestion and the dispersal of pear seeds by irritating the intestines of birds that eat the pomes.

Like apples, pears lent themselves to grafting, and their similarity tempted some to graft pears to apple trees, which rarely succeeded although the genera are closely related. Branches bearing unique pear varieties were readily grafted to mature pear trees, such as scions from an admired "sugar pear" that was planted in 1630 by Endecott near Salem (now Danvers), Massachusetts. The tree was still producing fruit 180 years later, when shoots from this sweet-fleshed pear were sent to President John Adams, who grafted them successfully to root stock in nearby Quincy. Since 1870, most cultivated pears have been grafted onto quince trees, which are a convenient shorter height. Familiar types also included Seckel pears, one of the relatively few to be developed in America from a chance seedling that appeared near Philadelphia. In 1807 Jefferson planted some (he called them "sickle" pears) and described them as "The finest pear I've tasted since I left France & equaled the best pear there." The old French 'Bon Chretien' variety became known as 'Bartlett' and still retains its popularity. The American Pomological Society was founded by Marshall Pincky Wilder in 1848 to assist growers of pome fruits, and by 1879, it listed one hundred fifteen distinct pear cultivars as useful in the United States. Cooks favored pears with fine, sweet flesh that could be stewed or baked, perhaps flavored with cloves, cinnamon, or lemon peel. Small pears were preserved whole in syrup or sieved to make marmalade, many of the same methods that were used to prepare quinces.

Quinces (*Cydonia vulgaris*) resemble pears in shape and in having stone cells in their pome flesh, and they were used interchangeably with apples in early recipes for pies, tarts, and other baked fruit dishes. Quince trees are native to Persia and perhaps as far west as Greece and the Caucasus region, overlapping the range of wild apples and pears. Many taxonomists long considered quinces a pear variety, the so-called Cydonian pear (*Pyrus cydonia*), but they will not hybridize, and quinces are now recognized as a separate genus.

Uncooked quinces are unpalatable and may have been the first fruits to be prepared into marmalades. Their pomes were valued in the early American colonies, and quince trees were among the earliest fruits planted in Massachusetts and Virginia. Quinces were also used in jellies, puddings, pies, and dishes of baked fruit layered with rice. Quince cheese was prepared by stewing the ripe fruit into a dense concentrate, which then was cooled in a mold and sliced. Eliza Leslie (1851) also prepared a potent quince cordial by steeping grated pears in brandy; flavoring the fruit with mace, cloves, or nutmeg; and filtering each fermented batch through muslin layers for clarity. Combined with apples and pears, quinces lent flavor and intensity, probably the result of their high malic acid concentration.

The genus *Prunus* includes the familiar stone fruits—peaches, plums, nectarines, cherries, and apricots—as well as almonds that are cultivated for their edible seeds. These drupes are defined by their inner ovary wall that differentiates into the hardened layer (endocarp) that surrounds the seed. The endocarp is composed of a dense layer of sclereids, stone cells with thickened cell walls that are similar to those that occur in the flesh of pears and quinces. The seeds of the *Prunus* species are known for the presence of amygdalin, a cyanide-releasing (cyanogenic) glycoside that imparts the characteristic almond taste and odor.

Amygdalin is a secondary compound that deters grazing animals, and it has had uses in folk medicine and as laetrile, an unproven cancer remedy. High concentrations occur in wild bitter almonds (*Prunus dulcis* var. *amara*), with lesser amounts in sweet almonds (*P. dulcis* var. *dulcis*) and the edible almond cultivars that have been selected over generations.

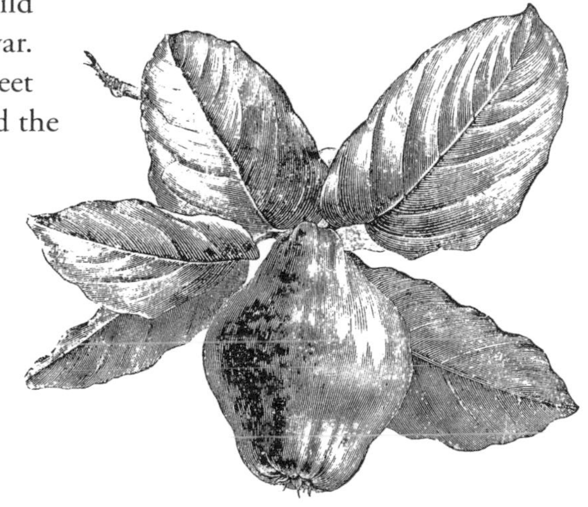

Quinces (*Cydonia vulgaris*) were once considered a pear variety, but now they are recognized as a distinct genus with pomes that must be cooked to become palatable.

When digested, amygdalin releases highly toxic prussic acid (hydrogen cyanide), a compound that is also present in apple seeds. Poisoning occurs at the cellular level when cyanide inhibits the activity of the enzyme cytochrome oxidase in mitochondria, the structures in which food breakdown occurs; affected cells can no longer use oxygen, resulting in their death. Of course, with the exception of almonds, the seeds of stone fruits are not usually eaten, and many prudent people decline to eat even a few bitter, cyanide-laced almonds for fear of toxicity.

Peaches (*Prunus persica*) were widely grown in America during the eighteenth century, cultivated varieties of the wild Chinese trees that during ancient times spread across Europe and were carried to the New World by Florida explorers and early colonists. The trees quickly naturalized in the South and could be easily raised from seed. Peaches were adopted by Native Americans who helped in the spread of the trees northward from Florida, and they were used fresh or dried by Cherokee, Iroquois, Navajo, and other tribes; in a 1663 letter, William Penn noted "not an Indian plantation without them." Colonists exploited their high sugar content to make peach brandy, a staple early beverage prepared by distilling peach juice that had been thoroughly fermented by yeast. The challenge for cooks faced with a generous peach crop was their preservation; this could be accomplished with quantities of sugar, alcohol (particularly brandy), or a combination of these methods.

Leslie (1851) suggested boiling some of the "peach-kernels" and using this extract (which contained some prussic acid) to flavor peach preserves, with the caveat that "a very little will suffice." Peach seeds also substituted for bitter almonds and were sometimes used to flavor custards, cakes, and creams. One such recipe was blancmange, a custard thickened with carrageenan from Irish moss (a marine red alga, see chapter 6), which could also be flavored with water or milk in which peach leaves were boiled. Methods for preparing peach marmalade, jams, and jellies took advantage of the abundant fruit in season but required sugar. Alternatively, unpared peaches were preserved by leaving

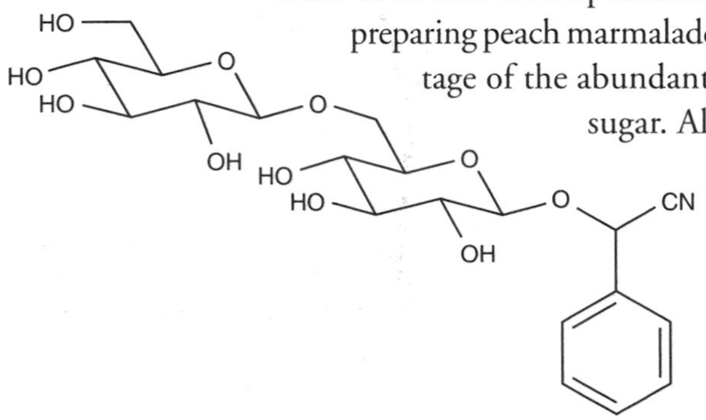

Chemical structure of amygdalin, the cyanide-containing glycoside in the seeds of various stone fruits (*Prunus* spp.)

them in the sun until dehydration rendered them the texture of leather, which was the method also used by many Native Americans. Beecher (1858) also recommended peach leather, made by drying peach pulp on plates or shingles, as a food "much relished by invalids." Peaches also had their critics; Hale (1841) blamed the "many bowel complaints" in southern cities on the consumption of peach skins, which she considered unwholesome.

The 1790 Prince Nursery catalog listed thirty-five peach varieties, with some possible duplicates, which included such appealing names as the yellow and green 'Catherine' peaches, as well as 'Old Newington,' 'English swalch,' and 'Scarlet nutmeg.' In 1802, Jefferson planted Italian peaches sent from Pisa by Philip Mazzei, a former neighbor; by 1811, the Monticello orchard included one hundred sixty peach trees, including American varieties such as 'Heath Cling' and 'Morris' Red Rareripe.' A few years later, Jefferson wrote to his granddaughter, "We abound in the luxury of the peach," and the fruits were both dried and used for brandy. Peach trees edged the fields at Monticello, and Jefferson considered their wood a useful fuel for fireplaces. He also planted nectarines (*Prunus persica* var. *nectarina*), drupes that lacked the familiar fine covering of hairs that characterize typical peach cultivars. His plan was to graft nectarine scions onto typical peach stock, on which nectarines often grow better than on their original trees. Nectarine-producing branches may occasionally appear as sports on peach trees, and their seeds may or may not germinate into trees with hairless drupes.

Like peaches, apricots (*Prunus armeniaca*) originated in China, but they have a higher carotene content, which accounts for their deep golden color. The trees may have arrived in the New World during the sixteenth century or perhaps as late as the eighteenth century, when they were grown in California missions. Apricots were particularly popular in the South, although during the 1790s, James Winthrop recorded apricots growing in his garden at Cambridge, Massachusetts. George Washington used apricot trees at Mount Vernon to border his "grass plats," and Jefferson planted trees described as

The hard endocarp of a peach (*Prunus persica*) contains the seed, and the soft mesocarp has a high sugar concentration and was used by colonists to make peach brandy.

producing large early fruits, one of the ten varieties sold by the Prince Nursery in 1790. Apricots were also often preserved in brandy or sugar syrups, and along with plums, they were often used interchangeably with peaches.

Many cultivated plums are actually hybrids of a few *Prunus* species, including some wild European species (*P. spinosa* and *P. cerasifera*) as well as American plums (*P. americana*). The 'Greengage' plums grown by both Washington and Jefferson may be a distinct Asian species (*P. italica*), which was introduced into England by Sir Thomas Gage in 1720 and eventually into America. New varieties readily appeared in American orchards, probably the result of both natural hydridizations and chance mutations. In 1790, the Prince Nursery catalog listed several varieties of "Plumbs," including 'Yellow Egg,' 'White Damson,' 'Red Perdigron,' and 'Brignole.' Like other stone fruits, plums were preserved whole, or they could be partially dried in the sun and layered in stone jars with plenty of brown sugar. Simmons (1796) recommended plum pies, prepared without spices by following her method for apple pies. Leslie (1851) described a plum charlotte made of stewed ripe plums poured over slices of buttered bread, cooled and eaten with cream. Dried plums were known as prunes, recommended in *The Family Nurse* (Child 1837) as "laxative and cooling," with juices that were "pleasant to moisten the parched lips of an invalid." Ironically, plums were absent from the various baked or boiled plum puddings and plum cakes often served at weddings; in those cases, the "plums" were currants and raisins that were mixed into the batter.

Cherries were gathered from the wild for centuries before they were cultivated, and both Europeans and Native Americans knew and used these fruits. The earliest American explorers discovered native cherries growing abundantly in North America, including the choke cherries (*Prunus virginiana*) and black cherries (*P. serotina*) that were both used as foods and medicines by Native Americans. Colonists adopted black cherry wood for excellent cabinetry, and the fruits were fermented with sugar and wild yeast to make cherry bounce. Black cherries were soon known as rum cherries, following the New England practice of combining them with West Indian rum to yield a highly palatable cherry liqueur. Wild American cherries disappointed English palates long accustomed to the domesticated varieties used in European cookery; garden cherries from England were introduced early to American gardens and orchards, including cultivars and hybrids of both sour cherries (*P. cerasus*) and sweet cherries (*P. avium*). Records of the Massachusetts Bay Company noted that sweet cherry stones were included with the seeds to be sent to colonists in 1629, and cherries were also cultivated in Rhode Island, New York, Maryland, and Virginia. A sour cherry known as 'Red Kentish' was cultivated by Massachusetts colonists, and these were likely the 'Kentish' trees being sold by the Prince Nursery in 1790, in addition to several other "English Cherries," including varieties known as 'Bleeding hearts,' 'Ox hearts,' 'Amber,' 'Duke,' and 'Honey.'

Those with enclosed kitchen gardens could follow the English tradition and espalier cherry trees against the walls, or they could be grown as standard trees in dooryards or orchards. Jefferson planted cherries along the long walk of his vegetable garden terrace at Monticello; he favored the variety known as 'Carnation' and considered it "so superior to all others that no other deserves the name of cherry." A surfeit of cherries demanded preservation, which included jelly and jam-making. Randolph (1824) suggested preserving stoned cherries in a brown sugar syrup, followed by sun-drying, sprinkling with additional loaf sugar, and packing the fruits in pots for later use in charlottes, pies, and boiled or baked puddings. The work of removing the stone from each drupe was tedious and time-consuming, but Leslie (1851) cautioned cooks against drying cherries whole, which "renders them so inconvenient to eat, that they are of little use." Sour cherries and wild black cherries were also combined to flavor brandy, and sour cherry juice was sweetened and bottled with brandy to prepare cherry shrub. Nineteenth century recipes for cherry bounce combined sour and black cherries with sugar and good whiskey in a closed vessel that was shaken daily for a month, followed by straining, bottling, and aging of the liqueur.

Almonds (*Prunus dulcis*) are the seeds of a stone fruit, a drupe that is grown for its edible seeds rather than its ovary wall. The shell of an almond is the endocarp of the fruit wall; some had completely inedible fruits, while others were "peach almonds" with a fleshy layer (mesocarp) that was edible, if not delectable. Cultivated almonds were selected from populations of wild bitter almonds that contained high amounts of amygdalin, a domestication that probably occurred in the Mediterranean region prior to 3000 B.C. Almonds were cultivated in England by 1548, and bitter and sweet varieties were carried to America. Some trees survived as far north as New York, and American cooks used the seeds commonly as a "dessert nut" and in various confectioneries. The Prince Nursery catalog listed "Hard shell almonds" and "Sweet shell almonds" under "Timber Trees and Flowering Shrubs." Almond trees are as hardy as other *Prunus* species, but their habit of early flowering renders them vulnerable to variable weather and spring frosts. They were probably an unreliable crop for many gardeners who attempted almond culture, but wealthy southern colonists could afford imported almonds to prepare rich dishes such as the almond pudding and almond ice cream described by Randolph (1824). J. H. Walden (1858) advised gardeners to "cultivate a few as a family luxury," and he suggested the variety with long, hard shells for cultivation in cold areas and the thin shell types for warm locations. Like peach trees, almonds were valued as ornamental trees, and dwarf and double-flowered varieties were cultivated for both their flowers and edible seeds.

Almonds were used whole or pulverized into small pieces or a paste in a mortar, using the hand tool known as a pestle. Almonds were added to various fruit cakes, such as the "plum" cakes containing raisins or currants; they were usually prepared for batters by blanching the seeds in boiling water so that the seed coats could be easily removed. Beecher (1858) recommended that a pound-and-a-half of pounded, blanched almonds be added to "Portugal Cake," and her method for almond cheese cake called for sweet almonds and rose water beaten together into a paste, with a few bitter almonds (or peach seeds, in a pinch) for additional flavor. The small cakes known as macaroons were also made of blanched, pounded sweet and bitter almonds combined with egg white, nutmeg, spice, and cinnamon and baked on buttered paper. Cyanide poisoning probably did not occur because people consumed relatively small amounts of various foods containing bitter almonds.

Strawberries, blackberries, and raspberries are a botanical conundrum that provide a useful footnote to a discussion of edible rose family fruits. These are all "berries" in the culinary sense, but they are excluded from the botanical definition of a berry as a soft-fleshed fruit without dehiscence (spontaneous opening to release the seeds). Blueberries, cranberries, and tomatoes are all berries in a botanical sense, but the red, fleshy portion of a strawberry is not a true fruit. In strawberries, the receptacle on which the floral parts develop differentiates into the tasty, juicy mass of sugary tissue; the true strawberry fruits develop from the pistils into numerous small achenes located on the surface of the red receptacle. The true fruits are the "seeds" on the outside of the berry, and the small, leafy "cap" is composed of the sepals that grow around the outside of each strawberry flower. The red receptacle is known to botanists as an accessory fruit. American botanist Asa Gray described a strawberry achene in *First Lessons in Botany and Vegetable Physiology* (1857) as "plainly a ripened ovary" that "shows the remains of its style or stigma" where the pollen arrives and germinates. He observed their similarity to the fruits of buttercups (*Ranunculus* spp.) and cinquefoils (*Potentilla* spp.), which are also members of the rose family.

Strawberries grew abundantly in eastern North America, where they were observed and collected by early explorers. These wild plants were *Fragaria virginiana*, a woodland species that was introduced to European gardens in the early seventeenth century; in the mid-eighteenth century it hybridized by chance with another American species (*F. chiloensis*, from the Pacific coast of North and South America) to produce the commonly cultivated *F.* ×*ananassa*. Mature plants colonize areas with horizontal stems know as stolons, which sprout at their tips to produce small clones or offsets of the parent plant. This asexual means of reproduction results in dense wild beds, and the offsets can also be used for cultivation. In his *Herball* (1633), John Gerard described the growth habit of European strawberries as "sending forth many strings, which disperse themselves far abroad, whereby it greatly increaseth," and a woodcut in the book clearly shows a stolon and offset. His text described strawberry leaves for treating "rednesse and heate of the face," which foreshadowed the American Shakers who later grew and marketed strawberries for the astringent and cooling properties of their leaves. The Prince Nursery sold four varieties in its 1790 catalog, including 'Chili' strawberries and wood strawberries, which were probably the two species that hybridized to produce cultivated types. Nearly one hundred years later, W. Atlee Burpee sold a few cultivars, including 'Belmont,' developed in Belmont, Massachusetts; these plants were described as disease and frost resistant, good cross-pollinators that produced fruits that ripened perfectly and were "very solid and sweet."

Strawberries were rare in markets. They were known as a thin-skinned dessert fruit, and most were grown in kitchen gardens for home consumption because of their fragile nature. Strawberries at their peak had to be eaten, pre-

Strawberries (*Fragaria* ×*ananassa*) form from the receptacle of the strawberry flower, which enlarges and differentiates as an accessory fruit; the true fruits are the small achenes on its surface.

served in syrup, dried, or transformed into jams or cordials or some other product. In her chapter on temperance drinks, Beecher (1858) described strawberry vinegar (ripe berries steeped in vinegar) and "Royal Strawberry Acid," prepared by steeping strawberries in spring water and citric acid, and then draining, boiling, and bottling the liquid for later consumption. Both were tart alternatives to hard liquor. Rural families could also gather wild fruit; Thoreau (1999) recorded that strawberries were the first edible fruit to ripen "as early as the third of June, but commonly about the tenth or before the cultivated kinds are offered." He criticized cultivated types when he wrote, "I do not think much of strawberries in gardens, nor in market baskets, nor in quart boxes, raised and sold by your excellent hard-fisted neighbor," preferring instead the wild *Fragaria virginiana* that was native to Massachusetts. Nevertheless, nineteenth century gourmands considered the cultivated fruit to be delectable; during his 1840 presidential campaign, Martin Van Buren was accused by Whig opponents of liberal indulgence in French cuisine, including strawberries and other foods of the wealthy. He lost the campaign to William Henry Harrison, who claimed to survive on "raw beef without salt."

Raspberries and blackberries are aggregate fruits composed of small stone fruits (often called drupelets) that each develop from a single ovary. Each individual fleshy "unit" of a blackberry or raspberry is structurally the equivalent of one peach, plum, or cherry, with a fleshy layer and inner hardened layer that surrounds the seed. Each "seed" is a pyrene, the equivalent of the entire pit of a peach or other stone fruit. We traditionally group blackberries and raspberries with other sweet fleshy fruits, probably because of their high sugar content and similar uses. Raspberries and blackberries are borne on prickly shrubs known broadly as brambles, and native species occurred on both sides of the Atlantic. European raspberries (*Rubus idaeus*) were first cultivated in the sixteenth century, but their pyrenes have been found in Neolithic sites, suggesting their long history as a human food. American colonists cultivated European stock but also gathered the wild fruits of New World raspberries (*R. strigosus*).

North American raspberries were among the wild fruits listed by Edward Winslow in 1621, and early settlers also soon discovered American blackberries (*Rubus occidentalis*), which occur in black, red, and yellow varieties. Black varieties of raspberries also occur, but pickers know that in true raspberries the cluster of drupelets always separates from the flower receptacle, while blackberry drupelets always cling to the receptacle, which forms a core in each fruit. In *A New Voyage to Carolina* (1709), Lawson recorded native purple raspberries as an "agreeable relish" but with the tendency to "run wild all over he country," an accurate description of the aggressive growth of various brambles. American raspberries brought into cultivation provided some new varieties that were superior to European raspberries, but many growers preferred the older, familiar cultivars; by the 1860s growers could obtain at least six varieties of American raspberries.

Birds are attracted by the red fruits, and various cultivated raspberries rapidly escaped kitchen gardens and naturalized into fields and along roadsides. European raspberries hybridized with New World *Rubus* species, which now poses a taxonomic challenge to botanists since individual wild plants often defy classification. Species have been described taxonomically that are probably hybrids between various cultivars and wild plants, and the result has been a confusing excess of scientific names and difficulties in identification. As a further complication, many brambles have life cycles that involve apomixis, flowering and fruiting without the sexual fusion of eggs and sperm. Plants produced in this way are actually clones of a single parental plant, but many of these clones originated as confusing hybrids of native, European, wild, and cultivated plants.

Regardless of their botanical difficulties, raspberries were a desirable cultivated fruit, and in 1790, the Prince Nursery sold four varieties, including both American and English types. Jefferson cultivated square beds of raspberries, similar to the practice of growing currants and gooseberries, and the fruits were typically used for preserves, cordials, and sweet vinegars that were diluted to use as a beverage. Raspberries were also used to

European raspberries (*Rubus idaeus*) were cultivated in colonial gardens and easily hybridized with New World species of *Rubus*.

make wine, which involved straining and bottling the juice and storing the bottles in sawdust to ferment and age. Thoreau gathered wild raspberries, which he preferred to cultivated fruit, and described them as "the simplest, most innocent, and ethereal of fruits." He also noted that the brambles produced prolifically during the wet summers and that they thrived in rich soil in areas that were recently burned or cleared. Many preferred blackberries to make wine and vinegar, bake into pies, or serve fresh with milk, but until the mid-nineteenth century, few gardeners considered cultivating blackberries because they grew abundantly in the wild.

Rose hips (each the mature cuplike hypanthium of a rose flower, *Rosa* spp.) also provided useful foods and medicines, valuable sources of vitamin C that helped to forestall scurvy during the winter months. Rose hips superficially resemble berries, but they are an accessory fruit in which the hypanthium enlarges to enclose the true fruits, which are small achenes. Rose hips were used traditionally in the preparation of jellies, jams, teas, and syrups, and they were fermented into wine and vinegar. In *Child Life in Colonial Days* (1898), Alice Morse Earle described tiny teapots fashioned for amusement from rose hips impaled with a rose thorn and twig. The Shakers grew several rose species and recommended them for their astringent and tonic properties; *Rosa gallica* was known as the apothecary rose, and the flowers were used extensively in preparing rose water, a common flavoring in cakes and confections. Leslie (1851) recommended a half glass of rose water to flavor small, rich "queen cakes"; it was also used to flavor blancmange, a pudding prepared by thickening rich milk with carrageenan harvested from Irish moss. During the nineteenth century, rose cordial was prepared by infusing rose leaves in water, with the addition of sugar, brandy, cinnamon, and coriander; rose leaves were also packed into crocks and covered with white vinegar to make a flavored salad vinegar.

Cranberries and Blueberries

Some New World fruits were more frequently harvested from the wild than cultivated, although cultivation eventually occurred for all fruits for which there was a demand. These include cranberries and blueberries, quintessentially American fruits that were native to many New England habitats. Cranberries and blueberries are all members of the genus *Vaccinium* in the heath family (Ericaceae), acid-loving plants with leathery leaves and symbiotic relationships with mycorrhizal fungi that invade their roots. Their fruits are true berries, fleshy fruits that do not dehisce to release their seeds. The plants are woody, but the species vary in growth habit from the prostrate, vining cranberry plants that colonize bogs to the highbush blueberries that can be twelve feet in height. Native Americans knew both fruit types well; they are credited with stewing and sweetening cranberries to make the first tart cranberry sauce, and they gathered native blueberries to use fresh, dried, or in preserves and cornbreads. Cranberries were combined with venison and tallow to make a northeastern version of pemmican, nutritious dried cakes that were preserved indefinitely by the high acidity of the berries.

The common name *cranberry* is derived from *craneberry*, coined by early German and Dutch settlers who saw a resemblance between the flower shape with the head of a crane. North American cranberries include the large or common cranberry (*Vaccinium macrocarpon*) and also the smaller European cranberry (*V. oxycoccus*), which has a range that includes north temperate regions of both Europe and North America. Lingonberry (*V. vitus-idaea*), also known as mountain cranberry, is a similar alpine species with a tart taste that improves after frost; these were probably the "Red Whortle" described by Gerard in his *Herball* as "rough and astringent" but yielding "the fairest carnation colour in the World."

Common and European cranberries colonize bogs, semiaquatic habitats defined by acidity and indigenous sphagnum or peat mosses, and their berries ripen from green to deep red late in the growing season. Many did not consider the fruit fully ripe and ready to eat until after a late autumn frost, and certainly cold weather seemed to decrease the natural tartness from the combination of citric, malic, benzoic, and quinic acids. A debatable point is whether cranberries were eaten at the first Thanksgiving celebrations; some food historians contend that the unsweetened berries would have been shunned by early colonists, and the tradition of serving cranberries with a meal of wild turkey was perhaps learned later from Native Americans. Simmons (1796) recommended "cramberry" sauce to serve with turkey, and her recipe for cranberry tarts specifies that "stewed, strained, and sweetened" berries be "baked gently" in a rich butter and egg pastry. The practice of serving a dish of cranberries with turkey also closely resembles the Elizabethan practice of accompanying meat dishes with barberries (*Berberis vulgaris*) preserved in vinegar, sugar, and salt. Barberries are also small, red, astringent berries, which may have suggested similar uses for native cranberries. During the 1700s, common cranberries were the first American fruits to be exported to England as fresh "Cape Cod bell cranberries," a testament to their resistant nature even when the berries are fully ripe. Cranberries could be kept fresh for weeks or even months, although some of the crop was also dried or made into jelly.

Wild cranberries were originally hand-picked, but efficient New Englanders soon crafted scoops that could be used to rake the berries from the lax stems. During the nineteenth century, bogs carpeted with wild cranberries transformed into cultivated sites that were raked systematically each fall. In the early 1800s, Captain Henry Hall of Dennis, Massachusetts, began fencing his bogs and spreading sand over the plants to mimic conditions that sometimes occurred in nature; he had observed that cranberries thrived when the plants became partially buried in blowing sands, and others soon copied his practices. Since the plants are perennials, wild cranberries did not require annual replanting, although improved cultivars like those developed by Eli Howes and Cyrus Calhoun during the 1830s and 1840s were soon preferred by many growers. Howes cultivated the 'Howes' cranberries in East Dennis, while Calhoun perfected 'Early Black' cranberries in Harwich, Massachusetts. By the 1850s, seafaring families knew the berries as a way to prevent scurvy among sailors who lacked access to fresh fruits and vegetables, and they were commonplace in northeastern kitchens. Nineteenth century recipes included cranberry pies and batter puddings, as well as a bright tea prepared by steeping the crushed berries in boiling water and then straining, sweetening, and flavoring the infusion with grated nutmeg. In *The Good Housekeeper* (1841), Thanksgiving crusader Hale barely mentions cranberries (only a brief note that stewed cranberries are usually eaten with turkey), but she does offer a detailed method for "Cranberry and Rice Jelly" to be eaten with cream, prepared by boiling a mixture of ground rice and cranberry juice until thickened.

Thoreau praised the cultivated cranberry bogs he saw in Harwich and Provincetown, Massachusetts; white sand was used to build up a bed a few inches above water level, which was then planted over with cranberry rows so that "what with the runners and the moss and so on between, they soon form a perfectly level and uniform green bed, very striking and handsome." Nevertheless, he favored the wild berries that could still be gathered in spring and consumed raw as "a refreshing, cheering, encouraging acid that literally puts the heart in you and sets you on edge for this world's experiences."

Blueberries include not only highbush and lowbush species, but also the closely related shrubs known variously as huckleberries, whortleberries, and bilberries. Their fruits are usually characterized by a bloom, a fine, whitish, waxy coating easily removed by rubbing, although some species have black-fruited varieties that lack a bloom. The name *huckleberry* probably originated in America from a mispronunciation of *whortleberry*, one of several common names applied to *Vaccinium myrtillus*, the common European bilberry, blaeberry, or winberry that colonizes acidic moors and heaths. These various common names have been used interchangeably for wild and cultivated shrubs, and American cooks substituted the berries for one another depending on their availability. Depending on location, homemakers could supply the table with various lowbush blueberries or bilberries (*V. vacillans*, *V. myrtilloides*, *V. cespitosum*, *V. uliginosum*, and *V. angustifolium*), highbush blueberries (*V. corymbosum*), and huckleberries (*Gaylussacia baccata* and *G. dumosa*). The cultivated shrubs were usually highbush blueberries, but most often blueberries and their relatives were harvested from wild populations. Since the plants were abundant and productive in nature, little was to be gained by cultivating them in fields.

Native Americans dried blueberries and used them in breads, porridges, and cooked mixtures of fruit and cornmeal. The earliest colonists learned to use them in similar ways; John Josselyn recorded in *New-Englands Rarities* (1672) that the English in New England purchased dried "Bill berries" from Native Americans and used them in gruel and in "boyled and baked" puddings. Berries were combined with flour, milk, and possibly eggs; boiled tied in a cloth; and served with a sweet or wine sauce. Various blueberries used in this way substituted for cherries or currants in traditional English puddings. When prepared with cornmeal ("Indian meal"), these were known as Indian fruit puddings. *The Good Housekeeper* (Hale 1841) counseled cooks to prepare blueberry pies flavored with "a good quantity of brown sugar, with very little spice or seasoning"; an inverted teacup submerged in the pie contained the excess juice that would otherwise seep from the edges. Child (1844) recommended whortleberry pie flavored with cloves and sugar as a good "common pie" for a family with many children, and a dose of tea steeped from the berries was in order "when the system is in a restricted state, and the digestive powers out of order." The nineteenth century Shakers in New Lebanon, New York, also cultivated European bilberries (*Vaccinium myrtillus*) for their astringent leaves that were used to treat dysentery.

Pumpkins, Squash, and Melons

Native American pumpkins and squashes (cultivars of *Cucurbita pepo*) were adopted by the Puritan colonists, and two centuries later, they were still used in ways that closely resembled the seventeenth century method of stewing pumpkin with spices. The New England "standing dish" of baked or stewed pumpkin described by Josselyn (1672) appeared in nineteenth century cookbooks in the form of pumpkin and squash puddings and pies. Food historians have described the evolution of pumpkin pies as beginning with inventive colonial cooks who hollowed out pumpkins; filled the cavity with milk, honey, and spices; and then baked them in hot coals. Leslie (1851) recommended selecting deep colored pumpkins, paring away the rind, and stewing the flesh in a small amount of water; the cooked fruit could then be flavored with various combinations of milk, eggs, cinnamon, nutmeg, mace, ginger, lemon, and molasses. Some cooks believed that the sweetest part was the flesh surrounding the seeds, so they sieved the softened, cooked flesh rather than scraping away the innermost layer of flesh and seeds.

Cooks used pumpkins and long-keeping squashes interchangeably in various dishes and in starting the wild yeast cultures known as "ferments" that were used in bread baking. Thrifty cooks saved the vegetable broth from stewed squash and used it in preparing brown breads of rye flour and cornmeal; stewed pumpkin and squash were also added to bread mixtures. Southern cooks prepared a simple bread of cooked, sieved pumpkin mixed with flour, which probably bore some resemblance to the cornmeal and pumpkin breads baked by the Iroquois and other Native Americans. Leslie (1857) recommended Yankee pumpkin pudding as a good farmhouse pudding "equally good for any healthy children." She pre-

Pumpkins (*Cucurbita pepo*) are a thick-walled fruit type known as a pepo, and they were valued as a versatile food that could be stored fresh for long periods.

pared her version of this pie without a crust, as a custard of stewed pumpkin or winter squash flavored with West India molasses and plenty of ground ginger; other methods called for additional spices and a deep pie plate lined with pastry.

The great advantage of many pepos, the fruit type of the gourd family (Cucurbitaceae), is the hard or leathery texture of the ovary wall that develops as the fruit matures. Whole, fresh pumpkins and squashes could be stored for several months in a dry spot, and seedsmen recommended varieties based on their sugar content, texture, and keeping potential. The so-called "seven years' pumpkin" was familiar to nineteenth century growers as a remarkable pepo that could be stored for years without decay. The fruits were also preserved by paring away the tough rind and slicing the flesh into strips, which were dried thoroughly in the sun. Cooks who could afford the ingredients also prepared pumpkin chips, a delicate sweetmeat in which pumpkin slices were boiled until transparent in a syrup made of loaf sugar and lemon. Leslie (1851) recommended a delicate dish of pumpkin chips served in a shell of puff paste, quite different from the subsistence dishes of stewed pumpkin and squash in the early Puritan diet.

Massive, fibrous field pumpkins nourished livestock, but kitchen gardeners grew the more palatable sugar pumpkins, which had dense, sweet flesh and smaller size (about ten inches in diameter). Not all early pumpkin varieties were the brilliant orange associated with the carotene pigments in modern cultivars. Pumpkins and squashes had variable shades and patterns of green, yellow, orange, and colorless cells. The color of the rind (exocarp) and flesh (mesocarp) is controlled by the presence of chloroplasts and chromoplasts, pigment-containing organelles that differentiate in the cells of the fruit wall. The variety 'Mammoth Tours' was developed in Tours, France, from American seed stock that was introduced into Europe, and it was advertised as a salmon-colored fruit good for both livestock and table use. So-called banana pumpkins were also known as custard squashes, a variety with fine-grained, cream-colored flesh that was used in pies. The dark green Nantucket or Saint George pumpkin was introduced from the Azores to New England farms about 1846, and its orange flesh and ability to remain fresh all winter made it another good pie pumpkin.

Pumpkins historically have stimulated competition among cultivators, perhaps because of their potential for enormous growth. Farmers and gardeners knew that the most productive vines grew in well-watered hills on new or enriched soil, planted alone or meandering freely on the ground among rows of corn or potatoes. To protect prize specimens, New England farmers spread old coats over pumpkins that had not been harvested by mid-October. In 1857, Thoreau described in his journal a yellow pumpkin weighing one hundred twenty-three and one-half pounds that grew from seeds that he obtained from the United States Patent Office, and he was not alone in marveling at the great size of cultivated pepos. Squash could also grow to enormous sizes, such as the two hundred thirty-one pound 'Mammoth chili' grown by A. E. Hitchcock of Weaver, Minnesota, described in the testimonial advertising of W. Atlee Burpee and Company. Of course, gigantic specimens were often cultivated as natural curiosities more suited to exhibition and decoration than kitchen use, although many huge field pepos were stewed with root crops and grains to fatten household pigs. Some were also carved into seasonal jack-o'-lanterns, a custom that began with Irish immigrants who repeated the legend of Stingy Jack, an unsavory character who tricked the devil and was then forced to roam the earth as a ghost with only a lantern lit by an ember. European peasants carved and illuminated turnips or potatoes, but Irish immigrants to the New World fashioned pumpkins into frightful faces to scare away evil spirits and "Jack of the Lantern." Nineteenth century Shakers at Mount Lebanon, New York, also grew pumpkins for their practical medicinal value; the seeds were valued for their mucilaginous and diuretic properties and were used in treating urinary infections and parasitic tapeworms.

Squashes selected from *Cucurbita pepo* varied in color, shape, size, and texture, revealing the vast genetic variation that occurred in the original wild populations. These included several clusters of cultivars, including scallop, crookneck, and straightneck squashes, as well as vegetable marrows and various winter squashes. Scalloped varieties were sometimes known as cymlings (or sometimes cimlins, cimblins, or simnals), presumably named for the similarity of their shape to simnel cakes, the English seed cakes baked in preparation for the Lenten observance of Mothering Sunday. In 1805, Thomas Jefferson sent a box of seeds including cymlings (which he called "cucurbita verrucosa") to Madame de la Tesse in Paris, with the note, "I recommend this merely for your garden . . . we consider it one of our finest and most innocent vegetables." Varieties that resembled Jefferson's cymling included early bush varieties with a "scolloped" shape. Late nineteenth century squash cultivars of *C. pepo* included 'Small Cocoanut' and 'Perfect Gem,' esteemed for their fine-grained flesh and sweet flavor. Other varieties cultivated in American gardens included the crookneck squashes that are classified as *C. moschata*, as well as *C. maxima* and *C. argyrosperma*, closely related species that include additional winter squashes.

Adding to the taxonomic confusion, the common name *pumpkin* has been applied to pepos produced by all four of these species. The situation is complicated by the tendency of the plants to hybridize when they are planted in common fields. Perhaps the greatest challenge to growers was keeping their pumpkin and squash seed lines genetically distinct; the cultivars of *Cucurbita pepo* and its close relatives easily cross-pollinated, with unappetizing combinations of parental traits appearing among their offspring. Decorative, inedible gourds were also classified as *C. pepo*, while large dippers and other household utensils were made from the calabash or bottle and drinking gourds (*Lagenaria siceraria*), which were introduced from Africa. These large pepos were cured by drying, and then, according to Earle (1898), "Gourd-shells made capital bowls, skimmers, dippers, and bottles." The water dippers used by slaves on southern plantations were made from drinking gourds, a traditional African crop that was cultivated commonly and probably introduced by African slaves. These were mentioned in the lyrics of "Follow the Drinking Gourd," a southern song that provided the geographic cues to travel north by following Polaris, the large North Star at the end of the handle of the Little Dipper constellation; the two stars of the bowl of the Big Dipper constellation point toward Polaris. Travel instructions were passed from plantation to plantation by word of mouth in the encoded message of the song, which helped runaway slaves walk north from Alabama and Mississippi with the help of the Underground Railroad; the trip typically took a year on foot. The last verse describes the Tennessee River and the Ohio River, where escaped slaves crossed the frozen Ohio River and met their helpful guides on the north bank:

> Where the great big river meets the little river,
> Follow the Drinking Gourd.
> For the old man in awaiting to carry you to freedom if you
> Follow the Drinking Gourd.

Melons (*Cucumis melo*) arrived in the New World with early colonists, who recognized Native American pumpkins as melon relatives and called them "pompions," the early English name for melons. "Pompion" seeds appeared on the list of seeds purchased for New World cultivation by John Winthrop Jr. in 1631, probably a reference to musk melons with their typical reticulated rind and sweet flesh. Musk melons (and closely related cantaloupe melons) are cultivars of wild vines that probably originated in Africa and thrive in bright, warm habitats and light, sandy soils. During antiquity they were introduced into Europe and later Asia, and as with squashes, the many cultivars available to nineteenth century American gardeners reflect the genetic diversity in wild populations. Growers could select melons such as the small 'Jenny Lind,' with pale green flesh; large 'Perfection,' with thick, salmon-colored flesh; or 'Japan Coral Flesh,' prized for their distinctive color, texture, and taste. Melons thrived in the middle and southern states, and in the northeast determined gardeners experimented with cultivating melons in boxes under glass or gauze, as protection against insect damage and the cold climate.

In *Soil Culture* (1858), Walden recommended that home gardeners start a few vines well-mulched with wood chips or leaf mould in their dooryards, where "They will grow luxuriantly . . . where hardly any other plant would flourish." Of course, problems could arise, such as excessive rains that caused melons to crack before they were fully ripe. He described cultivated melons as "among the most delicious of all the products of the garden. . . . A little use makes all persons very fond of them." They were favored for their sweetness and fresh texture, which enlivened many diets. But despite their appeal, not all melons were consumed as fresh fruit; some musk melons were preserved in sugar syrup, and following the English culinary tradition, some were transformed into "mango melons." This culinary fad began during the 1600s with the pickling of tropical mangoes (*Mangifera indica*) imported from India, but eventually other fruits were also commonly "mangoed." Eighteenth century American cookbooks provided instructions for packing hollowed peaches, peppers, cucumbers, and melons with potent herb and spice mixtures and pickling the prepared fruits; ginger, mustard, mace, cloves, black pepper, cinnamon, turmeric, onions, garlic, and horseradish were used to fill and flavor these mock mango pickles, which garnished meat dishes on festive occasions. Americans particularly favored mango melons, prepared from immature musk melons stuffed with a spicy mixture, packed in a stoneware firkin, and preserved in strong cider vinegar. These were put by in large quantities. Randolph (1824) provided instructions for preparing the fillings for forty melons, with the notation that a clove of garlic also be packed in the cavity of each fruit; she favored pickling melons "a size larger than a goose egg."

Elizabethans prized both Old World melons and cucumbers (*Cucumis sativus*) for their cooling properties, described by Gerard as an effective cure for "copper faces, red and shining fierie noses (as red as red Roses)." Cucumbers or "cowcumbers" were part of the traditional English diet that was carried to America, and colonial cooks pickled cucumbers in vinegar laced with strong spices such as cloves and ginger. Cultivars differed in flowering times, fruit length and color (creamy white to dark green), and the presence of surface spines; one garden curiosity was the "serpent cucumber," which matured into a peculiar curved pepo that seed catalogs claimed could reach six feet in length. New Englanders favored the fruit of 'Boston pickling' for their productivity and firm texture once pickled. In addition, gherkins (*C. anguria*), so-called "West India gherkins," were small cucumbers that were cultivated especially for pickling. Simmons (1796) probably had gherkins in mind when she noted "the prickly is best for pickles, but generally bitter." Nineteenth century seed catalogs listed them as "Jerusalem pickles" and described them as small, ovoid, spiny, and with small seeds. Gardeners found that the vines were weedy and prolific, but methods for preparing gherkin pickles called for some uncommon ingredients, so they were prepared less frequently. American editions of *The Art of Cookery* (Glasse 1805) included instructions for pickling a batch of five hundred gherkins with a mixture of white wine vinegar, cloves, mace, allspice, mustard, horseradish, bay laurel, dill, ginger, nutmeg, and salt; preserved, spiced gherkins were a condiment or side dish familiar to upper social classes, rather than a dietary staple like cucumber pickles.

A variation on pickling, cucumber vinegar was a flavored vinegar into which large cucumbers were sliced and steeped for a few days. Cucumbers had the potential to grow to enormous size and become tough and bitter, which destined them for preserving or cooking; one solution was to stew large cucumbers and flavor them with cayenne and catsup. Some even considered cucumbers potentially unhealthy, and *The American Frugal Housewife* (Child 1844) described soaking sliced cucumbers in cold water to remove "the slimy matter, so injurious to health." The bitter flavor of cucumbers and other members of the gourd family (Cucurbitaceae) results from cucurbitacins, terpene-based compounds that are known for both their toxicity and their medicinal applications, including use as folk cancer remedies. In fact, some cucurbitacins are now known to exhibit antitumor activity, although there have also been cases of livestock poisoning from consuming too many cucurbitaceous fruits. Nevertheless, nineteenth century agricultural texts generally referred to cucumbers as cooling and healthy in the diet, a crop that could conveniently share a field or an orchard with corn or fruit trees. The plants benefited from slight shading during hot weather, and they were often more productive if the vines were grown like peas, with cut brush for support.

Watermelons (*Citrullus vulgaris*) are native to tropical and subtropical Africa and are likely the most recent edible pepos to arrive in North America. The cultivated fruits may first have been grown in the Kalahari Desert region, where they served as a valuable water source. The massive, watery fruits were valued by the African slaves who first carried watermelon seeds to the American colonies; their coarse, climbing vines were productive in poor or sandy soils, and watermelons thrived in the garden plots cultivated by African slaves who had grown the fruit for generations. Along with eggplant, field peas, yams, and sesame (*benne*), watermelon typified southern food traditions that incorporated various African plants that were carried on slave ships. In his *Natural History of Virginia, or The Newly Discovered Eden* (1737), William Byrd listed in Virginia gardens "many species of melons, such as watermelons and fragrant melons . . . golden, orange, green, and several other sorts . . . several varieties of cucumbers, which are very sweet and good-tasting; four species of pumpkins; cashaws . . . simnals, horns, squashes are also very good." Simmons (1796) noted that watermelons are "cultivated on sandy soils only . . . if a stratum of land be dug from a well, it will bring the first year good Water Melons," and colonists also soon learned that the vines produced prolifically in heat.

By the end of the nineteenth century, watermelons were garden favorites, and in the words of Mark Twain (in *The Tragedy of Pudd'nhead Wilson*, 1893), "It is the chief of this world's luxuries, king by the grace of God over all the fruits of the earth. When one has tasted it, he knows what the angels eat. It was not a Southern watermelon that Eve took; we know it because she repented." Seed catalogs provided tempting descriptions of watermelon cultivars, with various rinds (green, white, orange, blackish, and variously striped or mottled) and pink, red, or scarlet flesh. 'Kolb's Gem' had a strong rind that resisted breaking on the way to market, while 'Hungarian Honey' was advertised as having the richest, sweetest flavor possible and fruit that matured uniformly into dark green spheres filled with succulent red flesh that melted on the tongue. The red mesocarp in which the seeds are embedded is more than 90 percent water, and its bright pigmentation is from lycopene, a carotenoid pigment that is now valued for its antioxidant properties; lycopene has the ability to absorb the charged oxygen atoms (free radicals), which are potentially damaging to cells.

Citron melons are a watermelon variety (*Citrullis vulgaris* var. *citroides*) in which the flesh is white and inedible, but the dense rind provided a substitute for the candied citron that was a favored European sweetmeat. True citron (*Citrus medica*) is Asian, and the rinds from the small, green citrus fruits were soaked in brine and then in a concentrated sugar solution; candied citron peels were so commonly used in European cakes and confections that citron melon became a desirable substitute in early American gardens. Thoreau (1999) noted that citron melons were harvested after the fruit was ripened by frost, and then they were destined for the preserving kettle.

Growers who saved seeds had to prevent cross-pollination between citron melons and watermelons, which resulted in watermelons that resembled the citron variety rather than those with edible red flesh. Agriculturalists knew that cross-pollination with citron melons could also cause genetic deterioration in the quality of other cultivated pepo varieties.

Citron melons were also known as preserving melons, and cookbooks provided detailed instructions for their preparation. The pared rinds were soaked in brine and alum water, stewed along with some leaves from the vine, and then placed in sugar syrup and dried to produce a sticky, candied sweetmeat. The leaves presumably helped to keep the fruit an appealing green color; alum (usually potassium aluminum sulfate) was used as an astringent to preserve the crisp texture, but it had to be used sparingly (no more than a half teaspoon per pound of sliced rind) to avoid a bitter taste and possible toxicity. Once prepared, candied citron rind was mixed into the batters of tea cakes and wedding cakes. Beecher (1858) recommended a that a half pound of sliced citron along with raisins and currants be used to enrich "black cake," which required four hours' baking. Some cooks trimmed the citron rinds into fanciful shapes, and the rinds of other melons or pumpkins could be preserved in a similar manner. Citron melons were also

Citron melons (*Citrullis vulgaris* var. *citroides*) were a watermelon variety cultivated for their rind, which was preserved in sugar as a substitute for true citron.

preserved in syrup, pickled, or even sliced and used like apples in pies. Randolph (1824) offered a method for citron cream, in which the sliced rinds were sugared to form a syrup that was combined with cream for freezing.

Tomatoes and Their Relatives

The nightshade family (Solanaceae) includes a number of large edible berries such as tomatoes, eggplants, and various bell and cayenne peppers—fruits that are frequently considered "vegetables" because they are not sweet. Like white potatoes, which are also members of the nightshade family, these are tropical edible plants that were introduced into temperate North America by explorers and immigrants. In their native habitats, the species are weedy perennials, but North American gardeners soon learned to grow them as warm weather annuals. Wild tomatoes thrive in the Peruvian Andes, which is the center of wild tomato diversity, suggesting that this is the region in which tomatoes were first cultivated and domesticated. Spanish explorers likely carried cultivated tomatoes (*Lycopersicon esculentum*) from Central and South America north to Florida settlements during the mid-sixteenth century. The plants thrived, seeds were shared, and tomatoes gradually appeared in gardens up the Atlantic coast. These first cultivated tomatoes bore small fruits that resembled modern cherry tomatoes and the wild tomatoes native to South America. Historians hypothesize that tomatoes may also have arrived in southern states with the slave trade; Portuguese explorers had already introduced tomatoes as a food crop into western Africa during the fifteenth century.

Spanish explorers introduced tomatoes to the Mediterranean region, where they became known as *pome dei Moro* (Moor's apple), which eventually may have been altered to *pomme d'amour*, which perhaps accounts for the notion that tomatoes have aphrodisiacal properties. Based on their flowers, tomatoes were recognized by early herbalists and botanists as close relatives of nightshades, mandrake, belladonna, and henbane, all notoriously alkaloid-laden members of the nightshade family (Solanaceae). Unlike henbane and mandrake, tomatoes do not produce hallucinogenic, highly toxic tropane alkaloids, but the plants do synthesize glycoalkaloids such as solanine and solanidine that can cause gastrointestinal upset and neurological effects. Glycoalkaloid concentrations in cultivated tomato fruits are extremely low, but tomato fears persisted for centuries, perhaps because of the known toxicity of the hairy, pungent vines. The botanical name *Lycopersicon* translates literally as "wolf peach," an ancient Greek name adopted by Linnaeus, reflecting the early notion that tomatoes (like henbane) were herbs tied to witchcraft practices and werewolf activity.

During the eighteenth and early nineteenth centuries, tomatoes were botanical and culinary curiosities, but by 1858, Walden noted that tomatoes had "recently come to be generally esteemed" and that they could be grown under any conditions and at less expense than potatoes. He described several cultivars that varied in size, shape (pear, round, and scalloped types), and color (red and yellow). In warm habitats, tomato plants have the potential for perennial growth, and the base of the vines can become somewhat woody. Nineteenth century gardeners soon learned to stake them for support; Walden described using "slats, or board fence" and training the "principal bunches," a cultivation practice that was easily accomplished by urban families with small garden plots. He also suggested placing unripe tomatoes on a shingle or stone to hasten their ripening.

Southern cooks experimented with tomato cookery during the late eighteenth century. In 1809, Jefferson recorded planting "tomatas" in one of his garden squares at Monticello; his daughter and granddaughters prepared tomato-based dishes such as gumbos, soups, and omelets, and they pickled green tomatoes and prepared tomato preserves. Gumbos were thickened with okra (*Hibiscus esculentus,* also known as gumbo, gombo, or gumbs), a fruit native to tropical Africa that arrived in the New World by way of the West Indies with the slave trade. Typical of members of the mallow family (Malvaceae), okra capsules have secretory cells that release the

The first cultivated tomatoes (*Lycopersicon esculentum*) had small fruits that resembled modern cherry tomatoes, and a variety of cultivars were developed from the genetic variation in wild tomatoes.

mucilage traditionally used to thicken dishes that would otherwise be watery. In Creole cookery, gumbos were instead thickened and flavored with filé powder made from finely ground, aromatic sassafras leaves (*Sassafras albidum*); sassafras trees (see chapter 1) are remarkable in producing four distinct leaf types on a single tree, and their distinct flavor results from terpene-containing volatile oils, primarily safrole, which is now recognized as a potential carcinogen. Native Americans gathered and used sassafras leaves, which were eventually incorporated into Creole cookery along with African and French influences.

In *The Virginia House-Wife*, Randolph described the preparation of tomato catsup, marmalade, and "tomata soy," a sharp sauce that was spiced with cloves, allspice, black pepper, and cayenne and allowed to ferment for a few days before boiling and bottling. "To scollop tomatas" according to her method, fully ripe tomatoes were skinned and layered with bread crumbs and butter. Adventuresome cooks also prepared "gaspacha," described as a Spanish dish of sliced, fresh tomatoes layered with biscuits or bread and seasoned with mustard, oil, and onion. Tomatoes were also incorporated into soups and stews or were cooked in syrup and dried to resemble figs. By the end of the nineteenth century, elegant dishes of "tomatoes a la mayonnaise" were assembled from chilled, peeled tomatoes served with thick mayonnaise sauce and garnished with fresh parsley. According to Mary Henderson in *Practical Cooking, and Dinner Giving* (1878), a dish of fresh tomatoes was good enough to eat every day during the tomato season.

Many may have grown their first tomato plants as curiosities, with no intent of eating the fruits. New Englanders were the most reluctant to adopt tomatoes, raw or cooked, into their diets; during the early nineteenth century, northern farmers often fed their tomato harvest to livestock. According to local legend, an Italian painter working in Salem, Massachusetts, introduced the

Okra (*Hibiscus esculentus*) capsules produce a characteristic mucilage that was used to thicken tomato-based dishes such as gumbo.

fruit in 1802, but he could convince few to venture even a taste. Some New Englanders adopted methods of tomato cookery that involved lengthy boiling or stewing; Beecher (1858) described an "Indiana Receipt" for tomatoes that "stew very slowly all the forenoon" and were thickened with bread crumbs and beaten eggs. Some conservative northerners may also have been suspicious of the name "love apple" or "apple of love," reflected in Gerard's description of "Poma Amoris," which he described in the *Herball* as "naught and corrupt." The names "love apple" and "gold apple" were still listed nearly three centuries later as a tomato synonyms in *Sturtevant's Notes on Edible Plants* (1919). According to legend, suspicions about tomato edibility were challenged early by Colonel Robert Gibbon Johnson, who in 1820 stood in front of the Boston courthouse and vowed to consume a bushel of tomatoes; accounts in early farm journals relate that many spectators were shocked to see him survive.

The turning point in tomato history may have come coincidentally with a 1835 essay by Dr. John Cook Bennett in which he described tomatoes as "the most healthy article in the Materia Alimentary." He urged consumption of both raw and cooked tomatoes as a treatment for bilious attacks, dyspepsia, and even cholera, and his advice was reprinted in home and farm journals ranging from the *Maine Farmer* to *Southern Agriculturist.* Bennett predicted that the medicinal benefits of tomatoes would soon be available in concentrated capsules, and soon various tomato-based "liver pills" appeared in the nineteenth century pharmacopoeia of patent medicines. Competition for market share and medical claims was keen among various purveyors of tomato pills and the fad eventually lapsed, but the result was a change in the public perception of tomatoes. By the mid-nineteenth century, tomatoes were more commonly viewed as palatable rather than as dangerous "love apples." Tomato varieties were relatively few prior to the Civil War, and their uses were correlated with the available cultivars. Fig-shaped tomatoes were used for confections, while cherry and yellow types were frequently pickled; large red tomatoes were cooked down into catsups, sauces, and stews. At the end of the nineteenth century, George Washington Carver actively promoted the dietary benefits of tomatoes to poor southern families, whose diets often lacked sufficient vitamins. He eventually published a detailed guide to tomato culture and one hundred fifteen recipes for their cookery.

Wild tomatoes are berries with two internal chambers divided by a central wall, a hollow seed-filled fruit that clearly reveals the two-carpellate pistil typical of the nightshade family. Selection and hybridization of fruit with desirable traits resulted in new cultivated types during the second half of the nineteenth century, including those with thicker and more abundant wall tissue that mostly filled the internal fruit chambers. During the 1870s, Alexander Livingstone developed 'Livingstone's Paragon,' marketed as the first perfectly round tomato, and 'Livingstone's Acme,' with fruit that were medium-sized and tinged with purple. He also bred 'New Dwarf Aristocrat,' tomato plants named for their "erect bearing and dressy appearance," and rot-resistant 'Livingstone's Large Rose Peach.' Farmers often favored thick-skinned fruits that survived shipping to markets, and all growers sought cultivars that resisted rot and cracking. Tomato canneries supplied Union troops during the Civil War, and canners desired thick-walled, few-seeded tomatoes. Variability among

cultivated tomatoes provided the genetic raw material for selection and development of numerous cultivars, but many early named cultivars were essentially duplicates of others being sold by competing firms. By the late 1880s, a study of one hundred seventy-one cultivars revealed that they represented about sixty different types, many distinguishable by only minute differences.

Tomato plants were prolific and fueled nineteenth century notions of self-sufficiency, even among city-dwellers with little arable land; they were a featured crop in publications such as "How to Support a Family on One-Fourth Acre of Ground," supplied during the 1890s by seedsman Amos Ives Root of Medina, Ohio. The notion was that anyone could grow a successful tomato crop, which was also the assumption in the communal gardens of Rugby,

An article titled "The New Rugby" in *Harper's Weekly* (16 October 1880) illustrated the vegetable gardens cultivated in the village of Rugby, Tennessee.

Tennessee. Rugby was a visionary community on the Cumberland Plateau that was founded during the 1880s by Thomas Hughes, author of *Tom Brown's School Days*, a fictitious account of his days at the Rugby School in England. With the support of some concerned Bostonians, Hughes's intent was to provide a New World home and livelihood for the younger sons of English gentry who did not benefit from primogeniture. The Rugby plan was to engage in farming, establishing a New World agricultural colony that would allow English gentry lacking inherited wealth to live dignified, productive lives. Hughes favored a southern site "to heal up the breaches in this country resulting from the Civil War," and the agricultural scheme relied on the premise that crops would thrive in a southern climate. Hughes invested much of his own fortune and served as president of the Board of Aid to Land Ownership; support also came from such New England luminaries as James Russell Lowell, an editor, poet, and diplomat. However, soil quality in the Tennessee tablelands was moderate at best, and the new arrivals knew little about farming.

Optimistic advertisements for the Tabard Inn in Rugby praised the fresh produce from the demonstration garden that was planted in 1880. Early crops of tomatoes, peas, melons, beans, and cabbages were a qualified success, and *Harper's Weekly* of 16 October 1880 proclaimed, "An English garden has produced vegetables this summer in quantity and quality which fully justify the belief that fruit-growing will prove highly remunerative." *The Rugbeian*, the colony newspaper, encouraged new arrivals to consult with Amos Hill, the British garden manager who had farmed in Tennessee for several years and served as an agricultural advisor to new Rugby farmers. The Board of Aid optimistically backed the Rugby Canning Company, founded in 1883, which would undertake to can the tomato crop cultivated by Rugby farmers on the Cumberland Plateau. Decorative labels were printed for "Rugby Brand" tomatoes, with their price in English currency (although presumably the product would be sold to American customers); however, the tomato plan ultimately failed because the thin plateau soil could not support crops large enough to supply a cannery. This was an unfortunate setback; within years, the utopian American community for English gentlemen dwindled, the result of unrealistic expectations, typhoid epidemics, and the reluctance of the new immigrants to engage in arduous agricultural enterprise.

Eggplants (*Solanum melongena*) are tomato relatives, and they are also large berries borne on coarse herbaceous plants that produce prolifically in tropical and semitropical climates. Eggplants were domesticated in India from wild populations, and they were introduced into Europe by Arabs and into Africa by Persians. Eggplants arrived in southern kitchen gardens along with yams, field peas, and watermelons, a cluster of food plants that were introduced, cultivated, harvested, and prepared by the slave women who did the cooking for southern families; these crops help to define traditional southern cookery. Some Virginians initially mistook eggplants for a squash or melon variety and knew them as "guinea squash" or "guinea melons." In fact, eggplants are members of the same genus as white potatoes (*Solanum*), but they are grown for their massive purple or white berries rather than starchy underground tubers. The partially fused calyx (sepals) is a trait that helps to reveal eggplant alliances; the sepals that remain attached to mature tomatoes, eggplants, and bell peppers are a common characteristic of the nightshade family.

White eggplants were considered by many to be strictly ornamental and possibly dangerous. Gerard probably referred to white eggplants in his *Herball* description of "Madde" and "Raging" apples, and he commented that "doubtlesse these apples have a mischievous qualitie, the use whereof is utterly to bee forsaken." According to Gerard, the plants bore fruit "of the bignesse of a Swans egge," and he likened the leaves to those of henbane. The old names may hint at the long association of eggplants with epilepsy, which may explain their infrequent early use among both Europeans and American colonists. By the early nineteenth century, purple eggplants were fried or stuffed with minced meat and baked, the result of southern influence and the slave trade on American food traditions. Leslie (1851) also recommended a breakfast dish of stewed eggplant, mashed and flavored with marjoram and nutmeg. Cultivars varied in their shape (roundish to elongate) and the darkness of their purple pigmentation; types with smooth, deep purple skin and dense, white flesh were the most desirable, and cooks knew that eggplants should be harvested before they reached maximum size and became bitter. The plants required little attention once they were established, and soil enriched with manure and leaf mould ensured an abundant crop.

Hotbeds enabled gardeners to anticipate summer and produce early warm-weather crops, and the nightshade family figured strongly in the mid-nineteenth century construction of glass-enclosed wooden frames. These could be filled with seedlings such as those of tomatoes and eggplants. Heat was produced by layering the floor with fresh manure, and the hotbed concentrated heat from the metabolic activity of decomposing bacteria. This was augmented by solar heat that was trapped inside by the glass, and care had to be taken on warm days not to overheat the delicate seedlings. Gardeners also used hotbeds to start the seeds of various cultivars of *Capsicum*, the genus that produces various "hot" and "sweet" peppers. These included the familiar green bell pepper (*C. annuum*) and the spicy red peppers (*C. annuum* var. *longum* or *C. frutescens*), although red cultivars of bell peppers and mild cultivars of the chili-type hot peppers were also grown. Like tomatoes, peppers originated in the New World tropics; seeds were carried by Columbus to Europe, and New World peppers spread rapidly into Asia and Africa, in some regions supplanting the Old World black pepper (*Piper nigrum*). Columbus apparently mistook powdered New World peppers for the Old World black peppercorns long known in Europe, but the pungency of hot *Capsicum* peppers is attributable to the presence of capsaicin and related phenolic compounds rather than the volatile oils in black pepper. Although both are known commonly as peppers, *Capsicum* fruits are completely distinct from black pepper; we will consider the various peppers used as spices in the next chapter.

Bell peppers were grown at Monticello, where Jefferson's garden notes described them as "a good salad the seeds being removed," implying that some of the fruit was consumed fresh. Nineteenth century cookbooks included recipes for pickled peppers, which began with soaking the fruits in brine for several days and then storing them in vinegar; the seeds and stems were sometimes left intact, but a slit had to cut in the side to permit complete infusion of the brine and vinegar. Bell peppers were also "mangoed," stuffed with a spice mixture in the manner of mango melons to resemble the pickled true mangoes that were prepared in seventeenth century English kitchens. Beecher (1858) suggested filling peppers with chopped cabbage "seasoned with cloves, cinnamon, and mace" before preserving them in cold vinegar, a modest version of mock mango pickles that stretched costly imported spices.

Some long red peppers were also pickled, usually before they ripened, including the bird peppers (*Capsicum annuum* var. *glabriusculum*) that were grown for their small, spicy, orange-red fruits. In 1812 and 1813, Captain Samuel Brown sent seeds from Texas to Jefferson, who noted his taste for "a pickle of the green Pods with Salt & Vinegar." In 1814, Jefferson recorded "Capsicum Techas" in his kitchen garden and sent some of the seeds to seedsman and nurseryman Bernard McMahon in Philadelphia. The plants were cultivated as ornamental pot plants and were soon cultivated on Philadelphia windowsills; the bright berries were also used to flavor vinegar and sauces. Cultivated bird peppers are likely a dwarf variety of the wild Mexican peppers known as *chiltecpin*, which were gathered and sold in markets.

Grapes, Raisins, Currants, and Gooseberries

Grapes (*Vitis* spp.) are Old and New World berries, borne in dense clusters on strong, woody vines and cultivated for wine-making and for table use as a fresh, dried, or preserved fruit. In *American Cookery*, Simmons extolled grapes as growing "spontaneously in every state in the union. . . . The Madeira, Lisbon, and Malaga Grapes . . . cultivated in gardens in this country, and are a rich treat or desert [sic]." To nineteenth century Americans, grapes symbolized a fruitful harvest and prosperity, an edible metaphor for optimism and the horticultural expression of Victorian domestic ideals. Perfect grape clusters were artistically displayed on elaborate epergnes, rendered on canvas in oil paints, and carved into walnut sideboards and parlor chairs.

Although they were a dining fashion during the nineteenth century, grapes were domesticated centuries earlier by both Native Americans and Old World peasants; grape seeds excavated from archeological sites on both sides of the Atlantic suggest that wine-making had its origins in prehistory. In 1616, Lord Baltimore introduced the European wine grape (*Vitis vinifera*) to America for wine-making, but early attempts at colonial viniculture failed. Jefferson collected a diverse array of both European and American grape varieties at Monticello, and he experimented with planting and replanting his vineyards, which suggests that he frequently experienced failure. The fibrous leaves of yucca leaves (*Yucca* sp.) were used to tie the vines to their arbors (see chapter 8). Many of the European grapevines introduced to America probably succumbed to black rot and the native aphidlike insect known as phylloxera (*Daktulosphaira vitifoliae*) that feeds on grape roots, even if they survived cool winter temperatures. By the mid-nineteenth century, a few European varieties were known that could reliably survive mild southern winters, but imported grapes usually required cultivation in glasshouses ("grapehouses"), while American grapes such as the fox grape (*V. labrusca*) and the muscadines (*V. rotundifolia*) and their various cultivars were suitable for outdoor culture.

Both of these native grapes were cultivated at Monticello, long before 1849 when Ephraim Wales Bull developed 'Concord' grapes in Concord, Massachusetts, by selecting from among twenty-two thousand fox grape seedlings for frost resistance and rich flavor. The so-called "slipskin" 'Concord' fruits had a rounded shape, dark purple-black color, and a pale, waxy bloom on the epidermis; they were well-suited to use as both a table grape and for jellies and wine. Bull won a first prize for his efforts from the Boston Horticultural Society in 1853 and subsequently charged one thousand dollars for a 'Concord' cutting. Cultivation was promoted by the belief held by many that fresh grapes were the most palatable of all fruits—a veritable health food. During the nineteenth century, both grapes and grape wine were esteemed for their medicinal properties. Patients at Dr. John Harvey Kellogg's sanitorium at Battle Creek, Michigan (see chapter 2), were kept on a grape diet to cure high blood pressure, consuming ten to fourteen pounds of fruit daily. Another enthusiast was Sarah Josepha Hale, editor of *Godey's Lady's Book*, who always kept a bowl of fresh grapes at hand, even when the out-of-season fruit was prohibitively expensive. Rumors of impurities, poisons, and "ardent spirits" in commercial wines motivated many to cultivate their own wine grapes and ferment their own pure grape wines at home (see chapter 5). The combined domestic efforts of grape cultivation and kitchen wine-making seemed to insulate wine (especially wine used in the sickroom) from the suspicions of intemperance that clouded whiskey and other distilled liquors.

The unfermented juice of 'Concord' grapes also became a familiar domestic and ecclesiastical product during the second half of the nineteenth century, a product first bottled by a New Jersey dentist from his cultivated garden grapes. Dr. Thomas Welch and his son Charles cooked the fruit briefly, pressed the juice through cloth bags, and pasteurized the unfermented grape juice in bottles; the heat killed any yeast or bacteria that naturally colonized grape skins. No alcohol accumulated from yeast fermentation,

American fox grapes (*Vitis labrusca*) were the wild plants from which 'Concord' grapes were selected and propagated for use in jelly, juice, wine, and as a table grape.

and the product remained juice rather than becoming wine while stored. Grape juice was adopted by local churches as a non-alcoholic communion "wine," and by the 1890s, Charles Welch established his large juice works in Watkins Glen and Westfield, New York.

Grapes lent themselves to cultivation even in small yards, since the vines were trained vertically on trellises, fences, or walls; Walden (1858) advised that even a city family could have an ample supply of fresh grapes in season by training a grapevine to grow over a trellis raised two feet above the roof in the rear of their house. Limestone or household wastes such as wood ashes, charcoal, and bones provided the high calcium requirements of grape vines. Cultivation of a few varieties planted in proximity resulted in hybridization, the genetic opportunity to produce new varieties that might yield particularly desirable traits; home gardeners were encouraged to experiment in this way, by growing seedling grapes and then propagating desirable new types asexually by layering or grafting. Even tender European varieties could be attempted in a south-facing glasshouse by planting the vines outside and then training the stems to grow through an opening into the interior. During winter months, the roots of European varieties had to be mulched deeply, and the vines had to be lowered to the ground and covered with straw; growers noted that even hardy varieties benefited from mulching and protection during winter.

Dried grapes were known variously as raisins, sultanas, and currants, depending on the grape varieties that were sun-dried to prepare these storable confections. Raisins were the traditional brown dried grapes, with or without seeds; their dark color in part resulted from melanin pigments that formed when phenols in the grape flesh react enzymatically with oxygen. Sultanas were small, seedless yellow raisins, while currants are an ancient Greek grape variety (probably derived from European wine grapes) that has been cultivated for two thousand years. These currants are distinct from the various gooseberry relatives (*Ribes* spp.) that were also known as currants and eaten fresh or in preserves rather than dried. Beginning in the mid-nineteenth century, grapes were grown in California specifically for raisin production. During 1851 and 1852, Colonel Agoston Haraszthy planted sweet muscatel and seedless sultana varieties in his San Diego County vineyard, and he imported vines and cuttings of choice varieties that yielded good dried fruit. Other growers followed suit in planting their vineyards with raisin grapes, and by 1873, California raisins were a routine market commodity.

Early raisins were not necessarily seedless, and Randolph described the process of preparing raisins (frequently referred to as "plums") for puddings and cakes by stoning the fruits, cutting each in half, and flouring them lightly so that they did not sink into the batter. Currants required careful washing and "picking" to assure that they were separated before they were combined with other ingredients. A recipe for "a good common wedding cake" described by Child (1844) required both currants and raisins, along with two dozen eggs, brandy, nutmeg, and mace, similar to nineteenth century recipes for dark fruit (black) cake. A recipe for plum pudding from *Godey's Lady's Book* (December 1885) called for raisins, sultanas, and currants, but no true plums. Mincemeat combined cooked, chopped beef with apples, currants, and raisins as a filling for pies; Leslie (1851) offered a Lenten version that substituted boiled eggs for the meat and specified sultana raisins and brandy, white wine, and rose water for flavoring.

Another grape product was tartaric acid, found as deposits of white crystals inside wine barrels. This is a natural by-product of grape juice fermentation that had been known since ancient times, sometimes found combined with potassium in the form of cream of tartar (potassium bitartrate). In fresh grapes, tartaric acid is dissolved in the sugar-laden juice, but in the presence of alcohol, it precipitates out of solution to form characteristic crystals. The association of the crystals with Tartarus, part of the mythological Greek underworld, dated from the tradition of medieval alchemy. Some of the earliest baking powders used tartaric acid as their acidic component; the acid reacts with baking soda to produce carbon dioxide bubbles, which make batters rise without the presence of yeast. Ancient Romans called these crystals *faecula* (little yeast), suggesting that they had a practical knowledge of tartaric acid chemistry and its use as a leavening agent.

Both tartaric acid and cream of tartar had medicinal uses during the Civil War as diuretics, laxatives, and skin coolants (so-called refrigerants), often prescribed along with jalap (*Ipomoea purga*) or senna (*Cassia acutifolia*). In her chapter "Cookery for the Sick," Hale (1841) recommended whey prepared by boiling milk and adding cream of tartar, which curdled the milk proteins with its acidity; the whey was decanted, sweetened, and used as a mild milky beverage for the infirm. Household uses included rubbing cream of tartar onto soiled kidskin gloves to whiten them. *Johnson's Universal Cyclopaedia* (1885–86) noted the use of cream of tartar and "acid tartrate" as mordants in dyeing wool and described the unfortunate practice of adulterating pure crystals with sawdust, clay, gypsum, flour, chalk, or alum. Tamarinds, legumes that are known for their tart astringency (see chapter 3), also contain tartaric acid and yield tartaric acid crystals as their juice ferments.

Common names, even as applied to familiar fruits, can lead to confusion; *currant* has been used for two distinctly different berries—one a type of seedless raisin and the other a close gooseberry relative in the genus *Ribes*. Currants (*Ribes* spp.) are members of the saxifrage family (Saxifragaceae) and are native to both Europe and America. Black and red currants were noted by John Josselyn in his seventeenth century descriptions of New England, and some American farmers domesticated the native black currant (*R. americanum.*). In 1790, the William Prince Nursery catalog listed red, black, and white currants (probably all European stock), while American black currants were being cultivated in England as a curiosity. Jefferson grew the golden currants (*R. aureum*) that were collected by the Lewis and Clark expedition along with raspberries and gooseberries in his garden squares, which were planted with berries and berrylike fruits.

Red, white, and black European currants (*Ribes sativum* and *R. nigrum*) were generally favored for cultivation. These were introduced early to New World gardens, and the shrubs soon naturalized into the American landscape. Nineteenth century agriculturalists considered currants among the best of all cultivated fruits both for their diverse uses (pies, jellies, and wine) and because the ripe fruit remained for several days on the shrubs without dropping or decomposing. As a result of interest and selection, currant varieties became abundant to the point of confusion; Dutch and French growers developed many of the cultivars that were grown in the United States, including the red and white Dutch varieties that were prized for their size, uniform growth, and low acidity. The shrubs grew well-mulched in various garden soils, and growers sometimes trimmed the shrubs to resemble small trees, but their natural shrubby growth form was more adapted to windy sites. By planting currants along a northern wall or partially shaded by trees, a gardener could prolong the fresh currant season until October.

Black currants from English stock were made into jam or jelly used medicinally for sore throats, and black currant wine was considered by many to be the most medicinal of any home-produced fruit wine; the fruits are now known to have a remarkably high concentration of both citric acid and vitamin C, which can vary with the season and cultivar. 'Black Naples' currants were larger and more abundant than the older varieties, and mid-nineteenth century agriculturists urged gardeners to replace their old black currant shrubs with this new variety, but the widespread cultivation of currants in America gardens did not continue for many more decades.

The demise of currants resulted from the arrival of white pine blister rust (*Cronartium ribicola*), a parasitic fungus introduced to the United States with imported nursery stock in 1900. Its complex life cycle required two hosts, pine trees and either currants or gooseberries (*Ribes* spp.); both hosts are necessary for completion of the life cycle and spread of the infectious spores, and the European black current (*R. nigram*) is particularly susceptible to the white pine blister rust. Since the rust is fatal to many pines (but only a minor infection in currants and gooseberries), beginning in 1918 various federal and state legislation forbid the planting or cultivation any *Ribes* species. White pines were essential trees for lumber, masts, and many other timber products, and their disappearance would have brought financial ruin to many. As a result, many cultivated currants and gooseberries were destroyed by the Civilian Conservation Corps, although plant pathologists now believe that wild populations of native currants are more susceptible to the fungus than most cultivars. With the possible exception of some resistant cultivars, currants or gooseberries are still restricted in several states including Massachusetts, Maine, Ohio, and Virginia, but the laws are enforced inconsistently. Prudence is still in order: during the 1930s, botanist Liberty Hyde Bailey of Cornell University recommended planting currants or gooseberries no closer than two hundred to three hundred yards from a stand of any five-needled pine, a rule of thumb that many still observe.

Gooseberries are close currant relatives, similar in having both American and European species and in their use as dessert fruits. European varieties have yellow, green, or red fruits, all cultivars of *Ribes grossularia.* The William Prince Nursery catalog (1790) listed "Large yellow oval," "Great amber," and green varieties, presumably all from European stock, and American interest in gooseberries seemed to arise only from the English taste for these fruits. Gooseberries were among the first fruits in Jefferson's garden; he planted twelve cuttings in 1767, and Randolph described gooseberry fool, prepared by stewing green gooseberries with sugar, straining, and serving the chilled juice with "rich boiled custard." Gooseberries were also used to fill tarts and make preserves, yet they remained a minor fruit in American gardens and markets, in contrast to England, where nineteenth century gardeners competed to grow the largest fruits. Perhaps this was the result of climate, since gooseberry shrubs thrive in cool, moist conditions and according to Walden (1858) "do not flourish so naturally south of Philadelphia." Many nineteenth century American gardeners considered 'Houghton's Seedling,' which originated in Massachusetts from European stock, the best gooseberry to cultivate for its good flavor, thin skin, productivity, and resistance to fungi. Wild American gooseberries remained undomesticated but could be gathered and used; the bristly gooseberry (*R. hirtallum*) produced red or purple berries that were collected by Thoreau in July and described by him as "wild tasted" and marked by "internal meridional lines."

Nuts

A nut is defined botanically as single-seeded, indehiscent fruit with a hard fruit wall (pericarp), and edible nuts contain large seeds with abundant, highly nutritious endosperm. Nuts are the typical fruit of wind-pollinated trees, including walnuts and butternuts (*Juglans* spp.), hickories and pecans (*Carya*), and chestnuts (*Castanea* spp.). Natural selection has favored single-seeded fruits, since wind pollination depends on the random dispersal of huge amounts of pollen by wind currents; if a flower is not pollinated, the loss of a single ovule represents only a minor investment for the tree. Dispersal is also linked to the large, single-seeded strategy. The nutritious nuts are frequently collected and stored by animals that effectively "plant" tree seedlings when stored nuts are forgotten. The endosperm of most nut seeds contains primarily proteins and fats, with the exception of chestnuts, which are about 70 percent starch; these nutrients support seedling growth and serve coincidentally as an important animal and human food.

American cookbooks, both manuscripts and published volumes, often refer to the nuts used in cookery only in a generic sense. This suggests that cooks used what was on hand, whether the larder held native black walnuts and hickories or imported nuts from the marketplace. The William Prince Nursery catalog (1790) listed several nut species and varieties, including "Long black walnuts," "Short black walnuts," "White walnuts, many sorts," "Madeira nut" (English walnuts), and "American hazle-nuts." It seems likely that more nuts were eaten directly out of the shell than were used in cookery, and they were a food especially familiar to children. Gathering, cracking, and eating nuts were typical youthful pastimes, and agricultural writers encouraged families with children to establish nut groves on their property. Chestnuts were particularly favored, perhaps because of the added adventure of roasting the nuts in an open fire; Walden (1858) asserted that "Five or six trees would afford the children in a family a great luxury, annually."

Chestnuts were part of the eastern rural economy; families sold chestnuts gathered from the wild as a cash crop and kept what they could use in a year. Chestnuts were stored dry, but pickling methods took advantage of a windfall crop by preserving whole butternuts and walnuts in vinegar and spice mixtures. Recipes advised scalding, scraping, or even soaking the nuts in lye to remove some of the outer shell. Cooks pricked each nut with a needle to allow complete infusion of the vinegar and various combinations of cloves, allspice, black pepper, ginger, nutmeg, mace, garlic, mustard, and horseradish. Similar spices were used in flavoring walnut catsup, a potent sauce prepared by mashing and infusing green walnuts in vinegar.

According to the Doctrine of Signatures, walnut seeds were useful in treating brain ailments because of the similarities in their shape and surface sculpting, although walnuts were probably not introduced to England until the mid-sixteenth century. So-called English walnuts are the fruits of *Juglans regia*, which originated from wild populations in eastern Europe and Asia. They acquired their name because they were frequent imports on English ships. English walnuts were among the first trees introduced by colonists in both Virginia and Massachusetts, but few of the trees survived eastern North American growing conditions. Colonists soon discovered black walnuts (*J. nigra*), a native American tree that grew to six feet in diameter and provided a reliable annual nut crop; eighteenth century furniture-makers valued black walnut wood, and it was exported to England from Virginia as early as 1610. Settlers in Pennsylvania selected sites with natural black walnut groves, and they followed the lead of Native Americans who mixed black walnuts with cornmeal in bread-baking and used the nuts to prepare soups. The trees required high soil concentrations of limestone, and presence of the natural black walnut groves suggested soil fertility. Although walnuts could serve as a dietary staple in times of need, they were sometimes also known as luxury food; Randolph (1824) recommended making a rich ice cream by freezing a quart of cream mixed with a pound of pulverized black walnuts. She also provided instructions for preparing walnut catsup made of pulverized nuts preserved in vinegar and flavored with cloves, mace, and garlic.

When walnuts arrive at the market, the outer layer (the husk) has been removed; most are now familiar only with the nutshell that surrounds each seed, but evolution of the outer husk remains a conundrum to botanists. If they are nuts, then botanists presume that their outer layer is derived from an involucre of bracts (modified leaves) or sepals that have fused and enclosed the fruits. In contrast, some botanists interpret walnuts as drupes, stone fruits with a pigment-rich mesocarp (the husk) and a hard endocarp (the nutshell) that surrounds the seed. Regardless of the interpretation, the husk can be impenetrably tough; it contains juglone, which is produced by several members of the walnut family (Juglandaceae), including butternuts and some hickories. Juglone is classified chemically as a quinone, and in nature it functions in allelopathy, the inhibition of competing plants. Soil impregnated with juglone from walnut husks or other tree parts will be toxic for about a year, depending on environmental factors. Colonists learned firsthand the effect of juglone inhibition, when apple seedlings or other crops wilted and died when planted in soil where walnuts had thrived. Based on the false notion that walnuts would also repel insects, some farmers planted their barnyards with walnut trees.

As their common name suggests, butternuts or white walnuts (*Juglans cinerarea*) had sweeter, more oily seeds, but some considered the nuts too rich for healthy eating. The oil was valued for lubricating clockworks, however, and a syrup similar to maple syrup was prepared by tapping and boiling down butternut sap. The hairy covering on their husks releases a resistant brownish stain, and various dyes were made the bark, roots, and fruits of both butternuts and black walnuts.

Hickories and pecans (*Carya* spp.) have an outer husk that splits into four distinct valves, revealing the nut inside, and abundant quantities were available for gathering from wild trees. In his *Travels Through North and South Carolina, Georgia, East and West Florida*, William Bartram (1791) described bushels of "shell-barked hiccory" stored by the Creeks and used in almost all of their cookery. These were the same shagbark hickories (*C. ovata*) that were gathered by pioneer children for nut-cracking parties and noted by Thoreau as growing commonly in New England. Child (1844) described the practice of using cracked shagbark hickory nuts as a bait for the red ants that were "among the worst

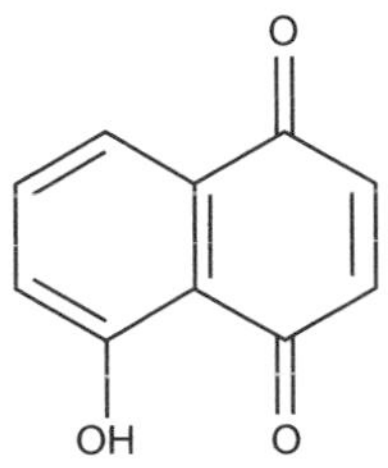

Chemical structure of juglone, produced by walnuts and their relatives (*Juglans* spp.)

plagues that can infest a house," and the inner bark yielded a yellow dye. The related mockernuts (*C. tomentosa*) were less desirable as a food, with small seeds encased in a hard shell.

Pecans (*Carya illinoensis*) had a wide range in central and southeast North America; despite their scientific name, Illinois is the northern limit of their range, and large natural populations occurred in Texas, Oklahoma, Kansas, and Missouri. Native Americans ground pecans to thicken stews, pressed the nuts for oil, and stored the nuts as an emergency food. Pecans were unknown to early colonists in Massachusetts and Virginia, although in 1541 they were discovered in present-day Arkansas by Hernando DeSoto, who mistook them for walnuts. Before pioneers settled the central United States, fur traders brought pecans east, where they became known as "Mississippi nuts" or "Illinois nuts." Jefferson planted pecans at Monticello and sent seeds to Washington at Mount Vernon, where the trees thrived. Settlers migrating into Texas relied on pecan trees for shade, shelter, lumber, fuel, and food, where some likened pecans to manna falling from heaven. By the mid-nineteenth century, horticultural varieties were selected and were grafted onto wild stock, but most pecans were gathered from wild trees until about 1900. Large trees were felled and picked clean by boys who scrambled among the branches, a common method that damaged wild pecan populations. Pecans are the only native American trees that farmers planted as cultivated orchards, primarily in Georgia.

American gardeners also occasionally cultivated various filberts, known also as hazelnuts or cobnuts, and sold in the Prince Nursery catalog (1790) as "American hazle-nut trees," "Barcelona nuts," and "Fill buds." Filberts are easily distinguished by a leafy bract that surrounds each nut, and they are related to birches, alders, and hornbeams (family Betulaceae). The original interest was in European filberts (*Corylus avellana* and *C. maxima*), but a closely related American shrub (*C. americana*) also yields edible nuts and hybridized with the European trees.

Chestnut trees (*Castanea* spp.) grew in forests across temperate America, Europe, and Asia, and the American species (*C. dentata*) was a dominant North American forest tree until most of the mature trees were destroyed by chestnut blight (*Cryphonectria parasitica*). This parasitic fungus probably arrived with Asian nursery stock of the Chinese chestnut (*C. mollissima*) imported from northern China about 1890. The fungal spores are wind dispersed and enter the bark of American chestnut trees through small cracks or openings; the blight then kills the trunks, leaving behind viable roots that sprout vigorously but succumb to the fungus before they flower and set fruit. The impact on eastern American forests was enormous, since one-fourth of all trees were chestnuts, but occasional chestnut trees avoided infestation and continue to produce the typical clusters of brown nuts surrounded by prickly valves that split apart at maturity. Some nineteenth century gardeners also planted chinquapin chestnuts (*C. pumila*) in their fruit gardens and ornamental landscaping. Chinquapin seedlings mature into shrubs or small trees, depending on their habitat; although its nuts are smaller, they were sometimes used like chestnuts.

Chestnuts in the genus *Castanea* are distinct from European horse chestnuts (*Aesculus hippocastanum*), which are relatives of native American buckeyes (*A. glabra* and *A. octandra*). The superficial, coincidental similarities in their fruits and seeds resulted in their confusing common names; the large, dark brown seeds of horse chestnuts occur in a spiny capsule, reminiscent of the appearance of true chestnuts (nuts) surrounded by spiny valves (bracts). The "spreading chestnut tree" mentioned in Henry Wadsworth Longfellow's "The Village Blacksmith" was in fact a horse chestnut that grew on Brattle Street in Cambridge, Massachusetts. When the tree was cut down to widen the road, Cambridge school children collected pennies to have the wood fashioned into a chair for Longfellow. Despite their similarity to true chestnuts, horse chestnut seeds contain the toxic glycoside aesculin and cannot be used as a food; however, small quantities were used medicinally during the nineteenth century to treat hemorrhoids and rheumatic pain.

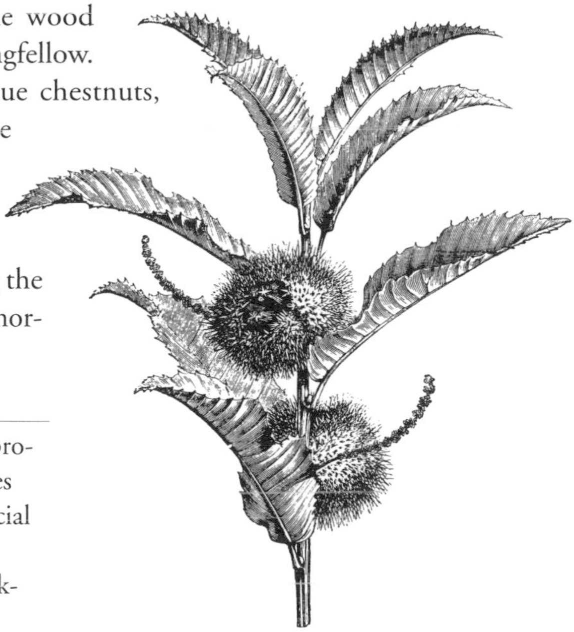

The spiny bracts and dark nuts produced by American chestnut trees (*Castanea dentata*) bear a superficial resemblance to the capsules and seeds of horse chestnuts and buckeyes (*Aesculus* spp.).

Exotic Fruits

The allure of the foreign is known to gardeners and cooks alike; exotic plants and foods brighten the landscape and the palate, and the same desire sparks curiosity for the unique or unfamiliar in both the kitchen and the garden. During the eighteenth century, imported fruits and spices were sold in coastal cities such as Portsmouth, New Hampshire. The markets, shops, and taverns of shipping ports were sites for the sale of imports, where inland shopkeepers could acquire a variety of goods for resale. Even frontier towns such as Deerfield, in western Massachusetts, received sporadic shipments carried by oxen teams from the coast, imports that included various tropical fruits. Fruits such as oranges (sweet oranges, *Citrus sinensis*, and bitter oranges, *C. aurantium*), lemons (*C. limon*), and limes (*C. aurantifolia*) were sold directly from sailing vessels as they unloaded their bounty onto the wharves.

The citrus fruit type is known botanically as a hesperidium, contained in a leathery rind with pockets of aromatic oils that characterize members of the citrus family (Rutaceae). The pulp is composed of internal hairs (trichomes) that differentiate into juice-filled sacs inside the maturing ovary. The outer rind is pigmented with chlorophyll and orange and yellow carotenoids, depending on the species, but many citrus fruits remain green even when they are fully ripe. Imports to America likely included shipments of green oranges and lemons that were fully palatable; ripening techniques using ethylene, a plant hormone, now produce the characteristic "ripe" pigmentation of various citrus varieties. Citrus juice is composed of sugars and acids, particularly citric acid, which is closely related to the ascorbic acid (vitamin C) in citrus fruits that prevents scurvy. Scurvy was known as "scorbutic fever," the first case of a disease known to be cured by a food, and seventeenth century naval surgeons realized the "antiscorbutic" properties of citrus fruits. The British navy adopted the practice of stocking limes on shipboard in 1795, and the American navy followed suit in 1812. Eventually, the spread of citrus followed ocean routes, and they were introduced in tropical regions worldwide to provide a reliable fruit supply for sailors.

Of course, it is unlikely that sporadic citrus imports prevented scurvy among American colonists, who relied on green vegetables as sources of vitamin C. In fact, the common colonial use of citrus fruits was in preparing a punch made from rum boiled with citrus and sugar, the preferred beverage for most occasions; early Americans generally imbibed citrus fruits with rum rather than using them as a food in the strict sense. Citrus fruits were also used as flavorings in confections, and oranges, lemons, and limes were often used interchangeably. *The Art of Cookery* (Glasse 1805) related methods for preparing creams, puddings, cakes, tarts, jelly, marmalade, and wine—recipes sufficiently diverse that when citrus fruits were available, cooks were not at a loss for ideas as to how to use them. Preservation methods often involved the use of refined sugar, also an imported luxury, but not even the citrus peels were wasted. Frugal cooks used lemon and orange rinds to flavor brandy, which was added to various pies, puddings, and cakes even by those who eschewed alcohol. Randolph (1824) also mentioned orange flower water that was used to flavor flummery, a molded mixture of jelly, cream, and wine. A few oranges were eaten as fresh fruit, but imports included both bitter and sweet orange species; Leslie (1851) instructed cooks to "strew powdered sugar" over peeled, sliced, and seeded oranges, hinting at their tartness. She also suggested a homemade version of curaçao, a liqueur made from grated orange peel, orange juice, sugar, and "the clearest rectified spirit." This was also made with shaddock (*Citrus maxima*), a primitive hesperidium ancestral to cultivated grapefruits.

Lemons were valued for their rind and juice, and during the nineteenth century they were used to flavor creams, puddings, and elaborate desserts such as floating island, sliced sponge cake spread with jam and submerged in lemon-flavored cream. Lemon peel was a substitute for true citron (*Citrus medica*) in cakes, and Child (1844) suggested to pie-bakers, "If your apples lack spirit, grate in a whole lemon." Lemonade (which was also made with oranges or limes) was sometimes frozen to resemble sherbet. A medicinal version of lemonade included

Lemons (*Citrus limon*) were used for the flavor of their juice and rind; whole lemons and other citrus fruits were often preserved in sugar syrup when they were available.

lemon juice, flaxseed, and gum arabic (a thickening agent derived from *Acacia* trees; see chapter 6) to yield a soothing, mucilaginous beverage for treating colds and coughs. Sporadic lemon imports were preserved for future use; lemon juice was bottled with sugar and brandy, and quartered lemons were pickled in vinegar flavored with nutmeg, mace, horseradish, garlic, and mustard. Whole lemons and other citrus fruit were packed in sugar syrup. Lumps of sugar were rubbed onto lemon rinds to prepare essence of lemon peel, and the flavored sugar was packed into an airtight china jar, another way to preserve lemon oil as a flavoring for confections. Child recommended a thrifty syrup that used citric or tartaric acid for tartness and a few drops of lemon oil for flavor, useful for cooks who lacked fresh lemons.

Even before their medicinal properties were appreciated, explorers carried citrus fruits from their native Asian habitats worldwide. Columbus brought seeds of bitter oranges, lemons, and citrons to Hispaniola in 1493, and by 1565 bitter oranges that were carried by Spanish and Portuguese explorers were growing on the Florida peninsula. Citrus fruits remained rare despite their spread, and until the last century citrus fruits were foods that most Americans rarely or never consumed. A few wealthy Americans followed European custom and built orangeries, including Andrew Jackson, who had one (now demolished) constructed at the White House during his presidency. Washington also cultivated lemons, limes, and shaddocks in a glasshouse at Mount Vernon. These glass and wood greenhouses provided a suitable habitat for cultivating mature citrus trees but were a maintenance burden from exposure to continual moisture, and outdoor citrus cultivation remained vastly more practical. After the United States took possession of Florida during the early nineteenth century, planters established groves of sweet oranges using bitter orange rootstocks for grafting, the beginning of the industry that still supplies American kitchens with citrus fruits.

California also acquired citrus crops, credited in part to an enterprising American missionary who in 1870 shipped twelve orange saplings to the United States Department of Agriculture from Bahia, Brazil. The USDA propagated the Brazilian variety and offered the trees to any Californian who would like to cultivate the fruits. Mrs. Luther Tibbets of Riverside, California, acquired two trees in 1873, and all modern navel oranges have been propagated vegetatively from her stock, passing to their descendants some curious floral anomalies. Navel orange pollen is usually sterile, and a second small abortive ovary forms on top of the normal ovary. The second ovary forms the "navel" when the normal ovary matures into a seedless orange through parthenocarpy, the occasional process in which a fruit develops without fertilization. Navel orange selection coincided with improved shipping, and these became popular table fruits. Manufacturers of domestic goods noted the interest in eating citrus fruits, and some perceived a market for a polite method of peeling the fruit. In 1879, Holmes, Booth, and Hayden patented a silver "orange peeler," a tool that performed the task of manual peeling. This was followed in the marketplace by a variety of specialized footed bowls, knives, and spoons designed for serving oranges at the table.

Historians have long considered pineapples (*Ananas comosus*) to be the quintessentially colonial fruit, although like other tropical fruits, they were rare or absent in most American homes. Columbus first found pineapples on the Caribbean Island of Guadeloupe, where they arrived with Native American migration from South America. He and others likened their rough, geometric exterior to a pine cone and interior solid flesh to an apple, the apparent origin of the common name. Pineapples reward the work of cultivation with extraordinarily sweetness hidden below a deceptively rough exterior; their taste was remarkable in an era when refined sugar was uncommon. Americans first knew pineapples as rare fruits that arrived on ships homebound from the Caribbean, although many decomposed before they reached colonial merchants in Boston, Philadelphia, Annapolis, and Williamsburg. According to local lore, sea captains impaled pineapples outside their homes to signal a safe return from tropical voyages. Rarity encouraged desire, and some colonists went to great lengths to obtain pineapples for decorative fruit displays that marked significant events and festivities; pineapples denoted social status, and the presence of a pineapple marked the occasion and guests as worthy of special provisions.

Although many pineapples were displayed rather than eaten, pineapple taste and texture appealed to the American palate. In 1751 during a trip to Barbados, Washington described in his diary the various new fruits that he tasted, noting "none pleases my taste as do's the pine." Pineapples evolved into a favorite nineteenth century dessert fruit, and the three-dimensional pineapple shape also lent itself to molds used for shaping gelatins and creams. The symmetrical pineapple form was also translated into carvings and other design elements; a stylized pineapple pattern was often rendered in stencils, weaving, needlework, plaster, pottery, wood, and metal. Pineapple designs were particularly common in Marseilles quilting, also known as stuffed work, in which cotton wool was used to pad outlined areas in quilted design. This technique for making decorative coverlets and petticoats was introduced to America in 1760, both as handwork and as machine-made imports; many colonists probably knew pineapples first as coverlet designs rather than as an edible fruit.

Pineapple plants are bromeliads (family Bromeliaceae) with large, spiky rosettes of fibrous leaves, and they are particularly adapted to dry conditions. A pineapple is defined botanically as a multiple fruit, a fleshy structure formed by the fusion of one hundred to two hundred flowers that grow on a single unbranched stalk. The flowers are pink to purple in color, adapted to attract hummingbird pollinators, and each is subtended by a small colored leaf (bract) that fuses into the multiple fruit. The plants have been variously described as monocarpic or biennial, flowering and fruiting one time. Most cultivated pineapples are seedless, the result of parthenocarpy in which the fruits develop without fertilization. Like many other bromeliads, pineapple plants reproduce asexually to establish genetic clones of the parental plants. Both stem suckers and the short, leafy stem that crowns the fruit can sprout adventitious roots when they are buried in sand, and these methods were used to establish cultivated pineapple plantations. Like citrus, pineapples have a high vitamin C content, and the fruits were carried on shipboard for their antiscorbutic properties. Sixteenth century Dutch, Portuguese, and Spanish sailors and traders planted leafy pineapple crowns on islands that they visited, and the ease of asexual propagation contributed to their cultivation in the Old and New World tropics.

Methods for pineapple preparation and preservation appeared frequently in American cookbooks by the mid-nineteenth century, as pineapples became more generally available. Of course, field-ripened fruit far surpassed those harvested for shipment while still green, but ripe

A pineapple is formed by the fusion of several flowers into a fleshy multiple fruit; its short, leafy stem is capable of rooting and cloning the parent plant if a pineapple becomes buried in sand.

pineapples decayed before reaching northern markets. Despite their natural taste, methods for pineapple cookery advised plenty of sugar, suggesting that many of the fruit that reached American markets were not always at their prime. Cookbooks recommended that slices of fresh pineapple benefited from a liberal sprinkling with powdered loaf sugar, and ice cream was made from sieved, sweetened pineapple mixed with cream and perhaps eggs. In *Practical Cooking, and Dinner Giving* (1878), Mary Henderson suggested serving pineapple ice cream molded and decorated with the crown of leaves. In her appendix to *The American Frugal Housewife* (1844), Child added a note about keeping pineapples fresh: "Pine Apples will be much better if the green crown at top be twisted off. The vegetation of the crown takes the goodness from the fruit. . . . The crown can be stuck on for ornament, if necessary." Leslie (1851) prepared "pine-apple-ade" from sliced ripe pineapples steeped in boiling water and powdered sugar, a concoction that was then strained and served with ice. Of course, those faced with a surfeit could preserve pineapples in syrup, either whole or sliced, the traditional method of fruit preservation. Cooks also learned by trial and error that pineapples contain bromelain, a protein-digesting enzyme that is denatured by cooking. Cooked or preserved pineapple was used in recipes including milk or gelatin to avoid the enzymatic breakdown of the milk or gelatin proteins.

Coconuts are the massive drupes (stone fruits) of the coconut palms (*Cocos nucifera*), a species that likely evolved in the western Pacific region and has been carried worldwide by ocean currents and human migration. The fibrous mesocarp and leathery exocarp are now usually stripped away before coconuts are brought to market, revealing the layers that surround the edible seed, the hard endocarp and papery seed coat. The water-resistant mesocarp keeps the fruit afloat, and occasionally coconuts arrive and germinate on island and continental shorelines. The mesocarp fibers, also known as coir, have long been used for matting and brushes, but the edible seed is the likely reason that by the mid-1500s traders and sailors carried coconuts to Spain, Portugal, and Brazil. The endocarp coconut shell had to be cracked to obtain the edible endosperm, and coconut shells had some uses as well. Earle (1898) noted that colonial craftsmen mounted coconut shells in silver or pewter to make decorative goblets, and she illustrated a "Silver-mounted Cocoanut Drinking-cup" that once belonged to John Hancock. Nineteenth century household manuals suggested using coconut shells to make hanging pots for plants.

Each massive coconut seed contains a small embryo and abundant white endosperm, which originates through double fertilization, similar to the edible endosperm in wheat and other grains. The endosperm begins its development with hundreds of divisions of the primary endosperm nucleus, formed from the fertilization of the central cell with a sperm produced by the pollen grain. This results in copious liquid endosperm tissue, the so-called coconut water, a liquid tissue that contains free-floating cellular structures, proteins, oils, and cytokinin, a plant hormone that promotes growth and differentiation of the coconut embryo. Eventually a cellular, solid endosperm layer forms from the liquid endosperm, and this is the familiar edible lining of a coconut seed that is grated to use in baking and confectionery. Coconut milk, used in cookery, is made by grating, moistening, and pressing fresh coconut endosperm to release a milky fluid; in many tropical areas, coconut milk is the primary liquid used in cooking. In tropical regions, local uses for coconut palms are diverse, including as a source of durable building materials, fibers, shell containers, and of course food and drink; its presence in a habitat almost assures human survival.

Coconuts arrived in New World markets as curiosities destined for limited use. Known in nineteenth century cookbooks as "cocoa-nuts," the endosperm was used exclusively in baking and confectionery, which required preparation: the shells had to be cracked and the endosperm removed and grated before it was mixed into ice creams, puddings, custards, and various cakes and macaroons. Typical flavorings for coconut confections and desserts were lemon, rose water, and nutmeg. Traditional Indian curries often called for coconut milk and were flavored with the mixture of turmeric and other spices known as curry powder; these dishes were incorporated into American cookery from England, where curried meats were adapted as part of the culinary tradition of the British Empire. In *The Virginia House-Wife*, Randolph described how "To Make a Dish of Curry After the East Indian Manner," but she omitted coconut milk as an ingredient, probably because of its rarity in colonial Virginia.

Bananas (*Musa*×*paradisiaca*, a hybrid of *M. acuminata* and *M. balbisiana*) were late arrivals in American kitchens. The fruit type is a berry with a leathery skin (exocarp), and like pineapples, domesticated bananas are seedless through parthenocarpy. Banana species evolved in east Asia and Australia, but both dessert (sweet) bananas and plantains (starchy cooking varieties) were carried worldwide and became an essential local crop in tropical countries where they were cultivated or naturalized. Early American colonists encountered few bananas carried to American port cities by trading vessels.

One historic anecdote relates the fate of the first bananas introduced to New England in 1690: the unpeeled fruits were stewed along with beef and vegetables in the traditional boiled dinner.

Bananas are delicate fruits that ripen when signaled by ethylene, a plant hormone that many ripening fruits release as a gas. Nineteenth century sailing schooners were too slow to provide reliable banana shipments from Central America before the bananas overripened and decayed from natural ethylene production, but the advent of steam ships during the second half of the nineteenth century increased the presence of bananas in American markets. By the 1880s, bananas could be widely purchased, and specialized serving pieces known as banana bowls or boats were marketed by crockery and glass manufacturers. Probably most bananas were eaten fresh; banana recipes were lacking in American cookbooks through most of the nineteenth century.

The Botanical Diet

The first colonists became competent agrarians who grew, harvested, and cooked with a diversity of fruits, vegetables, and grains, and early Americans consumed a variety of plants, from Native American corn, beans, and squash to European grains, coleworts, apples, and potatoes. Yet by the mid-nineteenth century, many working class Americans relied almost entirely on potatoes, cabbages, and turnips, augmented by preserves and pickled condiments, with most fresh fruits and vegetables limited to the summer months. Nor were dietary deficiencies limited to urban workers. Slaves cultivated garden plots on plantation land, but emancipated slaves and other poor southerners were often transient tenant farmers who lacked the stability to establish gardens for their households; their meager diets relied heavily on cornmeal, coffee, and molasses.

Nineteenth century social reformers focused particularly on the plight of the lower classes, which coincided with the first pioneering scientific studies of food chemistry and the first understanding of dietary proteins, carbohydrates, fats, and minerals. Some were also concerned that working classes were taking their meals at saloons and by the implications of that for temperance and family life. Ellen Richards, the first woman faculty member at the Massachusetts Institute of Technology, joined with Mary Hinman Abel to study "the food and nutrition of the working men and its possible relation to the question of the use of intoxicating liquors." Supported by Boston philanthropist Pauline

Agassiz Shaw, they fixed on the notion of opening up public kitchens to provide cheap, nourishing food "as a rival to the saloon," patterned on the *Volkskuchen* in Berlin. Part of their plan was to promote the Aladdin oven, a domestic cooker that provided the slow, steady heat necessary to cook inexpensive, tough meat. Rather than relying on costly beefsteak, their plan was that workers' wives could simmer inexpensive meat and vegetables all day in this early appliance, a tin-lined fiberboard or wooden box heated by a kerosene lamp. Edward Atkinson invented the prototype in the 1880s and promoted it as a way to improve diet and improve the plight of urban workers.

The first New England Kitchen opened in Boston in 1890, not as a soup kitchen but rather as an educational food laboratory with the cooking done in full view of the patrons who picked up meals for their families. The movement expanded to New York, Philadelphia, and Chicago, where a similar kitchen was established at Hull House, with the goal of assimilating recent immigrants by teaching them American cookery and eating habits. The fare consisted of long-simmered, vegetable-laden soups and stews prepared according to scientific standards, and wealthy benefactors were pleased with the notion that the kitchens were laboratories for food chemistry and preparation rather than soup kitchens. Well-meaning movement leaders ignored fresh fruits and vegetables in designing their menus because they did not understand the critical role of vitamins in metabolism; imperfect laboratory studies had concluded that most fresh fruits and vegetables consisted only of water and carbohydrates, which provided the "scientific" rationale for the omission of fresh fruits and vegetables. The outcome was a nineteenth century reprise of wintertime Puritan cookery that featured such dishes as baked beans, chowders, succotash, pea soup, and Indian pudding, and these were prepared and dished up year-round. Tomatoes were used for their flavor and only in small quantities, and leafy greens were thoroughly cooked; potato skins and bran from grains were removed and discarded. Of course, dietary calories often came from potatoes, which were consumed in huge quantities by many urban dwellers. An 1896 editorial in *Garden and Forest* magazine praised the visionary efforts of Detroit Mayor Hazen S. "Potato" Pingree in promoting potato cultivation by unemployed city residents; men with no agricultural experience grew sufficient crops to support their families for the winter, and soon a similar scheme was established in New York. Social reformers also valued the city potato plots as "providing the advantage of taking their children out of the heated houses and giving them a taste of rural life," while enabling "the superannuated and partially crippled" to supply their own tables.

Ethnic groups resisted dietary proselytizing and stuck with familiar foods, which in some cases were far more nutritious than the meals dished up by reformers; in addition, suspicions grew that the nineteenth century dietary reform movement meant to oppress working men by feeding them cheaply to justify low wages. By 1899, Atkinson ruminated that had he first introduced the Aladdin oven first to middle- and upper-class cooks for stewing venison and other game meats, the wives of workers would soon have clamored for his invention. Eventually, the New England Kitchens evolved into eateries that catered to middle-class workers, and nutrition reformers turned their attention to the teaching of nutrition and domestic science in public schools, as a way of promoting "melting pot" ideals. Jane Addams of Hull House noted that "an Italian girl who has had lessons in cooking at the public school, will help her mother connect the entire family with American food and household habits." Food traditions die hard, but perhaps immigrant children were more amenable to changing old ways and accepting the tide of Americanization.

Of course, it is ironic that the fruits and vegetables familiar to an Italian immigrant were nutritionally more sound than the meat and starch menu plans offered up by nineteenth century reformers. Yet reformers and welfare workers related some good lessons; thrifty meal preparation and domestic cleanliness were admirable goals, and children learned vegetable gardening in progressive settlement house kindergartens. Cultural tradition was the key influence in selecting food plants for cookery, more important than the dietary notions enforced by a few believers. Since the early seventeenth century, food traditions have influenced the diversity of food plants cultivated and imported to North America, which resulted in an early blending of Native American and European foods that continues to evolve with immigration and cultural changes.

CHAPTER FIVE

The Botanical Pantry I: Preservation, Wine, Vinegar, and Beverages

By the late fall, rural families transformed the bounty from kitchen gardens and orchards into an extensive larder stashed in pantries, storerooms, and root cellars. The practical goal was to store fruits and vegetables in ways that limited the growth of bacteria and fungi; depending upon the crop and local tradition, storage and preservation methods included burying, hanging, freezing, drying, pickling, candying, and jellying. Daily cooking also adapted to the harvest season, and some recipes reflected the thrifty ideal that nothing from the kitchen garden should be wasted. One such catchall dish was known as higdom, a spicy relish for meat that was prepared before the first frost to use the glut of tomatoes, peppers, cabbage, and onions; the method for preparing higdom appeared frequently in nineteenth century charity cookbooks.

Preservation methods were dictated by the nature of individual crops, and trial and error no doubt determined the best methods for various cultivars and climates. Fresh fruits and vegetables were kept from freezing in root cellars; root crops were buried in sand, celery stalks were earthed up in a corner, and apples and potatoes were carefully stacked in piles. Apples were sometimes packed in sand, sawdust, grain, or straw, or individual fruit were suspended by their stems. Walnuts and even lemons were also packed in sand for long-term keeping. Cabbages stored in dug pits stayed fresh until spring, but pumpkins and squashes required dry conditions and fared better in a cool, dry spare bedroom than in a damp cellar. Whole fruits and vegetables were still alive, respiring at low rates; their successful storage depended on low (not freezing) temperatures that kept their cells relatively dormant, with good air circulation, since living tissues use oxygen and release carbon dioxide. Small bruises or epidermal injuries provided an entry point for opportunistic bacteria and fungi, which explains the method used by some nineteenth century cooks to store tomatoes. Fresh, whole tomatoes were carefully placed in large glass jars and covered with lard, which solidified and effectively sealed the tomatoes against air and microorganisms.

Drying was another preservation solution, since dried foods last indefinitely if they are protected from dampness and insects. Apples, okra, pumpkins, and other fruits were sliced thin and threaded on string that could be hung in a storeroom to dry. Dried beans were stored in abundance and along with turnips and pumpkins formed the vegetable backbone of many diets, until spring arrived with new leafy, tender crops. Cold northern temperatures were also used to advantage; in *Oldtown Folks*, Harriet Beecher Stowe described the New England tradition of freezing fruit pies by storing them in an unheated "pie-room," where they often remained edible until spring. Typical colonial and nineteenth century winter diets also relied heavily on meat, and plants played a significant role in meat preservation. As an alternative to salting, meat was smoked with corn cobs or wood such as hickory and apple, which dehydrated the meat and produced antibacterial compounds such as cresols and organic acids. Early meat and fish cookery also used generous amounts of cinnamon, cloves, nutmeg, mace, sage, and garlic, spices with antibiotic properties that inhibit the growth of bacteria that would typically inhabit decomposing flesh (see chapter 6).

The bacteria and fungi that recycle nutrients in natural ecosystems can also decompose stored fruits and vegetables, and through the activity of a gaseous botanical hormone, one rotten apple could indeed spoil an entire barrel of fruit stored for the winter. The botanical effects of ethylene, a hydrocarbon gas, were first discovered in cities rather than rural root cellars, where shade trees dropped their leaves when they were planted near street lamps that burned illuminating gas. In addition to being a naturally occurring plant hormone, ethylene is a major chemical component of illuminating gas; trees interpreted ethylene in the atmosphere as a hormonal signal to defoliate. Ethylene is also released naturally by a ripening fruit and simultaneously promotes the ripening of nearby fruits. The presence of ethylene in the atmosphere signals a fruit to undergo changes in color, texture, acidity, and sugar concentration, the qualities that we associate with ripe, fully edible fruits.

Ethylene is also released in large quantities by damaged plant tissues that are beginning to decompose through the action of bacteria and fungi; fruits and vegetables that are bruised or injured hasten others nearby to overripen and become soft and watery. This is the botanical basis of the saying that "one bad apple spoils the barrel," the diffusion of ethylene from one wormy or bruised rotting apple and its unfortunate effect on essential winter stores. Fastidious housekeepers whitewashed their root cellar walls (and probably eliminated some microbial spores in the process) and routinely inspected the bins and shelves for signs of decomposition.

Pickling

Vinegar provides an acidic environment that retards bacterial growth, and preserving foods in vinegar is known as pickling, a method of fruit and vegetable preservation that was used by the ancient Romans. Colonial Americans made pickles in abundance to adorn meat dishes and platters, and pickling crocks lined colonial larders. Pickles of various sorts were often flavored with tropical spices and temperate herbs and were a mainstay of upper-class cookery, and New World melons and peppers substituted for the imported tropical mangoes that were pickled and served with meat in seventeenth century England. Then, as now, many pickling methods began with a step involving salt or brine, which dehydrated food tissues so that excess water did not dilute the vinegar used for preservation. Amelia Simmons (1796) described tiny gherkins as the best for cucumber pickles, and these were prepared in considerable quantity. Hannah Glasse (1805) instructed her readers to "Take five hundred gerkins" and combine them with vinegar flavored with cloves, mace, horseradish, bay leaves, ginger, dill, nutmeg, and salt. Fruits in particular were often pickled, since the price of imported sugar discouraged many from preparing fruit preserves.

Genteel cooks also pickled radish "pods" (botanically speaking, siliques, the fruits of radish plants), purslane, beets, walnuts, butternuts, peaches, cherries, asparagus, parsley, onions, cabbage, cauliflower, mushrooms, lemons, limes, nasturtium seeds, and barberries. Nasturtium seeds were pickled while still green, and some preferred them to imported capers served with boiled mutton; both contain mustard oils, the chemical basis of their sharp flavor. Pickled barberries (*Berberis vulgaris*) today may seem like an unlikely garnish, but their early appeal was thwarted when the shrubs were cleared from many wheat growing areas because the plants are alternate host for the wheat rust fungus (*Puccinia graminis*). Although the complex fungal life cycle was not understood until the 1860s, astute farmers observed that wheat rust seemed worse when barberries were present near their crops. By the mid-eighteenth century, Connecticut and Massachusetts both had prudent laws requiring the removal of barberries from grain fields; had the shrubs not been largely obliterated because of their association with a pathogenic fungus, perhaps pickled barberries would still be on American tables.

Nineteenth century cooks categorized pickles as kickshaws, delicacies or fancy foods that decorated tables and whetted appetites. Piccalilli mixtures, often made of chopped green tomatoes and peppers, also became popular by the end of the century, and many considered various pickles essential to a properly laid table. Homemade pickles were preferred over commercial products that were sometimes cooked in brass or copper kettles, which imparted bright but unnatural metallic hues to the finished product. Pickling spices and flavorings evolved into complex mixtures, which often incorporated whatever was on hand; Lydia Maria Child (1844) pickled cucumbers with allspice, mustard, pepper, and horseradish, "if you happen to have them." Her addition of "flag-root" (presumably sweet flag, *Acorus calamus*) suggests that perhaps she valued the uses of this aromatic plant to promote digestion and treat flatulence, the selfsame herb that was cultivated to treat colic and dyspepsia by the New England Shakers. Ambitious cooks could also use diverse garden produce to make "East India Pickle," described by Eliza Leslie (1851) as cabbage, cauliflower, carrots, cucumbers, beets, radish pods, nasturtium seeds, barberries, cherries, green grapes, string beans, onions, and various peppers "put promiscuously into a large earthen pan"; the mixture was soaked in brine and then pickled in vinegar spiced with ginger, garlic, turmeric, and mustard. Domestic manuals wisely advised storing pickles in wooden barrels or stoneware since vinegar reacted with the lead-based glazes that were used in making earthenware from porous clay.

Perhaps the most peculiar American pickles were made from martynias or "martinoes," the fruits of Brazilian devil's claw plants (*Ibicella lutea* and related species) in the martynia family (Martyniaceae). Child (1844) and

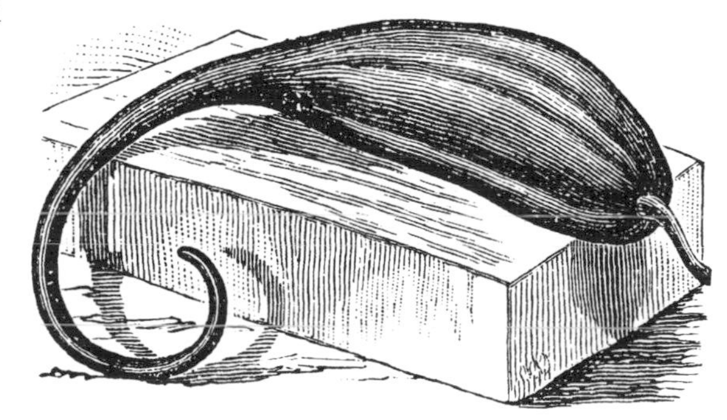

The immature capsules of martynias (*Ibicella lutea*) were pickled in spiced vinegar; the vines were a garden curiosity that self-sowed in the South and were grown as a hot-weather annuals in the North.

Catharine Beecher (1858) provided instructions for pickling "martinoes" in spiced vinegar, and some gardeners probably grew them out of curiosity rather than the need for another fruit to pickle. The plant's common name describes the elongated style that forms a woody hook on each mature fruit; these unique extensions clasp onto animals' feet and are an adaptation for seed dispersal in nature. The clusters of green capsules resembled okra or beans, and martynias thrived under the same hot conditions as okra. Northern gardeners learned to collect the seeds and start the young plants indoors as a annual crop, while gardeners in the middle Atlantic and southern states relied on the plants to self-sow. Martynias were harvested for pickling when they were immature and could still be pierced by a pinhead, before the capsules developed woody tissue. As coarse herbs with their bright yellow, catalpalike flowers and hairy, viscid leaves, these tropical plants also provided garden interest.

Catsup (or ketchup) originated as *kachiap*, a Malay term for the brine from pickled fish, and it was a thrifty way to use the flavorful dregs of vinegar left over from pickling. Sixteenth century English cooks developed spicy walnut and mushroom catsups, and two centuries later, these condiments were still prepared by American cooks. Tomato catsup first appeared in the early nineteenth century and at first was often confused with tomato soy, a sauce of tomatoes and spices that was permitted to ferment for a few days before bottling. Tomatoes were still relatively new to Americans, and many still distrusted them as poisonous "love apples" (see chapter 3). Mary Randolph (1824) instructed her readers to prepare tomato catsup in August from ripe tomatoes flavored with salt, pepper, and mace, but her recipe contained no vinegar. Nineteenth century cooks served tomato catsup with roasted meats and used it in cooking to improve "the richness of soup and chowder," as recommended by Child (1844). After the Civil War, Joshua Davenport of Massachusetts experimented with tomato catsup mixtures that contained both sugar and vinegar and were flavored with cinnamon and cayenne, and his combination of ingredients became the prototype for the familiar modern condiment.

Bottling and Canning

Bottling whole fruit in thick syrups and preparing fruit preserves were culinary traditions carried from kitchens in England to British colonies worldwide. High sugar concentrations dehydrate and inhibit the growth of microorganisms, but sugar was often a limiting factor in colonial kitchens because of its cost and availability. A cook faced with a surfeit of plums, pears, berries, and other fruit had to plan on approximately equal weights of sugar and fruit to produce a well-preserved product. When sugar was in short supply and less was used than advised, microbial contamination became a practical problem. Some cooks attempted to bottle fruit "dry" without sugar or additional liquid, a historical precursor to canning strategies. Simmons's method for keeping gooseberries and plums began with packing whole fruits in glass bottles, which were then corked and boiled until "they are a little codled and turn white." She advised sealing the corks with wax or pitch and then storing the bottles in the cellar. A similar method "To keep Green Gooseberries till Christmas" appeared in the 1805 American edition of *The Art of Cookery* by Hannah Glasse. She also recommended sealing fresh green peas in bottles and storing fresh green beans layered in salt, more preservation methods that provided a varied diet for the families of cooks who were willing to experiment. Beecher (1858) cautioned cooks preparing fruit preserves against skimping on sugar and recommended that all pots and jars be covered with brandy-saturated paper or sealed with "a split bladder," which would become air-tight as it dried. She noted that a thick covering of "leathery mould" was a help in preserving fruit, but she lacked modern knowledge of the dangers of mycotoxins; more wisely, she realized that "when mould appears in specks" (presumably bacterial colonies) the preserves needed to boiled again before they could be eaten safely.

Canning was first developed during the early 1800s by Nicholas Appert, a French chef and confectioner who perfected the method of a hermetical seal as part of a competition to provide food supplies for Napoleon's troops. He used glass bottles with wide necks, with the corks wired in place and sealed with pitch; the filled bottles were boiled in a water bath for several hours, which destroyed bacteria. Perhaps most significantly, he adhered to high standards of cleanliness, which at the time was rare. In 1810, Englishman Peter Durand patented tin-coated iron cans as an improvement over glass jars, but it was not until the Civil War and the need to supply troops that commercially canned foods became popular in the United States.

Nineteenth century cooks extended their bottling efforts beyond sweet fruit preserves and vinegary pickles, which both were kept relatively sterile with sugar and acidity. In 1858, John Mason patented his famous screw-neck jars that provided a mass-produced glass container for home-canned produce, but low-acid fruits and vegetables remained a potentially risky challenge. Jars of corn, pumpkin, beans, and other low-acid foods provided ideal conditions for bacteria that grow in oxygen-free habitats, such as *Clostridium botulinum*, a spore-producing bacterium that recycles organic matter in soil habitats. *C. botulinum* produces deadly nerve toxins, which cause paralysis and halt respiration; if these bacteria begin to grow in an oxygen-free jar of preserved food because a spore is accidentally introduced, they can cause botulism poisoning. Incorrect canning methods could indeed prove lethal, because the proper combination of heat and pressure are required to kill the bacterial spores that cause botulism poisoning.

Pressure cookers were not available for home use until after World War I, and these were necessary for the absolutely safe home canning of low-acid fruits and vegetables. Some nineteenth century cooks experimented with canning using only the hot water bath method, in which filled jars were covered with water and boiled intensively. Without pressure to kill spores and without sufficient acid, salt, or sugar to inhibit bacterial growth, this was high-risk food storage. The saving grace may have been the heat labile nature of botulism toxins, which break down when they are boiled for several minutes. When home-canned produce was used in preparing soups and stews, additional cooking likely prevented some cases of botulism.

Wine and Vinegar Production

While most modern consumers probably think of wine as fermented grape juice, early American wine-making was a broader enterprise that relied on a variety of sweet fruits or even vegetables such as carrots that store sugars in their root cortex layers. Some colonists envisioned thriving New World vineyards and wineries, but European wine grapes (*Vitis vinifera*) at first proved a difficult crop in colonial Virginia. The Jamestown Assembly required that all colonists plant grape vines annually, but viticulture was plagued from the start with fungi and the native aphidlike insect known as phylloxera (*Daktulosphaira vitifoliae*) that feeds on grape roots. Thomas Jefferson also hoped for successful Virginia vineyards, and he helped with the early experimental work done at Monticello by Phillip Mazzei of Tuscany.

During the first half of the nineteenth century, Virginia became an important wine-making state, well known for the production of a claret from the 'Norton' grapes that were first propagated by Dr. Norton of Richmond. Wines and many other fermented beverages were also made at home, using small-scale production and bottling procedures that would assure a pure product made from various wild and cultivated fruits. The method involved crushing ripe fruits to produce the must and then adding sugar and water that had been boiled thoroughly. Fermentation requires yeast, single-celled fungi that metabolize fruit sugars and convert them to ethanol and carbon dioxide through fermentation in the absence of oxygen. Fruit skins harbor natural yeast populations (*Saccharomyces ellipsoideus* and others) that live in the waxy bloom that coats the skins of grapes and other fruit. The cells multiply and grow while suspended in the fermenting juice, but eventually the wild yeast succumb to their own alcoholic wastes before the alcohol concentration exceeds the relatively low levels typical of wines.

In the case of grape wine, the skins of the crushed fruits (known as pomace) release anthocyanin pigments and impart the characteristic colors to red wines. An alternatives to grapes, various wild and cultivated fruits such as gooseberries and currants (*Ribes* spp.), elderberries and elderberry flowers (*Sambucus* spp.), and various raspberries (*Rubus* spp.) were also used in domestic wine-making. Sometimes sweet dessert wines, such as currant wine, were fortified with brandy before bottling. Bottles were sealed with corks cut from the outer bark of *Quercus suber*, a Mediterranean oak, and often sealed with rosin, the dried resin of various pines (*Pinus* spp.). The wines benefited from aging, and early wine-making instructions often mention storing the bottles in a sawdust bed, perhaps for a few years.

Colonists consumed large quantities of cider, whiskey, and rum, but during the eighteenth century, many Americans also acquired a taste for various wines, those fermented in both home kitchens and by European vintners. Many considered wine nutritious and medicinal, particularly when it was administered as wine whey, the liquid strained from milk that was boiled and curdled with the addition of sherry, Madeira, or

Gooseberries (*Ribes grossularia*) contained sufficient sugar for culturing yeast and were commonly used in making domestic wines.

currant wine. In her chapter "Receipts for Food and Drink for the Sick," Beecher (1858) described her method of mulling wine by boiling together a pint of wine with beaten eggs. Randolph's 1824 recipe for a "quire of paper pancakes" called for two gills (eight ounces) of wine to flavor the thin batter, and wine sauces were used on both roasted fowl and plum puddings. Wine jelly, sweetened wine congealed with isinglass (a gelatin protein from the air bladders of sturgeons, cod, and hake), was molded in shapes and eaten with cream.

During the 1840s, temperance reformers (see chapter 2) also turned their attention to wine as well as whiskey and encouraged ladies to provide water as an alternative to wine at dinner parties. Sarah Josepha Hale (1841) noted that "fermented liquors . . . if used at all as a drink, should be very sparingly taken . . . unless prescribed by the physician, it would be best to abstain entirely from their use." Nevertheless, home wine production and consumption continued unabated in towns such as Amherst, Massachusetts, where Edward Dickenson's family maintained a wine cellar with port, sherry, and sweet malmsey for themselves and callers, and their daughter Emily made currant wine for local friends. The prevalence of adulterated wine and vinegar encouraged their home production during the nineteenth century. Unscrupulous vintners sometimes colored pale wines with aniline magenta dyes, the red-purple pigment obtained from the heartwood of logwood trees (*Haematoxylon campechianum*), or the juice of beets, mulberries, or elderberries. The practice of cleaning bottles with shot led to wine tainted with measurable traces of lead, a particular problem in wines imported from France.

Vinegar was long known in the Old World as a product related to wine. The word *vinegar* means "sour wine" (Latin, *vin* for "wine" and *aigre* for "sour"), and wine is converted to wine vinegar through the metabolic activity of acetic acid bacteria. This chemical conversion was probably an accidental discovery, but through serendipity the high acidity (low pH) of vinegar made it an excellent fruit and vegetable preservative because acidity inhibits microbial growth. Vinegar became an essential domestic product because pickling relied on vinegar, which in turn can be made from any fruit juice that undergoes fermentation; most apples were converted into cider or cider vinegar, which became the essential preservative for pickling many other botanical foods. Cider vinegar was a New World invention and like wine vinegar was the product of acetic acid bacteria that break down the alcohol in fermented cider and release acetic acid as a waste product. It was easily made by adding sugar to apple cider and allowing the mixture to ferment for several months, or it could be purchased by the gallon or barrel. Child (1844) recommended that a barrel be purchased "when you begin house-keeping" and that leftover cider, beer, wine, and tea be added to replenish the contents and maintain the culture of acetic acid bacteria.

The acetic acid bacteria also benefited from the addition of a few beans or a small amount of molasses (both rich in carbohydrates), if the vinegar became weak or diluted. Aside from pickling, cider vinegar was also used as a flavoring (sometimes combined with cayenne, celery seed, horseradish, or cucumbers), and it was a familiar home remedy. In *The Family Nurse* (1837), Child noted the antiseptic properties of vinegar and its use in cases of ringworm and "putrid fevers and pestilential disorders." Poultices of vinegar and bran were applied to sprains and inflammations, and the vinegar scent was said to be good for patients prone to hysteria and fainting.

Tea, Coffee, and Chocolate

Tea, coffee, and chocolate are stimulating beverages, each derived from plants with potent secondary compounds that are the chemical basis for their unique appeal and properties. Tea is brewed from the leaves of *Camellia sinensis*, native to the hills between China and India and closely related to cultivated camellias. Tea leaves have been used in China for nearly three thousand years, but the beverage was introduced into Europe and North America only during the seventeenth century. Nevertheless, economic botanists now recognize tea as the most economically and culturally significant beverage plant worldwide. Evergreen tea trees are cultivated primarily in China and India, where they are usually pruned into a bush to simplify harvesting the terminal buds and top few leaves. The dried leaves contain caffeine (3 or 4 percent in fresh leaves, which decreases when the leaves are processed and brewed) and lesser amounts of the closely related alkaloids theophylline and theobromine. Caffeine stimulates the central nervous

Tea originated in China and India and was made from the terminal leaf pairs of *Camellia sinensis*.

system, while theophylline and theobromine dilate the smooth muscles that line the bronchial airways. Black tea also contains tannins from reactions that occur after the leaves are crushed. Leaf polyphenols combine with an enzyme (polyphenol oxidase), resulting in tannins such as catechin; the process is hastened by warming and is known as fermentation, although it does not involve any microorganisms. The characteristic aroma is the scent of theol, the volatile oil present in tea leaf cells, which is a mixture of organic compounds including the essential oils geraniol and linalool.

Tea first appeared in America about 1650, carried by Dutch traders arriving at the colony of New Amsterdam. Wealthy New Amsterdam colonists often served several tea varieties, sometimes accompanied by peach leaves or saffron (the styles of *Crocus sativus* flowers) for flavoring the brew. Colonists in Salem, Massachusetts, simmered tea leaves to make a bitter tea that was served with neither sugar nor milk, and the tea leaves were eaten with salt and butter. Despite its high cost, tea became increasingly popular with both Dutch and English settlers. Pehr Kalm, a Swedish naturalist who botanized in North America from 1748 to 1751, noted that most Albany residents consumed tea at breakfast, but most tea in the American colonies was served in punch bowls combined with citrus fruits, sugar, rum or brandy, and water. *Punch* is derived from *panch*, "five" in the Hindi language, which refers to the number of ingredients in the typical punch recipe. Most tea used in the American colonies for punch mixtures was green or hyson tea, for which the leaves were gathered and dried but not fermented to produce tannins; some economized on the expense of imported green tea by experimenting with alternatives, such as the southeastern hollies known as youpon (*Ilex vomitoria* and *I. cassine*). By the late eighteenth century, tea drinking was as much a social event as a way to quench thirst; in 1787, Benjamin Franklin entertained botanist Manasseh Cutler in his Philadelphia garden with tea served under a broad mulberry tree, and summer gardens were often places for strolling and tea service.

Bohea tea was the favored type of black tea in the American colonies, and both bohea and hyson tea were sold in port cities by eighteenth century merchants and importers, often directly from ships. The British East India Company supplied tea to New World colonies, but the company's trade was undermined by tea smugglers who sold their goods without taxes. Difficulties arose when the Tea Act of 1773 released the company from the tea taxes imposed on colonial tea merchants, promoting a monopoly by the British East India

Company. British taxation of any sort was unwelcome, and eighteenth century tea consumption was interrupted for many by the Boston Tea Party on 16 December 1773.

Indignant colonists disguised themselves as Mohawk Indians and boarded three ships at Griffin's wharf in Boston Harbor; Paul Revere, Sam Adams, and other patriots protested taxation on tea imported from England by destroying three hundred forty-two chests of tea. Although the Townshend Acts had been repealed, the tax was symbolic of the British control of the American colonies, and the colonists acted in defiance of Governor Thomas Hutchinson, who endeavored to uphold the law, pay the required taxes, and have the ships unload their tea imports. Similar acts of disobedience occurred in Charleston, New York, Annapolis, and Greenwich, New Jersey, and many Americans advocated a full-scale colonial tea boycott. With its start in 1774, the Continental Congress organized the Continental Association, which banned tea drinking and other British customs such as gambling and horse racing; local citizens watched over each other and threatened to publish the names of unpatriotic neighbors in the newspaper. Occasional mobs gathered outside of homes and intimidated known tea buyers into compliance with the boycott of tea from England.

A few colonial attempts to establish New World tea plantations were unsuccessful because the American climate proved unsuitable; although tea trees are generally hardy, they grow rapidly only in a warm climate with high rainfall. Zealots instead adopted "liberty teas," brewed from various native and cultivated garden plants that release pleasant essential oils when they are infused in hot water. To encourage the use of homegrown rather than imported teas, newspapers printed claims that Asian tea bred insect vermin and that the leaves were compressed into tea chests by barefoot Chinese pagans. Patriotic rhetoric blamed tea for nervous disorders and stomach troubles, which encouraged housewives to seek out palatable tea alternatives in their gardens and woodlands. In 1773, Susannah Clark wrote,

> We'll turn the tea all in the sea,
> And all to keep our liberty.
> We'll put on homespun garbs
> and make tea of our garden herbs.

A tea substitute was often brewed from Labrador tea (*Ledum groenlandicum*), a member of the heath family (Ericaceae) that contains tannins such as those found in the black varieties of Asian tea. C. S. Rafinesque, author of *Medical Flora or Manual of Medical Botany of the United States* (1828–1830), described Labrador tea as "very near to Chinese tea, but stronger, owing to fragrant resin." He was referring to the volatile oil present in *Ledum* leaves, including the compound known as ledol (also known as "*Ledum* camphor") that contributes to the characteristic flavor of these teas. Colonists may have learned about its use from Native Americans or Canadian colonists, who both brewed Labrador tea before it was known in the American colonies. Colonists may also have noted its similarity to marsh tea (*L. palustre*), long used as a folk remedy and household insecticide in Europe. Despite its familiarity and common use, Labrador tea may have caused ill effects among those who drank it to excess; anecdotal reports suggest its narcotic and headache-inducing properties. Other tea substitutes included the leaves of New Jersey tea (*Ceanothus americanus*), considered by botanist Asa Gray to be one of the best Revolutionary-era teas; raspberry (*Rubus* spp.); and sweet fern (*Comptonia peregrina*). All of these also had medicinal uses as astringents (suggesting the presence of tannins in their leaves) and were commonly used in tonics to treat ailments such as dysentery, stomach upset, and mouth cankers.

Savory herbs such as fennel (*Foeniculum vulgare*) and dill (*Anethum graveolens*) as well as flowers such as chamomile (*Chamaemelum nobile* and *Anthemis arvensis*), violets (*Viola* spp.), red clover (*Trifolium pratense*), and linden (*Tilia americana*) also appeared in dried tea mixtures. Tea was also brewed from combinations of familiar garden herbs such as peppermint (*Mentha* ×*piperita*), catnip (*Nepeta cataria*), lemon balm (*Melissa officinalis*), rosemary (*Rosmarinus officinalis*), sage (*Salvia officinalis*), thyme (*Thymus vulgaris*), lavender (*Lavandula officinalis*), marjoram (*Origanum vulgare*), and bee balm or Oswego tea (*Monarda didyma*). These are

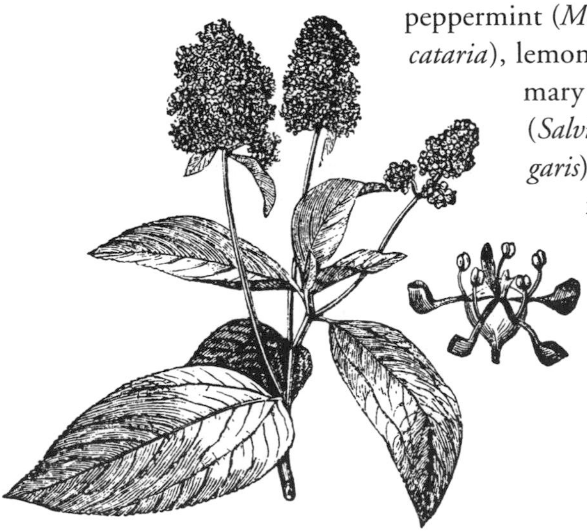

The leaves of the shrub known as New Jersey tea (*Ceanothus americanus*) were dried as a substitute for the imported tea that was subject to British taxation.

all members of the mint family (Labiatae or Lamiaceae), and some such as rosemary and sage may contain tannins that are typical of Asian tea. Their characteristic fragrances and flavors also reveal the presence of essential oils, including the pungent terpenes known as pinene, cineole, and limonene, and the leaves or flowers could generally be brewed into safe beverages to supply a household. One exception was tea brewed from pennyroyal (*Mentha pulegium*), also a member of the mint family, which produces a potentially lethal volatile oil containing the terpene pulegone; it was also used as a culinary herb (see chapter 6) and medicinally as an abortifacient (see chapter 7). Other mint teas, including sage, were also potentially toxic if consumed in large quantities. Sassafras bark (*Sassafras albidum*), licorice root (*Glycyrrhiza glabra*), and rose hips (*Rosa* spp.) also lent their own distinct flavors to specific mixtures.

A pleasant wintergreen-flavored tea was prepared from wintergreen or teaberry leaves (*Gaultheria procumbens*), a member of the heath family (Ericaceae) that contains the volatile oil of wintergreen that is composed of 98–99 percent methyl salicylate. Methyl salicylate is also potentially toxic if large amounts of the oil are ingested, but wintergreen tea infusions delivered safe amounts; larger quantities were known to cause vomiting. Thoreau (1999) knew wintergreen as "checkerberry" and noted that the scented plants and their red berries remained fresh despite frost and snow. He quoted botanist Manasseh Cutler who noted in 1785 that checkerberry commonly grew with oaks and pines and that New England children ate its red berries with milk.

The introduction of inexpensive Brazilian coffee supplanted much of the nineteenth century American tea market, but tea-drinking rituals continued to evolve as part of genteel society. After the Revolutionary War, the popularity of Asian tea revived, but never again to pre-war levels of consumption. Tea and other refreshments were handed around at ladies' teas, formal events at which participants sat in chairs drawn around a tea table or against the parlor walls. The awkwardness of precariously balancing cup, saucer, soup, and bits of food was illustrated in "The Tea Party," an etching by Edward W. Clay (Philadelphia, 1819), which illustrates the tedious traditions of the parties that became known as "tea fights." As its popularity developed, elaborate china services, trays, silver, and tables evolved with the custom of tea-taking. Tea

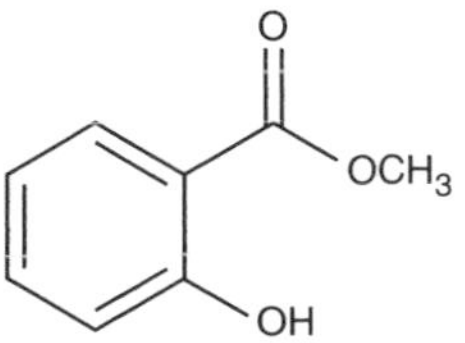

The chemical structure of methyl salicylate in oil of wintergreen.

varieties, brewing, and service were topics of considerable discussion and opinion; Leslie (1851) favored china teapots and devoted a full page to the proper methods for brewing and serving tea. She favored the pekoe and pouchong varieties of black tea and the young hyson, imperial, and gunpowder varieties of green tea.

Tea had to be stored with care to prevent the loss of flavor; some nineteenth century manuals recommended lead-lined tea boxes for long-term storage (plumbism, the effects of lead toxicity, were not yet completely understood), while others suggested that tea be kept in glass or tin containers. Tea also had a place as a flavoring in cookery. Randolph (1824) suggested using "one ounce of the best tea" to flavor cream, which was then curdled with enzymes from rennet (the stomach lining of a calf) and served with fruit. Iced tea was a latecomer, often credited to tea importer Richard Blechyden, who experimented with hot tea poured over ice at the sweltering 1904 Saint Louis Exposition. In reality, during the 1880s women were already serving tea with ice and lemons; in *The Cottage Kitchen* (1883), Marion Harland noted, "Great bowls of this, ice-cold and well-sweetened, are popular at fairs, church receptions, and picnics, and have become a fashion at evening parties where wines and punch are not served." We can assume that cold tea served elegantly and known as *tea a la Russe* appealed to proponents of temperance. Tea bags were also inadvertently invented during 1904 when New York importer Thomas Sullivan packaged small tea samples in silk bags; customers brewed tea without removing the contents and soon requested more tea packaged in small bags rather than the customary loose tea in tins.

By the mid-nineteenth century, some held peculiar notions about tea chemistry and its effects on physiology that centered on the volatile oil (theol). In his *Treatise on Food and Diet* (1843), London pharmacologist Jonathan Pereira likened the effects of tea to foxglove, which was a source of potentially toxic cardiac glycosides, and he incorrectly claimed that foxglove had the effect of slowing heartbeat. Of course, nearly a hundred years earlier, William Withering had worked out the effective dose of foxglove that regulated heartbeat and counteracted fluid retention in the body cavity, the condition (edema) that was known as dropsy (see chapter 7). Tea contains no cardiac glycosides, but Pereira argued that its volatile oil caused sedation in some but anxiety and tremors in those with a nervous disposition, while also functioning as a diuretic and diaphoretic. His work was cited in American cookbooks and household manuals and no doubt influenced some readers in their tea purchases and consumption. The potential danger of green tea in particular was a widespread notion, ironic in light of the current interest in green tea as an effective antioxidant. Hale (1841) recommended brewing hyson (green) and souchong (black) teas together to make "A pleasanter beverage than either alone . . . safer for those who drink *strong* tea, than to trust themselves wholly with green."

In *The American Woman's Home* (1869), Beecher also argued against the common nineteenth century practice of serving children tea and coffee, which she believed caused tooth decay and "gradually sap the energies of the feebler child, who proves either an early victim or a living martyr to all the sufferings that debilitated nerves inflict." She may not have understood caffeine chemistry, but she had a practical knowledge of the effects of this alkaloid on the young and old alike; in light of the possible mutagenic properties of caffeine, her caveat may have been prescient, indeed. Child (1844) noted that currant leaves were a useful substitute for green tea, which she noted was suitable only for the healthy: "Let those who love to be invalids drink strong green tea, eat pickles, preserves, and rich pastry." She also recommended green tea boiled in iron for washing silk, a method that may have revived old silks by slightly changing the hues of pH-sensitive anthocyanin pigments.

During the mid-nineteenth century, Americans still experimented with growing Asian tea. In 1858, federal agriculture officials provided young tea plants to Tennessee farmers who established a tea plantation near Knoxville. The plants survived frosts and responded well to the high rainfall. In 1870, the *Nashville Journal of Medicine and Surgery* described the leaves as comparable to imported hyson tea, but this was apparently a short-lived experiment. The 1885–86 edition of *Johnson's Universal Cyclopaedia* omitted any mention of tea plantations in the description of Tennessee agriculture; indeed, only China and India tea were mentioned in Asa Gray's brief article on the "Tea-Plant."

Coffee (*Coffaea arabica*) originated in Ethiopia and was domesticated by Arabs, and by the eighteenth century the trees were cultivated in Central and South America. The red fruits are drupes, each with a hard endocarp that usually encloses two seeds; the seeds are the coffee "beans" that are roasted, ground, and infused or boiled in hot water to produce the familiar beverage. Roasting converts seed starches to sugars, which then caramelize and turn the beans a dark brown color. The seeds also release oils such as caffeol as they are heated, which enhances their characteristic aroma, while breakdown of the cell walls from heat renders the seeds easier to grind. Coffee was sold by the same coastal merchants who provided other imports, but British rule and taxation kept colonial coffee consumption low prior to the American Revolution. Nevertheless, coffee houses did serve as public meeting places, and Daniel Webster referred to the Green Dragon coffee house in Boston as the "headquarters" of the American Revolution. After the Revolutionary War, Americans traded slaves for coffee from French colonies in Haiti and Martinique, which lowered import costs. Coffee prices fell from nine shillings per pound in 1774 to one shilling per pound in 1783, which laid the groundwork for the change from tea to coffee consumption by the 1830s. Coffee taxes were removed in 1832, reappeared during the Civil War, and then were finally abolished in 1872. Per capita consumption increased from five pounds a year during the mid-nineteenth century to nine pounds by 1882, probably the result of industrialization and the adoption of coffee by workers as a helpful stimulant. The alkaloid caffeine found in coffee seeds (1 to 1.5 percent) increases heartbeat and blood pressure and constricts blood vessels, which explains why coffee has long been used to combat fatigue, maintain alertness, and encourage endurance.

Merchants at first sold green coffee, which had to be roasted at home, but a single burned bean ruined the flavor of an entire batch. Once the demand for roasted coffee became apparent, Pittsburgh grocers John and Charles Arbuckle in 1865 patented a process in which each roasted bean was effectively sealed with a layer of sugar and egg. A 1872 color advertisement illustrated two women at a stove, one complaining, "Oh, I have burnt my coffee again," and the other responding, "Buy Arbuckle's Roasted, as I do, and you will have no trouble." Arbuckle's Ariosa Coffee with its recognizable "flying angel" trademark and yellow label became an American favorite, especially among cowboys, western pioneers, and Navaho fam-

Coffee was prepared from the seeds of a native Ethiopian tree, *Coffaea arabica*, which were roasted, ground, and infused or boiled in hot water to make the beverage.

ilies. "Ariosa" was a meaningless coined name, but the product distinguished itself with the marketing strategy of including a peppermint stick in each package, a rare sweet commodity on the American frontier. The coffee came packed in wooden crates, which in frontier towns were sometimes used as children's coffins, and each bag bore printed coupons that could be exchanged for practical premiums such as wedding rings, razors, and handkerchiefs. Coffee consumption was generally high, and many coffee drinkers described the "Arbuckle thumps," the effect of excess caffeine on heartbeat.

Eastern matrons also drank coffee, and they gleaned the correct techniques for storing, roasting, grinding, and brewing good coffee, which Hale (1853) described as a "wholesome, exhilarating, and strengthening beverage; and when mixed with a large proportion of milk, is a proper article of diet for literary and sedentary people." Beecher (1858) described the method for preparing coffee by crushing a whole egg into the ground coffee beans, combining with water, and then boiling the mixture; any small suspended particles sank to the bottom of the pot to produce clear coffee. If an egg were not available, she recommended substituting fish skin, which was cut into small squares, dried, and set aside for brewing coffee.

Nineteenth century cookbooks provided complex methods for preparing "French" and "Turkish" coffee and also suggested various coffee substitutes. In place of coffee beans, cooks roast and ground acorns, peas, rye grain, and okra seeds, and in desperate times even the dried crusts of brown bread were used to brew "coffee." Rural southern and midwestern families often drank true coffee only as a Sunday beverage, along with wheat bread, while they consumed corn bread and brewed rye grain during the rest of the week. Chicory (*Cichorium intybus*) and dandelion (*Taraxacum officinale*) taproots were also used as a coffee substitute or to extend or possibly adulterate a limited coffee supply. Both are European composites (family Compositae or Asteraceae) that were valued as beverage and vegetable plants, and both are now familiar naturalized weeds. Coffee and its various substitutes were favored by rich and poor alike, and by the end of the nineteenth century, etiquette writers addressed the various social rituals

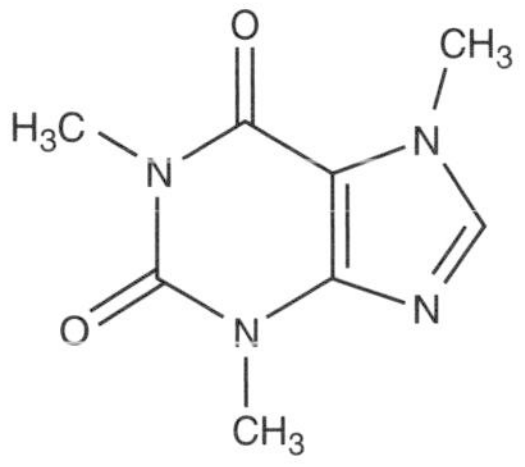

O
CH3
H3C
N
N
N
N
O
CH3

The chemical structure of caffeine found in the seeds of coffee (*Coffaea arabica*) and leaves of tea (*Camellia sinensis*).

of coffee drinking in polite society. In *Decorum: A Practical Treatise on Etiquette and Dress of the Best American Society* (1883), John Ruth insisted, "Tea and coffee should never be poured into a saucer," a criticism of the colonial custom of "saucering" hot beverages to cool, while the cups rested on small glass or ceramic cup plates.

Cacao trees (*Theobroma cacao*), the source of chocolate, were cultivated since ancient times in Central America. The trees exhibit cauliflory, meaning that their flowers sprout from buds found on the mature trunk and woody branches, and the large, pendulous yellow, green, purple, or red capsules stimulated early curiosity and experimentation. Classic Period Mayans (250–900 A.D.) were probably the first to select the trees for cultivation from the local rain forest flora, once they realized the value of the white seeds or "beans." Each cacao capsule contains up to sixty beans, which the Mayans prepared by fermenting, roasting, and then grinding into a paste; this was combined into a beverage with chili peppers, cornmeal, and water. Fifteenth century Aztecs traded with the Mayans for cacao, which they combined with vanilla and chili peppers to make *chocolatl*, a beverage consumed by rulers and priests and during religious ceremonies. According to Aztec legend, cacao was carried to earth by the god Quetzalcoatl, who was punished for giving a heavenly food to mortals. Aztec priests in return offered cacao to Quetzalcoatl, a custom later commemorated by Carolus Linnaeus who coined the botanical name *Theobroma* (Greek for "food of the gods") for the genus. Hernando Cortez brought cacao back to Spain, where cacao flavored with cinnamon and sugar gained favor among the wealthy; cacao became an valued import from Mexico, and knowledge of chocolate spread throughout Europe.

Elaborate customs and decorative pots and cups evolved for its use, even before Sir Hans Sloane of the Royal College of Physicians had the notion of adding milk to a cacao mixture to make the traditional beverage known as cocoa. Beginning in 1657, chocolate houses opened in England, and like coffee houses they were places for wealthy men to drink chocolate, congregate, and converse. Europeans also drank chocolate for its supposed medicinal benefits, as an aid to digestion, as a soporific, and as an aphrodisiac. American colonists as well consumed imported chocolate, which they had to prepare by grinding the beans in a hand mill or pulverizing them with a mortar and pestle. By the 1700s, mechanized commercial chocolate mills began to process cacao beans, which lowered the price of chocolate and made it more commonly available.

One such mill was established in 1765 by Dr. James Baker and John Hannon in Dorchester, Massachusetts, and was powered by water from the Neponset River. Cakes molded from the crushed beans sold well through word of mouth and handbill advertisements, but chocolate production was threatened during the Revolutionary War, when British naval ships cut off supplies of cacao beans imported from plantations in the West Indies. Smuggled beans kept the mill open and supplied colonists with chocolate for drinking. After the Revolution, Baker bought out another mill built by Hannon, who disappeared mysteriously while traveling to the West Indies in search of cacao seeds, and Baker Chocolate prospered. James Baker's son, Edmund Baker, later expanded operations at the Dorchester site, which by the 1830s was noteworthy for employing women as workers. Competing chocolatiers moved into the area, which became known as "Chocolate Village" because of the permeating scent; with the advent of refrigeration in 1868, the milling and shipping of molded cakes of chocolate continued year-round.

The large capsules of cacao trees (*Theobroma cacao*) contain the "beans" that were processed into unrefined chocolate for drinking.

Most nineteenth century cooks likely preferred factory-milled chocolate, although some cookbooks provided instructions for drying cacao beans in the oven and pounding them into a fine paste that was boiled in water, one ounce to a pint. Cocoa was prepared easily from milled chocolate by scraping the chocolate cake into a pint of water or milk, boiling the mixture for several minutes, and then adding sugar, milk or cream, and perhaps cinnamon and nutmeg to taste. Some cooks further enriched the mixture with beaten egg yolks. Volumes were approximate, but some chocolate was molded into squares that each flavored a pint of cocoa. Leslie (1851) mentioned making a foamy layer using "the small wooden instrument called a chocolate mill," a tool that was apparently similar to the *molinillo* used in sixteenth century Spain for stirring and whipping chocolate.

As the price of chocolate dropped during the nineteenth century, cocoa was marketed as a beverage particularly suitable for women and children. Trade cards for chocolate often featured young women or plump youth, such as the advertisements for "Phillips Digestible Cocoa, a delicious & highly nutritious beverage." By the end of the nineteenth century, nutritionists and educators alike argued that tea and coffee were too stimulating for children, but ironically the properties of chocolate as a stimulant arise from the presence of theobromine (2.5 percent) and caffeine (0.8 percent), the familiar tea and coffee alkaloids. However, unlike these beverages, chocolate also contains proteins, carbohydrates, and fats, even before the addition of milk and sugar. A thrifty alternative to purchasing chocolate was also employed: cacao shells (seed coats) were soaked overnight and then boiled in the same water to yield a weaker but chocolate-infused beverage. In *Every-Day Cookery for Every Family* (1868), Sarah Scott described roasting cacao shells and pulverizing them into a paste that was combined with boiling water, probably with a similar outcome. Hale (1841) observed about cacao shells, "They are considerably nutritious, allowed to be healthy, and are cheap." Those who enjoyed chocolate flavor also used the unsweetened milled chocolate bars to make custards or confections, but sweetened chocolate molded into candy bars was a late dietary arrival. Technical innovations, such as the conching machine developed in 1879 by Rodelphe Lind, helped to produce the smooth, consistent chocolate that lent itself to candy-making. Soldiers, rather than children, spread the word about chocolate; Queen Victoria had chocolate candy sent to her troops at Christmas, and during World War I chocolate bars were included in American military rations.

CHAPTER SIX

The Botanical Pantry II: Herbs, Spices, Sweets, and Miscellany

CONSUMPTION of herbs and spices evolved from the ancient use of aromatic plants in food preparation and preservation, medicine, perfumery and burial rituals. American colonists and their European ancestors found pungent flavorings wherever they could, which explains the remarkable overlap between culinary and medicinal plants and the secondary use of medicinal plants to flavor foods. Indeed, the epithets *officinale* and *officinarum* in the scientific names of some herbs and spices reveal their original use by apothecaries as medicinal plants; rosemary (*Rosmarinus officinalis*) and ginger (*Zingiber officinale*) are among the many culinary species with a medicinal history. The distinction between culinary and medicinal plants is often blurred, illustrated by sickroom care that relied heavily on teas infused from culinary herbs and on spices such as ginger, cinnamon, and cloves that were valued to promote digestion and relieve flatulence. Tansy leaves (*Tanacetum vulgare*) flavored puddings and cakes, but the plant was also applied as a medicinal poultice, brewed into teas and tonics, and used as a vermifuge to kill or expel intestinal worms. Tansy was one of many versatile European herbs that escaped from early colonists' gardens and naturalized in North America. Pennyroyal also had both medicinal and kitchen uses, Amelia Simmons (1796) listed "Penny Royal" under "Herbs, useful in Cookery," although European pennyroyal (*Mentha pulegium*) was probably better known as a tea herb, insect repellent, and abortifacient. Despite their early dual uses as kitchen herbs and medicinal teas, ironically both tansy and pennyroyal are now known to contain highly toxic essential oils with potential ill effects if the leaves are ingested.

The flavors and scents of aromatic plants arise from phytochemicals that evolved in response to environmental pressures; in most cases, these compounds deter or poison herbivores or function as natural antibiotics against bacteria and fungi. In light of their similar phytochemistry and uses, spices and herbs are impossible to distinguish, but botanically speaking, herbs are herbaceous (non-woody) plants, not necessarily species that lend themselves to food preparation or medicinal uses. Early European herbalists considered herbs to be "any useful plant," including medicinal, culinary, or household species with practical uses. In a culinary or medicinal sense, herbs are generally temperate plants, and most commonly their leaves are the part that is used. Spices are generally derived from the roots, bark, flowers, seeds, or fruits of tropical species, but these definitions break down under close scrutiny; for instance, caraway "seeds" are the small fruits of a temperate herb, and ginger leaves are used in addition to the rhizome that is the source of powdered ginger. Herbs and spices are impossible to separate botanically or practically, and from a historical perspective it is more useful to examine the properties that these various phytochemically-rich plants hold in common.

An explanation for the widespread use of a variety of herbs and spices in cooking lies with the antibiotic phytochemicals that most contain. The demand for exotic flavors may have motivated Columbus and Magellan to sail in search of spice-producing islands, but the same potential for food preservation was available through the common use of native European herbs such as basil and garlic, which were easily cultivated in kitchen gardens. Indeed, the evolution of herbs and spices use can be interpreted through Darwinian natural selection; families that used antibiotic plants in food preparation suffered fewer cases of poisoning from food-borne pathogens such as *Escherichia coli* and *Salmonella*, which put them at a selective advantage. Spice and herb use became a common practice in the most successful family groups in both temperate and tropical regions, and these culinary customs had a role in reducing illness and death caused by tainted foods. This contrasts with the common explanation that herbs and spices were used to mask the flavor of spoiled foods; in fact, many pungent phytochemicals kill or inhibit bacteria and fungi that decompose foods and in this way help to prevent the noxious odors of decomposition.

Paul Sherman and Jennifer Billings of Cornell University (1999) compiled data that show the antimicrobial properties of several herbs and spices, including those commonly used in colonial American kitchens. Allspice kills or inhibits all of the food-borne bacteria on which it has been tested and is among the most effective antimicrobial spices. Nutmeg, cloves, cinnamon, thyme, rosemary, marjoram, mustard, bay leaves, caraway, various mints, and parsley kill or inhibit at least half of the bacteria present in laboratory studies; black pepper, ginger, and anise seed have lesser but significant inhibitory effects.

When a colonial cook followed tradition and stuffed a turkey with marjoram and sage or seasoned pork with cloves and nutmeg, she unwittingly used herbs and spices as food preservatives. The methods for meat and fish cookery described in *American Cookery* (Simmons 1796) and in *The Art of Cookery* (Glasse 1805) relied heavily on herbs and spices, which in the absence of refrigeration kept protein-rich foods edible until they were consumed. For instance, Glasse's recipe for "beef alamode" called for cloves, mace, allspice, parsley, black pepper, cayenne pepper, onions, garlic, and bay leaves, which constituted a potent antimicrobial mixture. Bacteria not killed or inhibited by a single herb or spice might easily succumb to a mixture of antimicrobial compounds.

Meat that is chopped or ground provides increased surface area for bacterial decomposition, apparently reflected in the tradition of using generous quantities of cinnamon, nutmeg, and mace in meat pie recipes. Recipes for sausage and souse (meat trimmings preserved by pickling) made use of scraps of butchered pigs and relied on herbs and spices in their preparation. In "Minced Pie of Beef," Simmons recommended one ounce of mace and a whole nutmeg for four pounds of beef. Meat pies were sometimes made weeks in advance and stored; prior to serving stored pies, she advised cooks to "carefully raise the top crust . . . collect the meat into a bason [sic], which warm with additional wine and spices to suit the taste of your circle." This amounted to a last-minute "sterilizing" using alcohol and antimicrobial spices. Pickle recipes also included high concentrations of various herbs and spices that provided complex flavors and promoted effective preservation. Garlic (*Allium sativum*) was used more frequently as a medicine than as a herb, but garlic cloves appeared frequently in pickling mixtures; garlic and related onions (*A. cepa*) also kill or inhibit all of the food-borne bacteria on which they have been tested.

Culinary Spices

Until the end of the eighteenth century, the ginger, black pepper, cinnamon, nutmeg, mace, and cloves used in American kitchens were grown in Dutch-controlled plantations in countries such as Ceylon (now Sri Lanka). Ginger and black pepper were the first spices imported to America in large quantities; ginger was used in brewing beer and in flavoring metheglin, a type of mead made of fermented honey and water, and pepper had diverse uses in cooking and food preservation. Imported spices were costly commodities in colonial kitchens, but after the Revolutionary War, American ships traveled to Asia to participate in the lucrative spice trade. American port cities also began to participate in the spice trade, including Salem, Massachusetts, which became a capital of the black pepper trade when Captain Jonathan Carnes began to import pepper from Sumatra in 1795. During the same years, viable seeds or seedlings of cinnamon, clove, and nutmeg were smuggled out of Dutch colonies, and spice cultivation and the price of tropical spices was no longer controlled almost exclusively by Dutch growers. Spices became less expensive and more widely available, as reflected in their domestic use.

An exception to the Dutch spice monopoly was allspice (*Pimenta dioica*), which explains its apparent popularity in early American kitchens. Despite its similar taste to cinnamon and other Old World tropical spices, allspice trees are native to the West Indies; the dry, immature berries were probably traded along with molasses from the West Indies. Records from Deerfield, Massachusetts, indicate that in 1742 a half pound of allspice cost only a penny more than a half ounce of cinnamon, indicating that this was a thrifty New World alternative to the expensive spices monopolized by Dutch colonies. As the name implies, allspice resembles a mixture of cinnamon, cloves, and nutmeg, and it was used interchangeably with these in cooking and baking and as a medicinal carminative to relieve flatulence. Simmons's recipe for "A cheap seed Cake" used allspice in place of more expensive spices, and Sarah Hale (1841) recommended cooking beefsteak with allspice or cloves. All of these spices contain volatile oils with similar terpenes, including eugenol and geraniol, which provided the familiar aroma in spice cakes and gingerbreads. Typical early spice boxes provided storage for allspice, cinnamon, cloves, and nutmeg; to preserve their essential oils, prudent cooks pulverized whole spices and stored the powder in corked jars or small tins, which were fitted into a larger, airtight tin spice box.

Cinnamon trees (*Cinnamomum verum*) were native to Ceylon, where the Dutch established cinnamon plantations during the eighteenth century; the fragrant inner bark is stripped from the trees and cut into strips, which dry and curl into quills. Cinnamon quills were an essential ingredient in flavoring orgeat, a milky beverage described by Mary Randolph (1824) as "A necessary Refreshment at all Parties." The quills were also cooked in milk to flavor custards, and the thin bark was ground to make the powdered cinnamon used in baking; Catharine Beecher's (1858) "Minced Pie" of chopped apples, tongue, and suet was flavored with cinnamon, cloves, and mace. Apple butter, apple pies, and cider cakes used similar spices, as did pies and puddings made of pumpkin and squash. The combination of cloves and cinnamon also flavored traditional election cakes, the yeast cakes served at colonial New England election (or training) days, community events at which local militia troops paraded.

Cassia (*Cinnamomum cassia*) is a close cinnamon relative that was often sold as cinnamon, although the flavor difference is distinguishable; it may not contain the eugenol found in true cinnamon, but the two spices have historically been confused. The Food, Drug, and Cosmetic Act of 1938 allowed both true cinnamon and cassia to be sold as "cinnamon." In early American cookery, cassia was probably more commonly available than true cinnamon. While sassafras (*Sassafras albidum*) is not a tropical spice, it is a New World tree that is classified in the same family (Lauraceae) as cinnamon and cassia, and it was also used as a familiar spicy flavoring and spice substitute. The essential oil of sassafras contains a high concentration of safrole (now banned because of its toxicity to liver tissue) and several of the other terpenes such as eugenol, apiole, pinene, and anethole, which are found in familiar herbs and spices. Tea brewed from sassafras root bark was a home remedy, long believed to cure venereal diseases, and sassafras roots and the extracted oil were used to flavor beer and Godfrey's cordial, an opiate-containing syrup that was a familiar nineteenth century household medicine. Sassafras also contains water-soluble mucilaginous compounds (also important in filé powder; see chapter 4), and sassafras jelly was made by soaking the young branches in water until the water became glutinous. Beecher (1858) noted that it was "much relished by the sick, and is also good nourishment."

Cloves (*Syzygium aromaticum*) are the immature flower buds of Moluccan trees, which were widely known in medieval Europe and prized for their essential oil that contains a high concentration of eugenol. The dried buds resemble nails, and their name is derived from *clou*, the French word for "nail." Ground cloves were used as flavorings interchangeably with cinnamon, nutmeg, and mace, and cooks probably used whatever was available. In *The Family Nurse*, Lydia Child (1837) recommended that whole cloves be chewed as a stimulant or to relieve toothache, and she mentioned the use of cloves to mask the unpleasant flavor of medicines. Cassia buds resem-

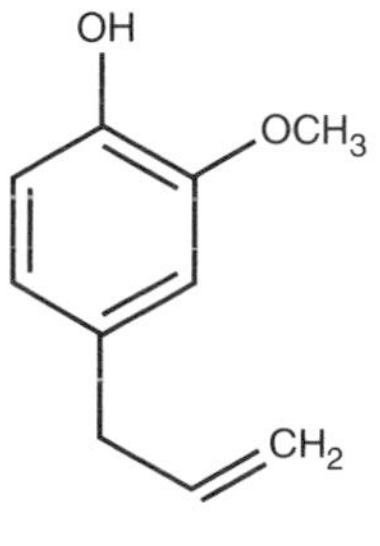

The chemical structure of eugenol, which occurs in cloves (*Syzygium aromaticum*) and other spices and spice substitutes.

ble cloves and were occasionally added to pickles in place of cloves. A clovelike scent and taste was also obtained from clove gillyflowers (*Dianthus caryophyllus*), also known as sops-in-wine, European wild carnations that were cultivated commonly in colonial American gardens (see chapter 9). Their essential oil also contains a high concentration of eugenol, as well as methyl salicylate (oil of wintergreen).

Nutmeg and mace come from the same source, the peculiar, apricotlike fruits of *Myristica fragrans.* These crack open to reveal a large seed (the nutmeg) that is partially covered by the fleshy aril (mace), a botanical anomaly because fleshy fruits are rarely dehiscent. The reddish aril contrasts with the dark seed, which botanists interpret as an evolutionary adaptation for bird dispersal.

The fragrance and flavor of nutmeg and mace arise from volatile oils that contain terpenes, including several of the same compounds that also occur in allspice, cinnamon, and cloves: eugenol, geraniol, camphene, and pinene. Nutmeg was a favorite spice for pies and puddings and had to be grated using a small nutmeg grater; mace was sold in blades that were easily crumbled into a powder. Simmons listed nutmeg or mace among the ingredients for various pies (including "pompkin" and tongue) and puddings based on rice, cornmeal, bread, almonds, winter squash, and potatoes. Nutmeg was used in remarkably large quantities, significant because of the presence of the psychotropic compounds myristicin and elemicin. The toxic dose of nutmeg is four or five grams; perhaps some suffered the hallucinogenic effects of myristicin, especially if cooks grated the entire nutmeg called for in many recipes. Large amounts of nutmeg over long periods can also cause liver damage, caused by the presence of safrole, which also occurs in sassafras.

Nutmeg fruits (*Myristica fragrans*) dehisce to reveal the black nutmeg covered by a reddish aril, which is powdered to make the spice known as mace.

The chemical structure of the myristicin that occurs in nutmeg (*Myristica fragrans*).

Nutmeg and mace were well known in Europe during the Middle Ages, and the Portuguese and later the Dutch defended nutmeg plantations in the Spice Islands. Wild populations of nutmeg no longer exist in the Moluccas, but Dutch growers cleverly protected their monopoly by trading nutmegs only after they were soaked in a calcium hydroxide solution, which killed the embryo. By the 1770s, French growers experimented with establishing colonial plantations with viable nutmeg seeds and seedlings smuggled out of Dutch plantations. As with other spices, nutmegs were carried on trade ships bound for American ports, but unscrupulous tradesmen perhaps sometimes peddled fake nutmegs carved of wood; school children learn that Connecticut is known as the nutmeg state, a name that may commemorate this common deception.

Ginger (*Zingiber officinale*) is derived from the robust, underground stem (rhizome) of a perennial Asian herb that is now known only in cultivation. Its pungent flavor is stronger than many other spices, and cooks know that ginger imparts "heat" to dishes. Ginger was used frequently in baking; gingerbreads made with powdered ginger were well known during the reign of Elizabeth I, and they became staples in colonial kitchens. George Washington spent electioneering funds for the House of Burgesses on ginger cakes and rum punch for prospective voters. Mary Randolph (1824) described "plebeian ginger bread" made with pearl ash, unlike her richer "sugar ginger bread" that was leavened with a dozen eggs and required "a large cup full of powdered ginger" and brandy. Child (1844) described two basic gingerbreads: a cakelike variety made with molasses, and "hard gingerbread" made with caraway and rolled out on pans for baking. Some cooks also preserved ginger rhizomes in syrup for later use as a sweetmeat; Leslie

The rhizomes of ginger (*Zingiber officinale*) were powdered to use as a spice or preserved in sugar syrup as a sweetmeat.

(1851) used eight pounds of "the best double-refined loaf-sugar" for each six pounds of fresh ginger. Preserved ginger was traditionally passed around after a meal to promote digestion and discourage flatulence.

Ginger beer was fermented by yeast mixed with molasses, powdered ginger, and water, and it was bottled for later use; ginger wine, a more elegant version, relied on sugar and lemons for flavoring. Medicinal teas were made by infusing powdered ginger or the whole rhizome in water, and these were used as stimulants and like cinnamon and cloves to relieve flatulence; ginger was also added to various quasi-medicinal tonics and bitters (see chapter 7). By the 1870s, extracts such as Lobstein's Quintessence of Jamaica Ginger sometimes replaced ginger rhizomes, advertised as having "all the active and valuable medicinal properties of pure white Jamaica ginger in its most concentrated form."

The nineteenth century Shakers at Mount Lebanon, New York, grew ginger as a medicinal plant for preparing tonics and treating stomach ailments, dyspepsia, and cholera; they called it African ginger, although the rhizomes that they cultivated apparently came from Jamaica. Shaker growers discovered that the rhizomes would sprout and multiply if the plants were cultivated as annuals. The Shakers also used the common name *ginger* for another plant that they cultivated, an American wildflower (*Asarum canadense*) in the birthwort family (Aristolochiaceae) with closely related species in Europe and Asia. This plant is now commonly known as wild ginger, and although it is not a relative of *Zingiber officinale* (family Zingiberaceae) from Asia, the plants both contain abundant essential oils and had some similar uses as flavorings and medicinal herbs. Wild ginger was used in meat and fish cookery, perhaps when imported spice supplies were short, but large quantities might have proven toxic due to the presence of aristolochic acid. The Shaker uses of wild ginger were similar to those of various Native Americans who used the plants to season fish, improve the edibility of meat, and treat gastrointestinal ailments.

The common name *pepper* also refers to unrelated Old and New World plants, black pepper (*Piper nigrum*) from India and cayenne pepper (*Capsicum annuum*) from tropical America. Black peppercorns were harvested from vines that were originally native to India, small drupes (stone fruits) that are dried and either ground or used whole. Food historians consider black pepper the most esteemed of all tropical spices, and the possession of pepper was equated with wealth; during the Middle Ages, the comment "he hath no pepper" implied low social status because peppercorns were used to provide dowries, pay taxes, and settle debts. Like ginger, black pepper became an essential imported spice in the New World colonies, with a long history of European use, and New England colonists actively participated in the pepper trade. Shipbuilders in Salem, Massachusetts, produced clipper ships, efficient vessels that carried imports worldwide and were responsible for establishing Salem as a trading capital, where black pepper shipped from Sumatra was distributed worldwide.

Black pepper was long valued for its ability to mask the taste of spoiled meat and rancid fat, properties that result from the alkaloids piperine and piperidine and a pungent volatile oil composed of several terpenes. Although its antibiotic properties are less than those of most other herbs and spices, black pepper has antifungal properties and has been used medicinally as a stimulant and to treat tumors, but medical evidence now suggests that pepper in large quantities might contribute to malignancies. Black pepper was ground and used as a seasoning in meat dishes, and whole peppercorns were used in many pickles. The soup known as "pepper pot" drew much of its flavor from black pepper; Leslie's 1851 recipe called for tripe and ox feet stewed with whole peppercorns and vegetables. Occasional recipes specified white pepper (particularly white dishes or pale sauces that would be marred by dark flecks), and this was prepared by removing the outer dark layer and then grinding the peppercorns in the usual way.

During the 1840s, food reformer Sylvester Graham warned against the dangers of pepper and other spices and condiments, which he considered to be exciting and exhausting to the human body. Hale (1841) suggested that tropical spices were most wisely consumed during summer because "they are productions of hot climates, which shows them to be most appropriate for the hot season." Nevertheless, Hale was suspicious of highly seasoned foods, which she insisted damaged stomach linings and should never be fed to children; she criticized Hannah Glasse, *The Art of Cookery* (1805) author,

The drupes of black pepper (*Piper nigrum*) were ground as a familiar spice and or used whole in pickling, but they were viewed as dangerous by nineteenth century food reform zealots.

"whose receipts seem little else than a catalogue of herbs, spices, essences, and all manner of flavors." Glasse did indeed cook with flavor, illustrated by her method for boiling and seasoning carp with a potent mixture of white and black pepper, mace, nutmeg, "a bundle of sweet herbs," horseradish, lemon peel, and an onion stuck with cloves.

On the island of Hispaniola (now the Republic of Haiti and the Dominican Republic), Columbus discovered red peppers (*Capsicum* spp.), relatives of potatoes and tomatoes in the nightshade family (Solanaceae). He called them *pimiento* (based on *pimienta*, referring to the black pepper that he hoped to find) and believed that their spicy nature justified his travels in search of exotic spices. Columbus introduced red peppers from tropical America to Europe, and their cultivation spread to warm regions worldwide. Mild bell peppers and some hot red peppers are both varieties of *C. annuum*, which was cultivated in Central America for a variety of uses since ancient times (see chapter 4). The spice known as cayenne is made of ground, dried red peppers, sometimes *C. annuum* and less frequently *C. frutescens*, the elongated red peppers cultivated in the West Indies since 1494 and also by Thomas Jefferson in his Monticello gardens.

The "heat" provided by the red peppers used as spices is from capsaicin and other capsaicinoids, colorless and odorless compounds that affect the mouth and throat pain receptors in mammals; capsaicinoids are amide-type alkaloids, released from the point of attachment of the seed with the fruit wall, the region known as the placenta. Exposure to capsaicinoids causes the selective loss of some sensory neurons, which explains the mild, temporary numbness associated with consuming this spice and suggested some obvious medicinal uses as well. In the early 1870s, J. and I. Coddington Apothecaries of New York patented and marketed "capsicum plasters," an external treatment for arthritic pains, lumbago, sciatica, pleurisy, and sore throat; the plasters consisted of muslin spread with *Capsicum* extract, which functioned as a counter-irritant that caused temporary superficial numbness. Red pepper

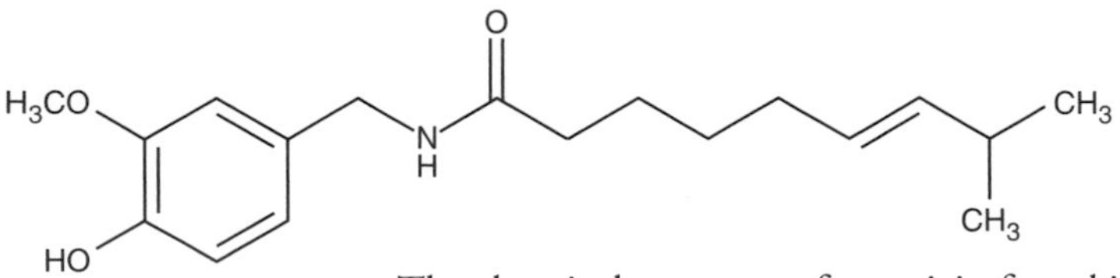

The chemical structure of capsaicin found in red peppers (*Capsicum annuum* and *C. frutescens*).

poultices probably originated as folk cures, since the pharmaceutical plasters were fundamentally similar to the cayenne poultices long used by both settlers in Appalachia and Cherokee Native Americans.

Maturity, growing conditions, and genetic variety all influence the capsacinoid concentrations and the "heat" of particular red pepper cultivars. Deep color and intense flavor seem to occur together; peppers with concentrated carotenoid pigments are also the most spicy. Cayenne was used particularly in southern kitchens to season gumbo and pepper pot, and Randolph's 1824 pepper pot recipe specifies "part of a pod of pepper" rather than the black peppercorns (*Piper nigrum*) used in northern kitchens. The gumbo (see chapter 4) prepared in colonial kitchens had an African origin and consisted of stewed okra flavored with salt and cayenne pepper, perhaps with the addition of more vegetables, rice, and meat. Cayenne pepper and black pepper also both appeared in recipes for the fermented sauce known as tomato soy.

According to legend, hot pepper sauces originated when an American soldier collected pepper seeds (*Capsicum frutescens*) in the Mexican state of Tabasco and planted them on a Louisiana coastal island owned by the Avery family; their cook used the peppers to prepare a hot, red sauce, which after the Civil War was marketed by their son-in-law Edmund McIlenny, a banker endeavoring to restore the family fortune. The sauce was made by pickling the pulp of hot peppers in barrels of vinegar and salt and then siphoning off and bottling the flavored juice. Beginning in 1868, he bottled Tabasco sauce in old cologne bottles and introduced it as a new condiment. E. C. Hazard and Company, wholesale New York grocers, helped to spread its popularity in the northeast, and by the late 1870s Tabasco sauce was being sold in England. McIlenny's product soon adopted slender glass sauce bottles, such as the one recently reconstructed from shards found at the site of the Boston Saloon in Virginia City, Nevada, an African American establishment that catered to miners from 1864 to 1875.

Curry was a spice mixture used in some American homes, particularly in southern colonial kitchens where British cooking traditions persisted. Various curry powders evolved in India and incorporated New World cayenne peppers after they were introduced worldwide. Curry mixtures were compounded according to regional and family tastes and availability, and they included various combinations and proportions of turmeric, ginger, black pepper, cayenne, mace, cloves, cardamom, cumin, coriander, and nutmeg, pounded into a paste or powder. A few of the ingredients were uncommon in American kitchens, including turmeric and cardamom, which are both members of the ginger family (Zingiberaceae). The characteristic yellow color of curry powder is derived from the ground rhizomes of turmeric (*Curcuma longa*), which contain the pigment known as curcumin, and turmeric was also used to flavor and color pickles. The fragrant seeds of cardamom (*Elettaria cardamomum*) have various uses in Indian cookery and were also used in baking. As with other umbellifers, the aromatic "seeds" of cumin (*Cuminum cyminum*) are the small fruits (schizocarps) of herbaceous plants related to parsley, caraway, and anise. Chickens stewed and flavored with curry powder became "East India curries" in the British colonial style, and Randolph also recommended currying rabbits, fish, beef, veal, and mutton. Hale (1841) described "currie soup," chicken cooked in veal broth flavored with onions, parsley, and curry powder.

Vanilla is less a spice than a mild flavoring made from the fruits of an orchid that is native to Central America and Mexico. Its capsules, or "beans," are typical of orchids containing thousands of minute seeds, and the fruit wall and seeds contain the volatile compounds vanillin and vanillic acid, which impart a characteristic flavor and scent. In nature, the flowers are pollinated by *Melipona* bees and hummingbirds, but a reliable crop of vanilla capsules requires hand pollination. These are cured by heating the capsules in the sun, followed by drying and then storing them for three months or more; since the green capsules lack distinguishable flavor or aroma, ethnobotanists have speculated as to how native people might originally have discovered vanilla and this process for curing the capsules. Vanillin accounts for 1.5 to 3 percent of the capsules by weight, and the concentrated crystals may appear as a slight surface bloom on the dried capsules. Aztecs learned to combine vanilla and chili peppers with chocolate to make spicy, bitter *chocolatl*; vanilla was encountered by sixteenth century French explorers and became popular both in France and England, but its flavor was unknown to eighteenth century American colonists.

Jefferson first tasted the vanilla flavor in France, and he sent for capsules after he had returned to the United States; his detailed manuscript notes for mixing, freezing, and molding vanilla ice cream are in the archives at the Library of Congress, and they specify a "stick" of vanilla. During the 1780s, he used a simple ice cream freezer with an inner canister (*sabotiere*) that was packed in a bucket layered with ice and salt. Randolph also described ice cream in *The Virginia House-Wife*. Her instructions were similar to her cousin Jefferson's method: boil a vanilla capsule in milk or cream to release its flavor, mix in eggs and sugar, and then freeze the mixture.

Vanilla was mentioned occasionally in nineteenth century cookbooks. In place of vanilla, rose water, peach water, and orange flower water were used to flavor delicate desserts; each was made by steeping or distilling flowers or leaves in water. These flavors were more typical of eighteenth and nineteenth century American cookery, but the recipes for elaborate desserts such as charlotte russe and Italian cream in *Mrs. Putnam's Receipt Book* (Putnam 1860) included vanilla. The cost and availability of vanilla were problems, and by the mid-nineteenth century, vanilla extracts became a popular alternative to using the whole capsules for flavoring. Extracts were prepared by soaking the crushed capsules in alcohol, and cooks could prepare or purchase them by the bottle. "Vanillo" was a vanilla extract made during the 1890s by B. F. Miner of Greenfield, Massachusetts, and sold throughout New England; the company's advertising boasted that it was "Made from Mexican vanilla beans" and of the "highest degree of strength."

Vanilla remained expensive, which encouraged nineteenth century chemists to experiment with cheap substitutes. By the 1880s, vanillin was synthesized from conifers, using the lignin that hardens the cell walls of wood cells, and it was also made from similar volatile molecules such as eugenol (from cloves) and guaiacol (from the resin of *Guaiacum officinale*; see chapter 7). Vanilla substitutes were sold as inexpensive "vanilla" extracts; in the years prior to the Pure Food and Drug Act of 1906, labels guaranteed little and promised neither purity nor authenticity. For instance, tonka beans (the seeds of the South American trees *Dipteryx odorata* and *D. oppositifolia*) contain coumarin, so-called "tonka bean camphor," a phenolic compound that has a taste and odor that resembles vanillin. Tonka beans were also used as an inexpensive "vanilla," but they were finally banned

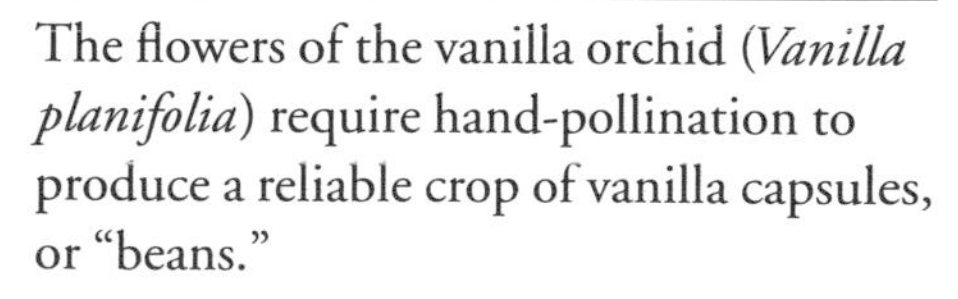

The flowers of the vanilla orchid (*Vanilla planifolia*) require hand-pollination to produce a reliable crop of vanilla capsules, or "beans."

in 1940 by the Food and Drug Administration because of the potentially toxic anticoagulant properties of coumarin.

Culinary Herbs

Sweet herbs are European natives, the traditional garden species that include a variety of easily cultivated, pungent plants that were both potherbs and medicinal plants. Traditional herb use originated in ancient European villages, when humans likely sniffed and sampled the aromatic field plants that were avoided by grazing animals. Herbs also have high concentrations of secondary compounds, the phytochemicals that impart characteristic flavors and fragrances to tropical spices. Humans are often attracted to salient plants, species with distinct traits such as pungency and indument (a hairy covering). As a result of informal handling and experimentation, some species evolved into culinary and medicinal herbs that were eventually cultivated in kitchen gardens. Useful herbs were introduced to New World Puritan gardens, the selfsame plants described in European herbals, and they were mentioned frequently in eighteenth and nineteenth century American cookbooks. Prior to the American Revolution, several of these herbs were also used as alternatives to imported Asian tea (see chapter 5), perhaps ironic because these herbs were originally brought to America from English gardens. Most importantly, because of their rapid herbaceous growth in temperate climates, herbs were easily cultivated as a reliable, inexpensive supply of food flavors and preservatives.

Several familiar herbs are members of the mint family (Labiatae or Lamiaceae), including sage (*Salvia officinalis*), rosemary (*Rosmarinus officinalis*), thyme (*Thymus serpyllum*), summer savory (*Satureja hortensis*), and marjoram (*Origanum onites*, *O. majorana*, and *O. vulgaris*). Most mints are Mediterranean natives that were introduced across Europe where they were grown as perennials, but rosemary shrubs often required winter protection. Mints characteristically have their stems and leaves covered with epidermal trichomes, microscopic hairs that contain the terpene-based essential oils that impart the characteristic odors of various mints. Other culinary herbs included several members of the parsley family (Umbelliferae or Apiaceae), which were cultivated for their small fruits that are known as "savory seeds." These included coriander (*Coriandrum sativum*), caraway (*Carum carvi*), fennel (*Foeniculum vulgare*), dill (*Anethum graveolens*), and anise (*Pimpinella anisum*). Botanically speaking, their fruits are schizocarps, mature seed-containing ovaries that split in half (each half is known as a mericarp) and superficially resemble single seeds. Umbellifers are chemically complex plants with essential oils that occur in internal canals; these oils include pungent terpenes such as anethole and apiole, and furocoumarins, which show antifungal activity and may cause photosensitivity.

New England Puritans chewed the intensely flavored fruits of various umbellifers as "meeting seeds," to help them stay alert at church services and perhaps suppress hunger. Various combinations of coriander, caraway, and fennel were baked into colonial cakes, including the New Year's cakes that were favored by Dutch settlers in New York and decorated with stamped designs using carved wooden molds that became popular in the early nineteenth century. Simmons's 1796 method for loaf cake required "a tea cup of coriander seed," along with twelve pounds of flour and six pounds of sugar, and like her recipe for a "cheap seed Cake" made with caraway, it was reminiscent of medieval European seed cakes. Dill was cultivated for both its leaves and its seeds, which were used for flavoring pickles and vegetables and to treat flatulence and colic.

Parsley (*Petroselinum crispum*) was cultivated for its leaves and flavored or adorned dishes that were brought to the table. Food historian Waverly Root (1980) has noted that although there is no written record of parsley in America prior to 1806, explorer Giovanni da Verrazano claimed to have seen parsley growing along the New England coast about 1524, perhaps a chance early introduction by Viking explorers or Basque fishermen who visited Newfoundland. Flat-leafed varieties were probably the ancestral type and are more flavorful, but Leslie (1851) recommended curled parsley for pickling in cider vinegar. Parsley sauce was made from the finely chopped leaves cooked in butter and served with fish, fowl, and other delicate meats. Fern-leafed parsley appeared as a cultivar in American seed catalogs in the 1880s, advertised as both a garnish and decorative garden plant. As with other culinary herbs, parsley had medicinal uses; Shakers grew parsley to treat urinary problems, edema (the condition once known as dropsy), and insect stings, and the small fruit were used to remove head lice, reminiscent of the European use of coriander to discourage bedbugs.

Like the mints and umbellifers, the mustards (family Cruciferae or Brassicaceae) were among the first families of flowering plants to be recognized, which explains why these familiar families have two names; the earlier name was coined before the suffix *aceae* became a standard nomenclatural practice for all plant families. Mustards are the primary source of the sharp-tasting mustard oils, glycosides such as sinigrin that are produced in specialized myrosin cells and have both culinary and medicinal uses. Mustard oils are found throughout the plants, including the leaves that were gathered and used as potherbs. The dark brown seeds of black mustard (*Brassica nigra*) and the pale seeds of white mustard (*Sinapsis alba*) were used both whole and ground for flavoring and condiments. Both were common European roadside and garden plants, and gardeners harvested the seeds from the siliques, the elongated mustard fruits. Shops offered "mustard flour," usually a combination of black and white mustard seeds ground into a fine powder and often adulterated with wheat flour and turmeric. Table mustard was prepared by combining the powdered seeds with boiling water or milk to make a sauce or paste, and sometimes flavored with horseradish (*Armoracia rusticana*), another member of the mustard family. Leslie (1851) recommended "French mustard" to serve with beef or mutton, prepared with tarragon (*Artemisia dranunculus*), which was uncommon; she noted, "If you cannot procure tarragon leaves, buy at a grocer's a bottle of tarragon vinegar."

Black mustard plants produce the mustard oil sinigrin, while white mustard yields sinalbin; both contain sulfur and are known chemically as volatile oils with both culinary and medicinal applications. Small amounts of mustard oils safely flavor food and provide the benefit of antibiotic activity, but larger amounts of ingested sinigrin are suspected to cause mutations. Early Americans relied on the external use of mustard plasters, poultices made of mustard seeds, as counterirritants that may have provided some antibiotic benefits as well. The *Druggists' Hand-book* (1870) described the use of white mustard plasters "to rouse the system to activity, relieve pain and mitigate inflammation." By this time, ready-made mustard plasters known as sinapisms were available from pharmacists; E. Fougera and Company advertised and imported "French medicinal preparations," including mustard plasters in two concentrations, one suitable "for children and delicate persons." Concentrated mustard caused severe stinging and possible blistering of the skin. *Johnson's Universal Cyclopaedia* (1885–86)

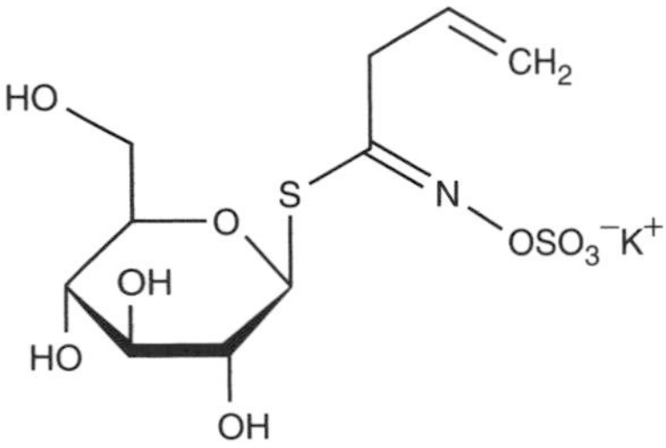

The chemical structure of sinigrin from black mustard (*Brassica nigra*), which had both culinary and medicinal uses.

noted the "peculiar pungent, irritant principle . . . which gives mustard its value as a food and medicine. . . . The moistened flour applied to the skin is a powerful irritant and vesicant . . . to relieve internal pains and spasms."

An historical footnote involves the use of allyl isothiocyanate (a portion of the sinigrin molecule from black mustard) variously as a flavoring, counter-irritant, and war gas. Sinigrin provided the chemical basis of the infamous mustard gas that was first used in 1917 by the German army against the Allies in Ypres. Known variously as "yellow cross" (the marking on the gas shells) and "yperite," mustard gas was actually a volatile oil that caused severe burns and blisters (both internally and externally) several hours after contact; mustard gas caused thousands of casualties in World War I, although it was invisible, odorless, and tasteless as it evaporated. Ironically, the synthetic production of lethal mustard oil was worked out by Fritz Haber, a pacifist who received the Nobel Prize in 1919 for making ammonia from hydrogen and nitrogen, which led to fertilizer production and staved off starvation in Europe. Mustard gas also provided the basic chemistry for the broad-spectrum anti-cancer drugs known as alkylating agents, which are nitrogen mustards (as compared to the sulfur mustards that compose mustard gas) that were used to treat lymphomas as early as the 1940s.

Mustard oils also flavor the massive taproots of horseradish (*Armoracia rusticana*), which were grated as a condiment or used to flavor sauces or pickles. Horseradish was grown in Puritan gardens (see chapter 1) and was valued for its antiscorbutic properties. It was used in tonics, syrups, and poultices, and by the 1870s, "elixir of horse-radish" was marketed as a diuretic, tonic, stimulant, and "Regenerator of the Blood." As with other members of the family Cruciferae, the mustard oils of horseradish are not released until the cells are damaged and enzymes break down the glycosides that contain the isothiocyanate compounds. Vinegar stops this reaction, so the horseradish vinegar recommended by Leslie (1851) had an appreciably milder flavor than grated horseradish used as a condiment.

Mustard oils also account for some confusing plant names. Nasturtium (as a common name) refers to the familiar garden flower *Tropaeolum majus*, a native of Peru and Chile that is also known as Indian cress. *Tropaeolum* is not a member of the mustard family, but coincidentally these plants also evolved the synthetic pathways for producing mustard oils as a defense mechanism. The flower buds and seeds of garden nasturtiums were pickled as substitutes for capers; capers are the pickled flower buds of *Capparis spinosa*, a shrub in the family Capparaceae that is related to the Cruciferae and also produces mustard oils. *Nasturtium* as a scientific name includes watercress (*N. officinale*), a species in the Cruciferae that was cultivated informally by sowing seeds in wet areas and streams and harvesting the plants as needed (see chapter 3); as with the other mustards that were introduced from Europe for cultivation in American gardens, watercress is now widely naturalized. Apparently, the coincidental similarities caused by the presence of mustard oils caused an overlap in common and scientific names and uses alike. Colonial cooks understood the practical chemistry of mustard oils as flavorings; Simmons (1796) recommended caper sauce as well as horseradish, watercress, turnips, and "colliflowers," all mustard relatives, to accompany a dish of roast mutton. Years later, Leslie (1851) reminded cooks to substitute "nasturtians" or pickled radish pods (also in the Cruciferae) in place of capers in preparing a sauce for boiled mutton.

Sweets

Humans have a taste for sugar. Long before the arrival of Europeans in North America, some Native Americans concentrated the sap of maples and gathered caches of wild honey; colonists adopted maple sugar and honey and introduced imported sugar cane, so various sugars were available in colonial kitchens. Maple sap is a mixture of water and dissolved sugars that travels in the phloem (food-conducting cells) that compose the inner tree bark, and it is the source of maple sugar, the only food produced exclusively in North America. Sap flows abundantly in sugar maples (*Acer saccharum*), but it can also be collected and concentrated from other maple species. Butternut (*Juglans cinerea*) sap was also collected and concentrated as a source of sugar, and the sap-rich legumes of the native North American honey locust (*Gleditsia triacanthos*) were used locally by pulverizing the whole fruits into a sweet meal that was used in cookery. These sap sugars are mostly disaccharides that consist of two simple sugars bonded together; for instance, sucrose comprises a glucose bonded to a fructose molecule. Some simple sugars may also occur in sap, along with proteins, amino acids, organic acids, vitamins, and minerals; these compose the nutrient mixture that trees transport from their roots to the branches during the late winter and early spring.

Since sugars constitute only about 8 percent of maple sap, the taste is slightly sweet, but it was appreciated by Native Americans who probably tasted drops that dripped from bark wounds. Early colonists observed the Native American practices of concentrating the sugar by freezing the sap and removing the ice or evaporating the water by dropping hot rocks into troughs of collected sap. Once concentrated and crystallized, the sugar was stored in bark boxes and traded with colonists. In 1706, Governor William Berkeley of Virginia observed of these practices: "The Sugar-Tree yields a kind of Sap or Juice which by boiling is made into Sugar. This Juice is drawn out by wounding the Trunk of the Tree, and placing a Receiver under the Wound."

"Sugaring-off" also became a colonial tradition, and the early practice of collecting sap involved "boxing," in which a large piece of the trunk was cut away to release sap from the wound. This was soon replaced by a less invasive method that did not kill the valuable trees; small spouts were carved from basswood, and these were inserted into small notches cut into the bark four or five feet from the ground. Sap dripped into a wooden trough, often hewn from a butternut log, and was collected for boiling. Eventually, buckets replaced the use of troughs, and augers drilled the holes for spouts. Boiling was done in iron kettles at the site, a maple grove known as a "sugar bush." Settlers sometimes encamped for several nights, particularly if the ideal weather conditions of cold nights and warm days resulted in abundant sap. Variation occurred in the availability and quality of the sap; depending on the weather, sap ran for a few days or several weeks and even varied in its sugar concentration. In the American edition of *The Art of Cookery*, Glasse added "A receipt to make Maple Sugar," instructions for tapping trees in mid-February and boiling down the sap to make granulated maple sugar and "maple molasses." A small amount of butter or fat kept the pots from boiling over, and a long period of boiling was necessary to concentrate the syrup. Once the syrup was supersaturated, it crystallized easily. She recommended cooling a small amount of syrup to check for granulation and then pouring the solution into boxes or bags in which the sugar crystals would form.

Tales of sugar maples reached Europe. In 1663, Robert Boyle, an English chemist known to chemistry students for his contribution to the gas laws, had described "a kind of tree . . . whose juice that weeps out of its incisions . . . doth congeal into a sweet and saccharin substance, and the like was confirmed to me by the agent of the great and populous colony of Massachusetts." Nearly a century later (1751), Swedish naturalist and explorer Pehr Kalm recorded gifts of maple sugar: "When we reached the villages of the savages we received more than anything else gifts of large pieces of sugar which stood us well in hand on our trip into the wilderness." His party ate cornmeal mush flavored with maple sugar, which became a traditional dish especially in New England, where antislavery sentiments persuaded many to use maple sugar in place of molasses and sugar imported from the West Indies. The "sugaring off" process yielded both syrup and sugar, depending on how much water was boiled away from the sap in the kettles, and both were used in many colonial homes as the only or most common sweeteners. Jefferson also considered the possibility of maple sugar production in central Virginia, and he cultivated groves of sugar maples at Monticello.

Cakes of maple sugar were the only sweetener philosophically acceptable to nineteenth century New England Transcendentalists such as Bronson Alcott, recorded by his daughter Louisa May Alcott in her short story "Transcendental Wild Oats," which parodies life at the Fruitlands farm. The Alcotts, along with Englishman Charles Lane and several others, formed the "Con-Sociate Family," a communal group that endeavored to live a life close to nature on a ninety-acre farm known as Fruitlands, thirty miles west of Boston, in Harvard, Massachusetts. Their philosophy demanded that they live without commodities that "caused wrong or death to man or beast," and maple syrup was a convenient substitute for the molasses and refined sugar that were part of the slave trade triangle. Regardless of philosophy, Americans used maple sugar commonly until the end of the nineteenth century; forty gallons of maple sap yielded one gallon of syrup or four pounds of sugar. Children sometimes poured the hot, concentrated sap over snow to make a hardened "maple wax" candy, and some brewers tried to ferment the syrup to make maple beer or vinegar; maple beer was sometimes flavored with essence of spruce.

Honey was another sugar source, a unique botanical product produced by social bees that are adapted to collect, process, and store the nectar of flowering plants. Nectar is a mixture of water, amino acids, and sugars (including sucrose, glucose, and fructose) produced as a floral reward for pollinating insects. Nectar is collected and carried in a bee's honey sac (crop), where it mixes with the enzyme invertase that breaks down sucrose into equal amounts of glucose and fructose. The nectar is then stored in the wax cells of the hive, where it evaporates and thickens into honey. Both male and female social bees use nectar and sometimes pollen as food, and adult females gather and store nectar in the form of honey that nourishes the larvae. Honey contains about 80 percent sugar, and a pound of honey represents thousands of bee visits to flowers, depending upon the amount of nectar that the flowers produce.

Ships traveling from England during the winter carried dormant honeybees (*Apis mellifera*), and hives were brought to Virginia before 1622. By 1634, the Puritans established European honeybees in Massachusetts, but in *Home Life in Colonial Days* (1898), Alice Morse Earle noted that colonists also gathered "wild honey" from hollow trees. In fact, this honey was not produced by bees native to North America but rather by escaped honeybees that colonized hollow tree trunks in natural habitats. Honeybees spread rapidly in North America, sometimes arriving before colonists; Native Americans called them "English flies" or "white man's flies" but willingly adopted honey as an ingredient in traditional Native American foods. Hives flourished in forest clearings, where the bees collected nectar from North American wildflowers, shrubs, and trees. Both escaped and domesticated bees provided honey as a source of sugars for cooking, baking, and brewing; diluted honey was used to ferment mead, also known as metheglin and hydromel, described by Randolph as a "pleasant and wholesome drink."

In 1851, Reverend Lorenzo L. Langstroth invented a hive design with movable frames, which was widely adopted by beekeepers; he authored *The Hive and Honey-Bee: A Beekeeper's Manual* (1853), a classic text on apiculture, in which he argued for the better management of hives. Farmers often kept hives, and honeybees became important pollinators of agricultural crops and fruit trees with nectar-producing flowers, a function that far exceeded the economic importance of honey. Bees can also maintain hives in populous areas, and by the 1870s, city dwellers were encouraged to take up bee-keeping in their gardens or on rooftops as a way to augment the family income; an article in *The Druggists' Circular and Chemical Gazette* (March 1876) described a New York beekeeper whose insects used "the waste of a sugarhouse and such flowers as the public parks might afford" with great success. Nevertheless, honey production was labor intensive. By the end of the nineteenth century, charlatans

Honeybees (*Apis mellifera*) collect and store nectar from flowering plants in the wax cells of beehives, where it evaporates and converts into honey.

produced clever fake honey and honey combs using inexpensive sucrose syrup, glycerine, essential oils, and paraffin, with occasional dead bees added for authenticity. The prevalence of adulterated or artificial "honey" encouraged legitimate beekeepers to press for the Pure Food and Drug Act of 1906.

Sugar from sugar cane (*Saccharum officinarum*) was a refined product in two senses; the crystals were refined from the crude molasses syrup, and granulated sugar was sought for tea and delicate baking. Sugar cane is a perennial, tropical grass (family Gramineae or Poaceae) that was propagated asexually by planting pieces of stems known as seed pieces, each with a node that sprouted adventitious roots and a shoot. Sugar cane has C_4 photosynthesis, meaning that the plants grow efficiently even under harsh environmental conditions; the plants accumulate and store carbon dioxide as part of a four-carbon compound, allowing the plants to keep their stomata closed during the hottest times of the day and avoid dehydration. Cultivated sugar cane is unknown as a wild grass but probably originated in the Pacific region, perhaps Papua New Guinea, where it was selected and hybridized from wild species of *Saccharum*. During his second voyage, Columbus introduced sugar cane to the West Indies, where the plants thrived. The tall stems or canes were cut and harvested by hand and then pressed to release their sucrose-rich sap, processes that required the cheap labor provided by African slaves on colonial Caribbean sugar plantations. The unrefined sap was evaporated to make the dark, concentrated syrup known as molasses. Despite its desirable sweetness, sugar was avoided by those who abhorred slavery because of the complex trading triangle that revolved around slaves, molasses, and rum.

Molasses and raw sugar were traded to New England rum distillers and sugar mer-

Sugar cane (*Saccharum officinarum*) is a tropical grass that was used to make white and brown crystallized sugar and molasses, which fueled the New England rum industry.

chants, rum was traded to Africa for slaves (a large amount was also consumed in the American colonies), and these slaves were traded in the West Indies for molasses and raw sugar. Antislavery pamphlets illustrated cruel sugar plantation practices, where slaves were tethered to weights to prevent their escape and prevented from eating sugar cane by wearing heavy head frames. Nevertheless, the early American economy was deeply tied to sugar production; in eighteenth and early nineteenth century New England, the sugar trade promoted shipbuilding and spawned a rum industry with serious social ramifications. Colonies also traded lumber, grains, meat, livestock, and horses to supply the sugar plantations in the West Indies, where the owners concentrated exclusively on sugar production. By the mid-1700s, Massachusetts had sixty-three distilleries that produced rum by fermenting imported molasses followed by distillation and aging. Rum was used along with tea in punch mixtures (see chapter 5) and to prepare arrack, rum flavored with spices or the pungent flowers of the spicebush (*Lindera benzoin*), a native North American shrub classified in the same family (Lauraceae) as cinnamon. The Sugar Act of 1764 levied British taxes on sugar imported into the American colonies, which resulted in sugar smuggling and the burning of the HMS *Gaspee* by colonists, events that predated the Boston Tea Party (1773) and eventually led to the Revolutionary War.

Despite the slavery controversy, both molasses and sugar were used in many colonial American homes. The sap from sugar cane was boiled and evaporated into sugar syrup from which sugar was crystallized; molasses was the unrefined liquid waste from the sugar-making process, a less expensive sweetener that was used both domestically and for rum production. Sugar was an expensive commodity often used in the colonial ritual of tea drinking, and in the words of Earle (1898), "housewives of dignity and elegance desired to have some supply of sugar, certainly to offer visitors for their dish of tea." Various grades of brown and white sugars were imported from the West Indies. White sugar was the most desirable for table use, while brown sugar and often molasses were used in cooking, and exasperated housekeepers sometimes burned small amounts of brown sugar to rid their houses of mosquitoes. White sugar was sold as loaves or cones that were shaped in clay molds that drew away water as the sugar crystallized. Specialized sugar cutters were used to break loaf sugar into lumps of uniform size, or the sugar was pounded into a powder. The loaves and cones of sugar weighed from ten to thirty pounds, and each was wrapped in deep blue paper that colonists often saved to dye wool; the paper contained blue pigments from indigo and woad plants (see chapter 8), and similar blue papers were used to wrap needles and other sewing supplies.

"West India molasses" was often specified in recipes for gingerbreads, spice cakes, election cake, puddings, candy, and beer. Molasses was often purchased by the barrel for household use, and Beecher (1858) mentioned molasses from "Porto Rico" as the best for cooking, while sugar house molasses (probably produced at sugar refineries in the United States) was preferred for table use. Ardent abolitionists also avoided molasses because of its dependence on slavery, and they substituted honey for molasses in their baking. Sugar molds were abandoned during the second half of the nineteenth century when refiners began to use the vacuum pan and centrifuge to produce crystals. Brown sugar, known as "muscovado," replaced maple sugar and molasses as the common sweetener in homes; it was sold in large lumps, a mixture of coarse crystals and molasses, and required grinding before it was used. By the 1880s, the price of imported sugar was comparable to American maple sugar, and cooks used large quantities of refined sugar in preserving fruits. Molasses was also used as a food preservative; Child (1844) suggested using molasses to cover stored suet and to preserve barberries. The transition from brown sugar to refined white crystals as the most common sweetner occurred during the 1890s, when whiteness became an admired quality that was associated with purity; sugar merchants also perhaps implied that brown sugars were impure and insect contaminated, to promote sales of white sugar.

Sugar, molasses, and rum had medicinal uses as well. *Eau sucre*, sugar dissolved to taste in boiling water, was recommended for weak nerves and fatigue, and according to Hale (1841) it was "much used by French ladies." Loaf sugar was used to sweeten various teas and wheys used as sickroom beverages, and Child (1837) recommended warm molasses mixed with water or milk as a mild laxative for children. Following the wisdom of *The American Frugal Housewife* (Child 1844), molasses was also applied to cuts, used on brown paper as a poultice for inflammation, and mixed with milk and the mucilage from mallows (such as marsh mallow, *Althaea officinalis*, and related plants in the family Malvaceae) as a drink for those with piles (hemorrhoids). A warm posset of milk flavored and curdled with molasses, butter, ginger, and lemon juice was used to treat colds. Rum was also used to treat colds, but nineteenth century temperance adherents urged that its medicinal uses be curtailed, a reaction to the excess rum consumption of the prior century.

Garlic mustard (*Alliaria officinalis*) was introduced from Europe as a medicinal mustard; it escaped from early American gardens and naturalized as a weed that can thrive in deep shade.

Groundnut or potato bean (*Apios tuberosa*) is a wild legume with potatolike enlargements of the roots; colonists learned of its use, and groundnuts were among the earliest American species introduced to European gardens. Courtesy New England Wild Flower Society / Frank Bramley.

Horseradish (*Armoracia rusticana*) was cultivated as much for its medicinal properties as for its culinary use, both based on potent mustard oils; traditional medicinal uses included horseradish plasters and poultices to treat sciatica and gout and the juice of fresh roots to treat intestinal worms.

Wild ginger (*Asarum canadense*) was used in meat and fish cookery, perhaps when imported spice supplies were short, but large quantities might have proven toxic due to the presence of aristolochic acid.

Common milkweed (*Asclepias syriaca*) was an alternative source of bast fibers similar to those of hemp, and the hairs on its seeds were sometimes spun into candlewicks.

Pickled barberries (*Berberis vulgaris*) were a colonial garnish, but the shrubs were cleared from many wheat-growing areas because the plants are an alternate host for the wheat rust fungus.

Carolina allspice (*Calycanthus florida*) was listed by the Prince Nursery as "the sweet scented shrub from Carolina"; its crushed leaves release essential oils with a strong, spicy scent.

Tea substitutes included the leaves of sweet fern (*Comptonia peregrina*), which also had medicinal uses as an astringent and was used in tonics to treat ailments such as dysentery, stomach upset, and mouth cankers.

Kenilworth ivy (*Cymbalaria muralis*) was introduced from western Europe and cultivated for its unique property of colonizing rock walls; the stalks that bear the fruits elongate and "plant" the seeds in crevices.

Yellow lady's slipper (*Cypripedium calceolus* var. *pubescens*) was one of several wild orchids known as nerveroots because of their deeply impressed veins, which according to the Doctrine of Signatures recommended them as a treatment for nervous ailments.

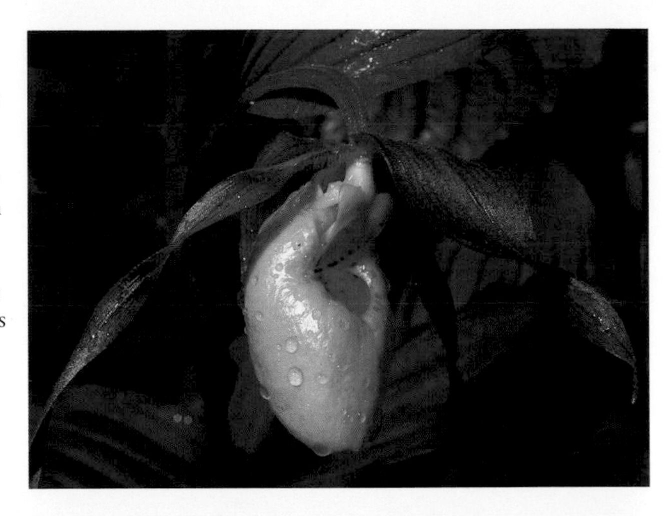

Clove gillyflowers (*Dianthus deltoides*) were also known as sops-in-wine; they were named for their early use as a folk substitute for cloves and as a flavoring for cordials.

Bleeding heart (*Dicentra spectabilis*) was introduced from China in 1846 by Robert Fortune, and within a few decades it was a favorite garden plant in both England and America.

Yellow foxglove (*Digitalis grandiflora*) contains cardiac glycosides, which were the chemical basis of the use of foxglove to relieve the edema (once known as dropsy) associated with heart failure.

Fullers used the spiny inflorescences of teasel (*Dipsacus fullonum*) to raise the nap of the clean, dry woolen cloth; several of these mature heads were attached to a wooden handle to make an implement for finishing woolen textiles.

Coneflower (*Echinacea purpurea*) was used by Plains Native Americans as an almost universal remedy, and early settlers soon adopted the herb; by the late 1800s, several proprietary tonics and blood purifiers were available that contained coneflower, including Meyer's Blood Purifier, which became popular in the 1870s.

Trailing arbutus (*Epigaea repens*) was cultivated in early Wardian cases, but the plants eventually became uncommon because of over-collection to supply the late nineteenth century florist trade.

Horsetails (*Equisetum hyemale* and related spp.) became known as scouring rushes because they have a rough texture that made them useful for scouring pots and pewter; their epidermis is covered with silica in the form of opal, a mineral that strengthens the stems and probably protects them from herbivores and desiccation.

Settlers adopted boneset (*Eupatorium perfoliatum*) as a familiar cure for colds and malaria, but the common name probably originated from its use to treat break-bone fever, another term for the virus known as Dengue fever.

American holly (*Ilex opaca*) was a native North American substitute for English holly, after the hanging of Christmas greens became customary during the second half of the nineteenth century.

Thomas Jefferson cultivated twinleaf (*Jeffersonia diphylla*), a native wildflower, at Monticello, and Benjamin Barton named the plant in Jefferson's honor in 1792.

Mountain laurel (*Kalmia latifolia*) is also an evergreen shrub with leaves that are similar to those of bay laurel; the old common name "calico bush" referred to the markings on the pale pink petals that resemble a printed design.

Coral honeysuckle (*Lonicera sempervirens*) is a native North American vine that was cultivated for its bright, hummingbird-pollinated flowers. Courtesy New England Wild Flower Society / John Lynch.

Bee balm (*Monarda didyma*), had medicinal properties attributed to the presence of the antiseptic compound thymol, and it was used as a substitute for imported tea after the Boston Tea Party on 16 December 1773.

Purple loosestrife (*Lythrum salicaria*) was cultivated as a medicinal herb, a nectar flower for bees, and an ornamental garden plant, but during the 1800s it began to colonize North American wetlands as a troublesome invasive weed. Courtesy New England Wild Flower Society / Jean Baxter.

Wax produced by the small fruits of bayberry (*Myrica pensylvanica*), above, was used to make candles and soap, and the shrubs were grown as aromatic garden plants.

May apple (*Podophyllum peltatum*) was used as a strong emetic and cathartic herb, and it was also used variously to treat jaundice, hepatitis, fevers, and venereal disease; the roots contain a resin composed in part of podophyllotoxin, which is now used to synthesize the anti-cancer drug etoposide.

The use of Senega snakeroot (*Polygala senega*) was promoted by John Tennent as a cure-all to treat snakebite, gout, rheumatism, dropsy (edema), and nervous disorders. Courtesy New England Wild Flower Society / Dorothy S. Long.

The Royal Botanic Gardens, Kew, distributed Japanese knotweed (*Polygonum cuspidatum*) as a unique garden ornamental during the mid-nineteenth century; the plants thrived under environmental extremes such as heat, dry soils, and salinity and during the 1890s had already naturalized as an invasive weed near Philadelphia.

Shakers grew damask roses (*Rosa damascena*) for use in treating hemorrhages and bowel complaints, and rose petals were combined with the pith of sassafras branches in an infusion for treating eye inflammation.

Rosemary (*Rosmarinus officinalis*) shrubs were tender perennial mints introduced by early colonists for a wide range of culinary and medicinal uses.

Bouncing bet (*Saponaria officinalis*) was known as soapwort or fullers' herb, the source of saponins that were lathered in water to remove grease and dirt from wool or fleece.

House leek (*Sempervivum tectorum*) has succulent, astringent leaves with a gelatinous texture that resembles aloe; they were once used to treat burns, stop wounds from bleeding, and combined with plantain (*Plantago* spp.) into an ointment to treat skin eruptions.

Pinkroot (*Spigelia marilandica*) contains the potentially toxic alkaloid, spigeline, and it was an effective but dangerous home remedy for intestinal worms. Courtesy New England Wild Flower Society / William Larkin.

Tansy (*Tanacetum vulgare*) was used as a potherb, tea herb, and culinary flavoring; to kill intestinal worms; and as an abortifacient, but its essential oil is potentially lethal—violent spasms, convulsions, and paralysis were observed in women who ingested tansy oil, now an illegal ingredient in any food or medicine.

Coltsfoot (*Tussilago farfara*) contains potentially toxic pyrrolizidine alkaloids, but the plants were once valued for their mucilaginous properties in treating sore throats and coughs.

Puritans learned about cranberries (*Vaccinium macrocarpon*) from Native Americans who stewed and sweetened cranberries to make the first cranberry sauce; cranberries were also combined with venison and tallow to make nutritious dried cakes that were preserved indefinitely by the high acidity of the berries. Courtesy New England Wild Flower Society / Dorothy S. Long.

Heartsease (*Viola tricolor*) was used to treat heart ailments according to the Doctrine of Signatures interpretation of its heart-shaped petals; the linear petal markings are visual guides that direct pollinating bees to the nectar deposits, a reward for floral visitors that collect and carry pollen.

Fox grapes (*Vitis labrusca*) had a rounded shape, dark purple-black color, and a pale, waxy bloom on the epidermis; cultivated varieties such as 'Concord' grapes were well-suited to use as both a table grape and for jellies and wine. Courtesy New England Wild Flower Society / Catherine Heffron.

The wood of white pine (*Pinus strobus*) in cross-section reveals water-conducting tracheids of a uniform size and occasional resin canals.

The wood of white oak (*Quercus alba*) in cross-section reveals small-diameter, thick-walled fibers and large-diameter vessel elements; the vessel elements are blocked by tyloses, which form when nearby living cells protrude into the vessel elements and prevent the passage of water through the vessels.

Some women used rum cosmetically to wash their faces and hair, but Hale (1841) argued that rum resulted in wrinkles and gray hair, more evidence to support her plan for "the entire exclusion of distilled spirits" from American homes.

Nevertheless, New England rum production continued into 1919, when the Eighteenth Amendment to the Constitution prohibited the manufacturing and selling of alcoholic beverages. Earlier that year, on 15 January, a bizarre accident involved an uncontrolled flood of molasses from a tank owned by Boston's Purity Distilling Company, perhaps caused by the rapid expansion of the stored molasses caused by a warm winter day. Air temperature in Boston reached 40 degrees Fahrenheit on 13 January, which may have caused rapid expansion of the stored molasses. More than two million gallons surged down Boston streets in viscous waves, destroying buildings, injuring several, and killing twenty-one people. Years before and on a considerably smaller scale, Beecher (1858) had warned cooks of the problems of storing household molasses; she noted that "No vessel should be corked or bunged if filled with molasses, as it will swell, and burst the vessel, or run over."

Botanical Miscellany

Storeroom and pantry shelves also housed a variety of other botanical ingredients used in cookery. Puddings and similar desserts required thickening agents, which came in a variety of forms. Tapioca was made from manioc (*Manihot esculenta*), also known as cassava, a plant native to Brazil and valued for its massive, starchy roots; as a member of the spurge family (Euphorbiaceae), manioc is related to castor beans (*Ricinus communis*, see chapter 7) and poinsettias (*Euphorbia pulcherrima*, see chapter 10). Some manioc plants contain high concentrations of cyanogenic glycosides (linamarin and lotaustralin), which release hydrogen cyanide when they are damaged or digested; the roots require soaking in water to remove these toxic secondary compounds. Manioc plants can grow in poor or acidic soils and became an important staple crop in the New World tropics, the "venomous roots" mentioned by Italian historian Peter Martyr Anglerius in his description of the first voyage of Columbus in *De Orbe Novo* (the *New World Chronicles*, 1516). Native Americans learned to grind manioc into bread flour, and the roots also served as cheap, starchy food for African slaves laboring in Latin America and those in transit to America, until abolition in the mid-nineteenth century.

North American settlers knew manioc not as a dietary staple but rather as the source of tapioca for occasional puddings; tapioca consisted of small "pearls" of precipitated manioc starch that were cooked on hot plates and sold as a dry ingredient for cooking. Tapioca required soaking in water and then was mixed and baked with milk, eggs, sugar, and nutmeg to make puddings. The translucent quality of tapioca was appealing; the proteins in dissolved wheat flour reflect light and cause its opaque nature, but tapioca is almost entirely starch that absorbs water and becomes somewhat transparent. Cassava bread (or "cassada") may have arrived with African slaves in Virginia and perhaps was made from manioc roots cultivated in slave gardens; Randolph (1824) noted that a flat bread made of rice is "nearly as good as cassada bread."

Puddings and custards were also thickened with starch from the fibrous tubers of arrowroot (*Maranta arundinacea* and closely related species), which was cultivated in Jamaica, Bermuda, Georgia, and Florida. Milk thickened with arrowroot starch was an infant food, but arrowroot pudding was an elegant dessert when flavored with peach leaves or almonds and decorated with citron and other preserved fruits. Gruel prepared with arrowroot or tapioca was bland sickroom fare, while puddings were flavored with spices and often included raisins or currants.

Sago is a another tropical starch, harvested from the central stem (pith) of palms (*Metroxylon sagu*, *M. rumphii*, and related species) native to Papua New Guinea and the Moluccas. The trees were introduced by way of southeast Asia and the Pacific islands, where sago as used as a source of food, fibers, and medicines; in Indonesia, Marco Polo noted, "Meal is procured from a certain tree." Sago evolved into a starch used in western cookery, and cooked sago flavored with sugar and cinnamon was another example of nineteenth century sickroom fare. Like manihot, it was prepared as small beads of starch by drying droplets

Manioc or cassava (*Manihot esculenta*) roots were the source of tapioca starch, a crop that was possibly cultivated in some slave gardens as a source of bread flour.

on a hot surface, and Hale (1841) described sago as one of the easily digested "farinaceous vegetables," along with rice, tapioca, and potatoes, all foods that she recommended for children. Beecher (1858) suggested keeping eight or ten pounds of tapioca, arrowroot, and sago as staples in the family storeroom; these starches were often used interchangeably.

The pudding known as blancmange was sometimes thickened with Irish moss (*Chondrus crispus*), a marine alga that was easily collected by those who lived in coastal areas. Beecher recommended using "a teacupful of Carrageen, or Irish moss, after it has been carefully picked over," soaking it to remove the salt, and then cooking the alga in milk. The cells of red algae produce carrageenans, cell wall compounds that are water-soluble carbohydrates (hydrocolloids); in nature carrageenans accumulate water and help to protect the algae from desiccation. The carrageenan is released from Irish moss and absorbs water from the milk, resulting in a slightly glutinous texture that was considered a good food for the infirm. Blancmange was flavored to taste with various combinations of almonds, nutmeg, cinnamon, mace, vanilla, rose water, and peach water. Nineteenth century druggists could also often supply Irish moss, but cooks without a source of algae thickened their blancmange desserts with arrowroot or isinglass.

Pale food sometimes needed color, and blancmange was occasionally tinted green with spinach juice or yellow with saffron, the stigmas and styles of saffron crocus flowers (*Crocus sativus*). Cooks used beet juice for red and pink hues, and indigo (from the leaves of *Indigofera tinctoria*) provided shades of blue. Alkanet (from the roots of *Alkanna tinctoria* and a few other members of the Boraginaeae, the borage family) was another source of red color. Randolph described tinting and molding blancmange into fruit shapes, and Beecher suggested a remarkable "Variegated Blanc Mange" molded with red, yellow, blue, and white layers and then cut into decorative shapes.

Nineteenth century cooks obtained gum arabic and gum tragacanth for occasional kitchen use; both are hydrocolloids produced by Old World legumes. Gum arabic is extracted from the northern African tree *Acacia senegal* and related species, and the product is harvested by cutting the tree trunks and collecting the hardened drops of gum that exuded from the wound. Gum arabic dissolves slowly in water and imparts viscosity; Beecher (1858) noted its use in candy-making, and Leslie (1851) added it to flax seed, sugar, water, and lemon to prepare a thickened, demulcent beverage for treating colds and coughs. Gum tragacanth was collected from bark incisions made in thorny

Astragalas shrubs (*A. gummifer* from Asia Minor and others); Leslie (1851) added a small amount of gum tragacanth to cake icing recipes made of "finely powdered loaf-sugar." Both of these gums are water-soluble, odorless, and nearly tasteless, and they were used only for the texture that they imparted. In compounding medicines, local druggists used gum tragacanth and gum arabic as emulsifying agents and to bind insoluble ingredients. Gum tragacanth was also used to stiffen and size calico fabrics.

Most cooks used animal fats in cookery, but occasional references to "sweet oil" or "Florence oil" referred to olive oil that was pressed from the flesh of olives (*Olea europaea*). Olives are drupes that grow on small Mediterranean trees, and they are among the oldest domesticated fruits. The skin (exocarp) and flesh (mesocarp) change from green to black with ripening, and the hard inner endocarp encloses the seed. Olive oil is pressed from the fleshy mesocarp, and it was valued both for its texture and taste. Jefferson considered olive trees a gift from heaven and "the most interesting plant in existence." He favored salads of mixed greens dressed in oil and vinegar and sought sources of imported olive oil when he was minister to France. Jefferson struggled to establish olive plantations in the United States and planted olive tree cuttings at Monticello in 1774, but many of the olive seedlings did not survive cold weather. He encouraged friends in North Carolina to plant olives, coffee, sugar, and sesame, but again olive cultivation met with no satisfactory outcome. During these same years, olive trees were already planted and flourishing in the interior valleys of California, cultivated by Franciscan missionaries; the trees require bright sunlight and warm growing conditions but also need a period of winter dormancy.

Discouraged with his failure in cultivating olives in the south, Jefferson turned to sesame (*Sesamum indicum*), an African crop plant that was introduced to America by slaves (see chapter 3). The herbaceous plants were easily cultivated as annuals. Africans used sesame as a grain, but Jefferson realized its potential as an oilseed because the seeds contain about 50 percent oil. Jefferson grew sesame at Monticello from 1809 to 1824, and he struggled to extract oil from the seeds using three different presses. Oil yields were probably disappointing, and operating the presses was also difficult, but nineteenth century agricultural writers encouraged the cultivation of sesame for its medicinal uses (see chapter 7). Sesame was also cultivated in the gardens planted by slaves, and the whole seeds were used in traditional African American cookery.

CHAPTER SEVEN

Herbs, Herbalism, and the Practice of Domestic Medicine

With New World exploration came great interest in American medicinal plants, but as ships laden with sassafras and sarsaparilla sailed across the Atlantic to Europe, colonists also imported European herbs to America. Waves of Puritans arriving in America were often unaccompanied by physicians, but they came supplied with cuttings and seeds of the familiar herbs that had been used medicinally since ancient times. Colonial medicine had its roots in English herbalism, and during the seventeenth century it resembled the traditional practice of medicine in England. Medical care was often conducted primarily by women, "doctresses" and midwives, who also planted and tended the herb plots that were the New World equivalent of European physic gardens. When care was needed, ladies and housewives alike tended to their families and neighbors; some clergy and perhaps their wives and daughters also nursed local parishioners, either as charity or as paid work. Women were experts in obstetrical and gynecological matters, and colonial diaries reveal that ministers and even physicians consulted with midwives regarding remedies for their female patients. Apothecary shops were sometimes opened by entrepreneurial women, such as the early eighteenth century business operated by Elizabeth Greenleaf in Boston. She relied on Nicholas Culpeper's *Pharmacopoeia Londinensis: or the London Dispensatory* (1683) as a source for medical information. Plants occupied a critical role in healing, and aside from a few mineral medicines, the most reliable drugs came from botanical sources; of course, colonists practiced such dubious remedies as blood-letting and tongue-scraping, and superstition and empiricism also influenced seventeenth century medicine.

Herbal medicines that were wisely administered were often the key to survival in the face of serious illness or widespread epidemic, which probably explains the continued colonial use of familiar European herbs rather than hasty adoption of New World medicinal plants. Indeed, many of the plants familiar to colonists, such as mullein, roses, wormwood, nightshades, and violets, had been illustrated and described by the Greek herbalist Dioscorides in *De Materia Medica* during the first century A.D. Colonial herb gardens were the practical equivalent of reliable apothecary shops, plots that provided medicines that had been known, grown, and used since ancient times. The gathering and drying of herbs was part of domestic lore for millennia, traditional methods written by hand in seventeenth century stillroom books and printed in nineteenth century household manuals. Herbal preparation was still part of the domestic training of a nineteenth century housewife and practical nurse; in *The Family Nurse* (1837), Lydia Child reminded her readers to "remove the sticks and pods from senna; other seed from flax-seed; dirt from roots; and all mouldy or musty parts from herbs."

Herbal knowledge was acquired as part of the oral tradition and to a lesser extent by trial and error, and some printed books were carried from Europe by immigrants to America. *The English House-wife* by Gervase Markham was first published in 1615, and new editions appeared for about a century; copies of his book and similar household manuals were used in American homes as valued references for the preparation of simple medicines. Markham's cures were evidently effective and required only a knowledge of common botany. For example, for dysentery (the so-called "bloody flux"), Markham recommended shepherd's purse (*Capsella bursa-pastoris*) and knotgrass (*Polygonum aviculare*). Both herbs were introduced to American gardens as botanical cures and were among the earliest medicinal plants naturalized in the New World; we now think of them as weeds rather than medicinal herbs. Shepherd's purse and knotgrass were listed by John Josselyn in *New-Englands Rarities* (1672) as "Plants as have sprung up since the English Planted and kept Cattle in New-England."

Like Markham's dysentery cure, most medicinal preparations were simples, medicines based on one

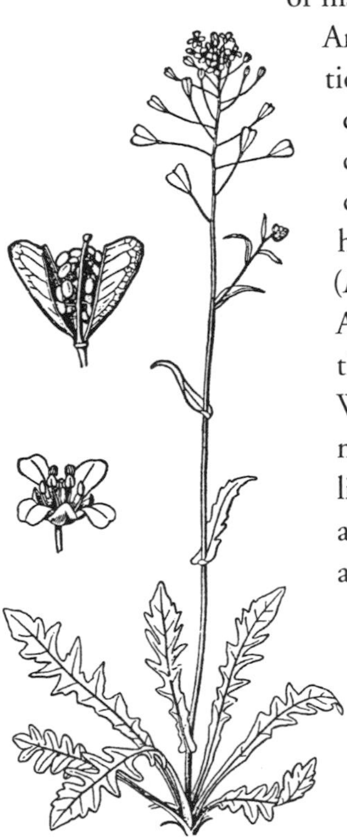

Shepherd's purse (*Capsella bursa-pastoris*) was cultivated as a traditional cure for dysentery and is now a familiar naturalized weed.

or a few herbs that were easily grown, collected, and stored for household use. Gardens were essential for cultivating a supply of useful herbs, but stillrooms were the domestic repositories for drying and storing herbs and the laboratories for preparing medicines, including various salves, waters, plasters, and oils. Stillroom books were manuscript records of useful lore and practical procedures for herbal chemistry; as their name implies, some stillrooms also included the equipment necessary for distilling essential oils. Wine, itself a botanical product (see chapter 5), was valued as a medium for mixing medicines because it was believed correctly that alcohol extracted many of the active principles from plant tissues. Indeed, this was the chemical basis of tincturing, in which macerated herbs or their fresh juice were combined with alcohol to prepare effective medicines. The women who toiled in stillrooms had considerable knowledge of fundamental natural product chemistry including solubility, extraction, and distillation, the skills necessary for converting plants into effective medicines.

The Doctrine of Signatures originated in ancient China, but it was embraced by Europeans during the Middle Ages and Renaissance as a method to sort out botanical diversity and interpret the uses of various medicinal plants. Theophrastus Bombastus von Hohenheim (also known as Paracelsus) was a fourteenth century physician and mystic who later revived the notion that plants bear "signs" that suggest their practical uses; the signs were presumably revealed by shape, color, odor, or habitat. For example, heartsease (*Viola tricolor*) was used to treat heart ailments because of its heart-shaped petals, and mandrake (*Mandragora officinarum*) was known as a panacea because its branched taproot resembles the entire human torso. The black, shiny stalks of maidenhair fern (*Adiantum* spp.) suggested its use for treating head and hair ailments, while various pigmented latexes or juices suggested body fluids such as bile or blood. Vernacular names such as heartsease, maidenhair, liverwort, and eyebright recall the signs and medicinal uses that were interpreted through the Doctrine of Signatures. Some also predicted that botanical remedies should grow near the afflictions that they cure; willows, which grow in wet habitats, were used to treat rheumatic pains that are caused by damp and cold. Although the Doctrine of Signatures is based on fallacy, occasional coincidences kept the notion alive. Mandrakes produce potent tropane alkaloids, which cause hallucinations and may be analgesic, and willow trees contain salicylic acid, the molecular precursor of the semi-synthetic analgesic drug known as aspirin.

The Doctrine of Signatures was part of the Old World herbal tradition that was carried to the colonies; many colonial physicians embraced the empirical teachings of Paracelsus, and the Doctrine was eventually used to sort out practical uses for the vast diversity of American medicinal plants. The use of parasitic plants to treat cancer, thought of as a parasitic disease, was an extension of the Doctrine of Signatures; one such plant was beech drops (*Epifagus virginiana*), a parasite on the roots of beech trees (*Fagus* spp.). Puritan minister Cotton Mather communicated to the Royal Society about the use of beech drops (he called the plants "Fagiana") to treat cancers, a practice that he reportedly learned from Native Americans.

Herbal Botany and Chemistry

Herbs are now often thought of as culinary plants, but historically these species were a diverse botanical assemblage that served as medicines, flavorings, insecticides, and other domestic products. In many cases, the selfsame plant that was ingested as a food, tea, or medicine was also valued for its preservative or insecticidal qualities. Tansy (*Tanacetum vulgare*) was used variously as an early potherb, tea herb, culinary flavoring, vermifuge, and household insecticide. Tansy had long been linked to the ancient superstition that eating fish during Lent resulted in intestinal worms; tansy cakes prepared as a vermifuge were traditional Lenten foods, which we now recognize as potentially toxic. Rue (*Ruta graveolens*) was valued for its bitter flavor, but large amounts were used as a vermifuge and to treat colic, epilepsy, and hysterics. Gardeners learned to use care in cultivating rue plants because the plants contain psoralens, which in bright light cause dermatitis and blisters; in nature, the

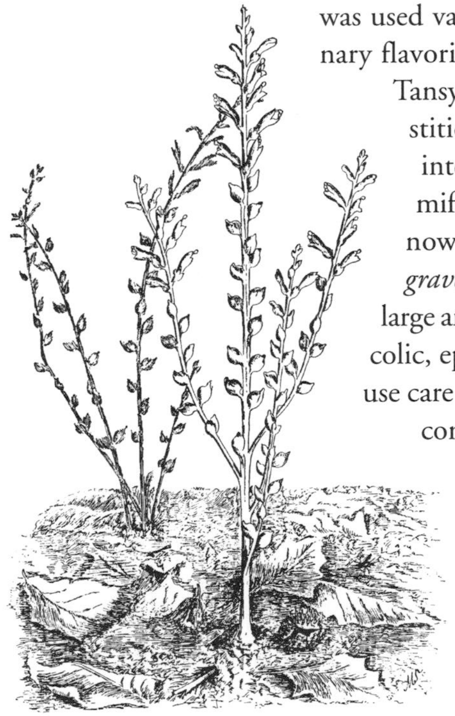

According to the Doctrine of Signatures, the parasitic growth habit of beech drops (*Epifagus virginiana*) suggested their use as a remedy for cancer, which was considered a parasitic disease.

psoralens in rue also function as phytoalexins that protect plants from fungal infections. Southernwood (*Artemisia abrotanum*) was used to repel moths and as a medicinal astringent, while horseradish (*Armoracia rusticana*) was used as a condiment, diuretic, and cough syrup ingredient; both were also potentially toxic, depending on dose and manner of use.

The phytochemical basis of herbal properties lies in the secondary compounds produced by many families of flowering plants; these include alkaloids, glycosides, and terpenes, which are all now interpreted as protective molecules that evolved in response to environmental pressures. Given the structural complexity of many secondary compounds, it is unlikely that they are merely metabolic wastes. They are synthesized at an energy cost to plants, so it is reasonable to expect that they have one or more functions that are essential or beneficial to plant survival. In fact, many secondary compounds deter or poison plant-eating insects and larger herbivores, and some protect plants from microbial parasites including fungi, bacteria, and viruses. In some cases, these protective functions overlap, just as the ethnobotanic and economic uses of many herbs overlap. On a more subtle note, once they reach the soil through decay or leaching, some secondary compounds are involved in allelopathy. Allelopathic plants chemically inhibit the growth of seeds and seedlings of neighboring plants that are competing for space, light, water, and soil nutrients. The natural arsenal of allelopathic compounds include alkaloids, terpenes, and other phytochemicals that may also have properties as medicines or microbial antibiotics.

Alkaloids are compounds that are bitter and basic (as opposed to acidic), contain nitrogen, and cause physiological reactions in animals. More than five thousand alkaloids have been isolated, many of them from herbs and other medicinal plants: morphine from opium poppies, aconitine from monkshood, and caffeine from coffee are all alkaloids with varying physiological effects and levels of toxicity. Toxicity is often a function of dose; small amounts of secondary compounds may be medicinal while larger doses are potentially lethal, as illustrated by pinkroot (*Spigelia marilandica*), a member of the logania family (Loganiaceae) that contains the potentially toxic alkaloid spigeline. Pinkroot was a useful vermifuge especially for children, but overdoses risked the life of the patient, and Lydia Child (1837) cautioned against administering too much of the fresh root until its effects were known. Settlers learned about pinkroot from Native Americans, and caution about dose was likely part of its herbal lore. Similar situations arose with homemade opium syrups of unknown concentrations, which were made from the capsules of opium poppies (*Papaver somniferum*) cooked with water and sugar. The soothing, analgesic effects of morphine and other poppy alkaloids were tempered by the possibility of overdosing a teething or fretful small child; Child (1837) wisely suggested using the mixture sparingly when dosing children because "this syrup is an opiate of uncertain strength."

Glycosides consist of a sugar molecule bonded to another, variable molecular component, which is released upon digestion and may have a variety of effects depending on its chemical structure. For instance, some species in the rose family (Rosaceae) contain amygdalin, a potentially toxic glycoside that releases cyanide upon digestion (see chapter 3). Cardiac glycosides such as digitalin from foxgloves and calactin from milkweeds affect heart muscle, while the sulfur-containing glycosides known as mustard oils cause blistering and pain (see chapter 4).

Terpenes make up the essential oils of many herbs, including fragrant oils of various mints, and some terpenes such as camphor from *Cinnamomum camphora* had both medicinal and insecticidal uses. Many terpenes are now known to have antibiotic properties as well (see chapter 6). More complex terpenoid molecules such as artemisinin from wormwood (*Artemisia* spp.) are based on smaller ten-carbon monoterpene subunits.

Many medicinal herbs fit the ethnobotanical definition of salient species, plants that have notable traits that may include remarkable growth form, color, indument (a coating of plant hairs, known botanically as trichomes), scent, or pungency. These plants are prominent in their habitats and have historically received attention and encouraged experimentation because of their obvious or unique traits. Foxgloves such as *Digitalis grandiflora* and *D. purpurea*, with their large basal rosette of leaves and tall inflorescences of tubular flowers, are clearly salient species. Experimental nibbles of foxglove leaves or flowers revealed the presence of potent cardiac glycosides,

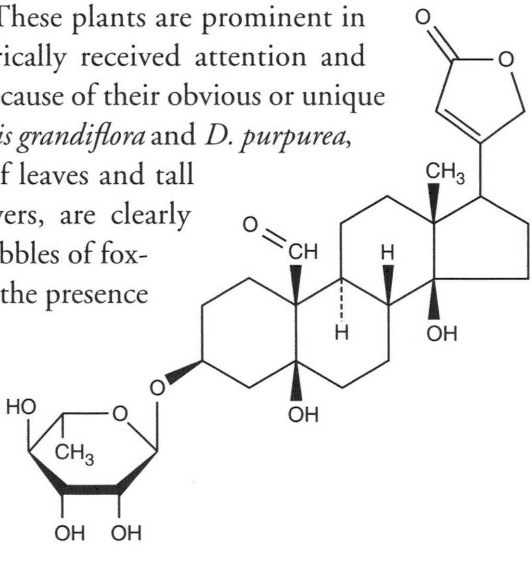

The chemical structure of convallotoxin, a cardiac glycoside produced by lily of the valley (*Convallaria majalis*).

which were the chemical basis of the use of foxglove to relieve the edema (known historically as dropsy) associated with heart failure. Foxglove was a risky medicine for folk use because of its toxicity, but it was nevertheless introduced from English gardens and grown by colonists. Working in a kitchen with primitive equipment, British physician William Withering worked out the foxglove doses that were effective in regulating heartbeat. He published *An Account of the Foxglove* in 1785, long after the plant had been introduced as a medicinal herb to the American colonies. A similar cardiac glycoside known as convallotoxin occurs in lily of the valley (*Convallaria majalis*), a fragrant plant that colonizes large areas; this species was also used by colonists as a diuretic and heart stimulant.

Mullein (*Verbascum* spp.) is in the same family as foxglove (Scrophulariaceae) and also exhibits salient traits: a large basal rosette of leaves, a dense covering of trichomes, and a prominent inflorescence. Mullein leaves were used externally as poultices and were used internally to treat asthma and coughs, but the plants contain both coumarin (precursor of the anticoagulant dicoumarin) and the potent insecticide rotenone, both potentially toxic to ingest.

Mints (family Labiatae or Lamiaceae) are usually herbaceous plants with leaves and stems that are covered with trichomes containing characteristic essential oils. The essential oils of mints are composed of mixtures of terpenes that are released if the leaves are crushed; in many cases the distilled oils are potentially toxic and even lethal, even if consumed in relatively small amounts. Medicinal mints overlap with culinary mints (see chapter 6) and include many of the same plants used as flavorings and tea substitutes, including sage (*Salvia officinalis*), clary (*S. sclarea*), rosemary (*Rosmarinus officinalis*), peppermint (*Mentha* ×*piperita*), spearmint (*M. spicata*), pennyroyal (*M. pulegium*), catmint (*Nepeta*

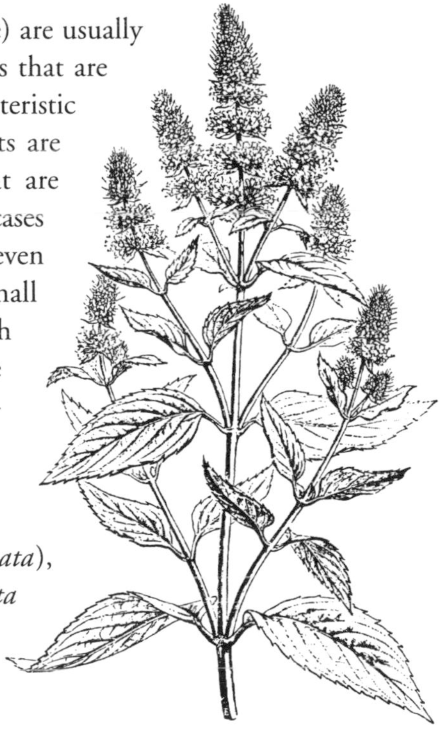

Peppermint (*Mentha* ×*piperita*) had varied uses as a medicinal and tea herb.

cataria), thyme (*Thymus vulgaris*), marjoram (*Origanum vulgare, O. onites,* and *O. majorana*), lavender (*Lavandula officinalis*), lemon balm (*Melissa officinalis*), hyssop (*Hyssopus officinalis*), and horehound (*Marrubium vulgare*). These mints had diverse uses, from brewing medicinal teas for sore throats and improved digestion to treating specific complaints such as eye troubles (clary), liver problems (sage), stomach troubles (peppermint and spearmint), and falls and bruises (catmint).

Perhaps the most notorious component of several essential oils is thujone, which occurs in several mints and causes symptoms similar to epilepsy if consumed in large quantities. Sage contains high concentrations of thujone, camphor, and other terpenes, and it was long used to promote longevity and wisdom and to improve brain activity. Clary, a related species, contains a pleasant-scented mixture of terpenes including thujone, pinene, and cineol and was used in treating eye inflammations. Pennyroyal was once a familiar remedy, but the plants contain pulegone, structurally related to thujone, which causes abortion and death. Catnip contains nepetalactone, which resembles the valepotriates found in valerian (*Valeriana officinalis,* family Valerianaceae) and has similar sedative properties; catnip and valerian were both used commonly for nervous conditions. Bee balm (*Monarda didyma*) was found growing in North America and was obviously a relative of Old World mint. As discussed in chapter 5, it was used as a tea substitute, with the added medicinal benefit of being useful in treating sore throats; we now know that bee balm contains the antiseptic compound thymol. Of course, many other plant families contain terpenes, and the essential oil of sage resembles the oil found in wormwood (*Artemisia* spp.), which also contains thujone.

Comfrey (*Symphytum officinale*) was long valued as a herb with healing properties, and its compound allantoin does promote the healing of wounds; however, the alkaloid lasiocarpine found in comfrey leaves may cause liver damage, and the tannins in its roots may cause cancers.

Similar pyrrolizidine alkaloids occur in coltsfoot (*Tussilago farfara*), a composite that was once valued for its mucilaginous properties in treating sore throats and coughs. Other introduced medicinal composites (family

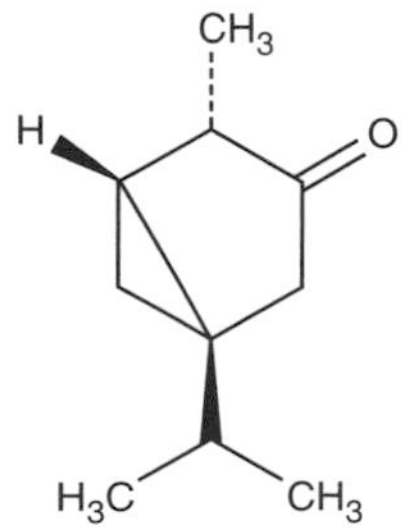

The chemical structure of thujone, found in the essential oils of sage (*Salvia officinalis*) and many other aromatic species.

Compositae or Asteraceae) included yarrow (*Achillea millefolium*), chamomile (*Anthemis nobilis*), German chamomile (*Matricaria recutita*), lavender cotton (*Santolina chamaecyparissus*), feverfew (*Chrysanthemum parthenium*), tansy (*Tanecetum vulgare*), and wormwood (*Artemisia absinthium*) and related species. All are related, strongly-scented members of the composite tribe Anthemidae that were frequently used as vermifuges, antiseptics, and insecticides. Yarrow was also known as an effective hemostat, sometimes known as "nosebleed" for its ability to staunch the flow of blood; according to legend it was used by Achilles to treat his soldiers who were wounded in battle. Santonin, a well known antihelminthic compound, was derived from the dried, unopened flower buds of wormwood; it was the active principle in dragees or troches of "santonine," the wormwood-infused candies and lozenges sold as a common nineteenth century vermifuge.

Chamomile brewed soothing teas, and lavender cotton and tansy were used to treat both jaundice and intestinal worms, as noted earlier. Tansy in larger doses was also used as an abortifacient, with potentially lethal effects; in *Medicinal Plants* (1892), Charles Millspaugh described the violent spasms, convulsions, and paralysis observed in several women who ingested tansy oil (see chapter 3), now an illegal ingredient in any food or medicine. The genus *Artemisia* includes tarragon (*A. dranunculus*) as well as the medicinal species wormwood (*A. absinthium*), and southernwood (*A. abrotanum*). Wormwood contains thujone as well as the bitter compound absinthin and related molecules. Its most notorious nineteenth century use was in flavoring absinthe, a green liquor flavored with a distilled mixture of wormwood, hyssop, nutmeg, lemon balm, and other herbs and spices. The chlorophyll-green color of absinthe was often deepened by the addition of adulterants such as copper sulfate or aniline green, but its narcotic and addictive properties were a greater concern. Absinthe consumption resulted in cases of toxic addiction, and by the end of the nineteenth century, the narcotic dangers of absinthe were recognized. *Johnson's Universal Cyclopaedia* (1885–86) noted, "Trembling, vertigo, fearful dreams, and epiletiform

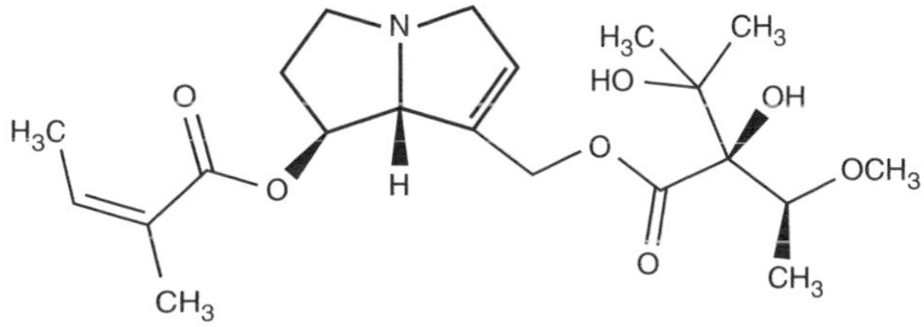

The chemical structure of lasiocarpine, a pyrrolizidine alkaloid found in comfrey (*Symphytum officinale*).

convulsions are among its severer consequences. Absinthe-drinking is one of the most dangerous forms of stimulation yet invented—the more so because its immediate consequences are usually more agreeable than those of alcohol." Because of its potential hazards, in 1912 absinthe was banned in the United States.

Umbellifers (family Umbelliferae or Apiaceae) also varied from deadly to benign. Poison hemlock (*Conium maculatum*) contains coniine and other alkaloids that cause respiratory paralysis. This is the hemlock used in the death of Socrates and others at Athens and should not be confused with the coniferous hemlock (*Tsuga*) that was used to brew antiscorbutic teas in some frontier regions. Millspaugh described cases of cancers cured by doses of poison hemlock, and the Shakers marketed it for treating nervous conditions, asthma, and rheumatism, with the caveat that it be used only by a skilled physician. The fruits (schizocarps) of wild carrot or Queen Anne's lace (*Daucus carota*) were used as emmenagogues to bring about abortions, and the leaves of the plant made poultices for ulcers. Caraway (*Carum carvi*), angelica (*Angelica archangelica*), fennel (*Foeniculum vulgare* and *F. dulce*) and anise (*Pimpinella anisum*) were umbellifers grown primarily for their small, flavorful fruits that were also used for their carminative properties in promoting digestion.

By the second half of the nineteenth century, seed catalogs usually included listings of "pot, medicinal, and sweet herbs" following the descriptions of vegetable seeds. Advertisements for the Hiram Sibley seedsmen of Rochester and Chicago (1882) described "medicinal herbs . . . as domestic medicines, several kinds are held in high repute. A very small space in the garden will give all of the herbs needed in my family." Sibley provided seeds of traditional European herbs for American gardens, including horehound, hyssop, rue,

Wormwood (*Artemisia absinthium*) was used in tonics and worm medicines, but its most notorious use was in flavoring absinthe.

sage, and wormwood; he was a forward-thinking entrepreneur who was a founder of Western Union, but he (or his firm) had faith in traditional medicine and cures that originated in the Old World. *Burpee's Farm Annual* (1888) noted, "A few pot and sweet herbs are indispensable to every garden . . . some of the medicinal herbs will also be found very useful." *Nigella sativa* appeared in the Burpee listings, a European and Asian member of the buttercup family (Ranunculaceae) that was known as "All Spices" but now commonly known as black cumin. The flavorful seeds were used as a food seasoning but were also used medicinally to treat jaundice and intestinal worms; it is a close relative of love-in-a-mist (*N. damascena*), a commonly cultivated garden annual that was also grown from seed. Burpee also marketed 'Holt's Mammoth' sage, presumably a variety of *Salvia officianalis*, advertised as being "perfectly hardy, even in New England." This cultivar had large leaves and spreading growth; it was probably grown more often as a curiosity than as an important domestic herb, although it was a descendant of the plants prized since ancient times for imparting wisdom and longevity to those who used them.

Many annual and perennial garden plants once supplied the domestic pharmacopoeia, but now they are grown as ornamental cultivars rather than as medicinal herbs; this transition from medicinal herbs to cultivation as garden ornamental began in the second half of the nineteenth century, as many families obtained medicines from druggists or physicians and no longer necessarily grew a wide diversity of medicinal herbs. Lily of the valley (*Convallaria majalis*), periwinkle (*Vinca minor*), hollyhocks (*Althaea rosea*), heartsease (*Viola tricolor*), primrose (*Primula veris*), poppies (*Papaver somniferum*), foxglove (*Digitalis* spp.), pot marigold (*Calendula officinalis*), and house leek (*Sempervivum tectorum*) are all garden plants with medicinal histories. Periwinkle, now a common ground cover, was cultivated for use in tonics and skin ointment, while hollyhocks like other members of the mallow family (Malvaceae) contain a demulcent mucilage that was useful in preparing cough syrups. The succulent, astringent leaves of house leek have an gelatinous texture that resembles aloe; they were once used to treat burns, stop wounds from bleeding, and combined with plantain (*Plantago* spp.) into an ointment to treat skin eruptions. The yellow color of pot marigolds suggested their use for treating jaundice, and primroses were used to treat headaches and supposedly had a mild narcotic effect. Even in the case of opium poppies and lily of the valley, which supplied medicines with strong analgesic and cardiac effects respectively, traditional medicinal uses were gradually forgotten and the plants were known for their ornamental value alone. Patent medicines provided opiates, and physicians prescribed compounded medicines that supplanted many risky folk remedies such as the cardiac glycosides from foxglove and lily of the valley, which were used to treat congestive heart failure.

American Medicinal Plants

During the seventeenth century, the promise of new botanical cures from the American flora attracted European attention, and imports from the New World included medicinal plants; during the next century, a uniquely American pharmacopoeia evolved through the combination of European herbalism with the practical knowledge of North American medicinal plants. Domestic gardens still included familiar European herbs; 1794 records from Monticello reveal that the Jefferson family continued to grow (and presumably use) tansy, rue, wormwood, southernwood, and various mints including marjoram, sage, balm, rosemary, and hyssop. Nevertheless, New World plants presented new medicinal opportunities, and the knowledge acquired from Native Americans and perhaps through individual experimentation revealed the medicinal uses of several North American plants. Some individuals learned Native American cures and revered "Indian medicine," but in many cases physicians incorporated American medicinal plants in compounding their medicines before these new plants were used commonly in home cures.

In *The American Gardener's Calendar* (1806), Bernard McMahon listed forty-two "Plants Cultivated for Medicinal Purposes, & c.," which included for the most part familiar European plants such as comfrey, poppy, rue, and yarrow. Few families would likely have cultivated forty-two medicinal plants, but it is interesting to note the inclusion of some American medicinal species into an otherwise Old World list of recommendations for domestic physic gardens. A few North American plants appeared on the 1806 list, including pinkroot (*Spigelia marilandica*) and boneset, also known as ague-weed or thoroughwort (*Eupatorium perfoliatum*), both plants with a long history of medicinal uses by Native Americans. Cherokees used pinkroot as a vermifuge, and Osages used it as a sedative for its sudorific effects; combined with senna (*Cassia* spp.), pinkroot evolved into an early folk remedy for intestinal worms. Native Americans collected boneset to treat a vast number of ailments, ranging from fever and snakebite to its use as a spring tonic, cathartic, and emetic. Settlers adopted boneset as a familiar cure for colds, fevers, and malaria; the common name seems to originate from its use to treat break-bone fever, another term for the virus known as Dengue fever that was once common in the southern states. It was also used as a poultice for broken bones (perhaps suggested by the fusion of its opposite leaves across the stem), another possible origin for its common name.

The closely related species white snakeroot (*Eupatorium rugosum*) became known as the source of a lethal toxin rather than as a medicinal plant. When white snakeroot was grazed by cows, they developed a condition known as trembles. Farmers often noted that cows with trembles had grazed in woodlands, but they did not realize that the animals had consumed white snakeroot, which thrived in moist, deep soil. Humans who drank the milk of these cows developed a potentially fatal condition known as "milk sick" or milk sickness, with symptoms of nausea, thirst, and extreme weakness; according to legend, Nancy Hanks Lincoln, mother of Abraham Lincoln, died from the effects of tremetol poisoning from tainted milk. During the nineteenth century, severe "epidemics" reduced the populations of many towns and caused villages to be abandoned, with blame placed on toxic miasmas, insects, and metals such as arsenic and lead. The connection between milk sickness and the tremetol in white snakeroot was not understood until the 1920s. Chemical analysis now reveals that tremetol contains tremetone, a white snakeroot toxin that is similar to the rotenone (a powerful insecticide produced by some composites) and often deadly when concentrated and consumed in milk.

There was particular early interest in American plants with potential for curing venereal diseases. Wild sarsaparilla was believed to work with mercury in curing syphilis and restoring health; the French physician Michel Sarrasin sent wild sarsaparilla or angelica tree (*Aralia spinosa*, which he called "Aralie") from Quebec to the botanist Joseph Pitton de Tournefort in the early seventeenth century, and soon the plants were collected and imported for use abroad. Sassafras (*Sassafras albidum*) was valued and exported during the seventeenth century as cure for venereal diseases (see chapter 1), and curative claims were also made for great lobelia (*Lobelia siphilitica*). Similar recommendations were made well into the nineteenth century; the *Druggists' Hand-Book of American and Foreign Drugs* (1870) recommended wild sarsaparilla (*A. nudicaulis*) for "venereal complaints," lobelia for gonorrhea, and sassafras for "scrofula and eruptive diseases." Wild sarsaparilla species were also used as a flavoring for colonial beers and ales (see chapter 2) and in various patent medicines and tonics.

Ginseng (*Panax ginseng*) was long valued in China as an herbal panacea, which like mandrake was suggested by the resemblance of its branched taproots to the human torso. Asian ginseng populations had already nearly disappeared from demand and over-collection, but the New World offered a new ginseng resource. William Byrd I, a seventeenth century Virginia grower, discovered the similarity of American ginseng (*P. quinquefolius*) to Asian ginseng, and morning and night brewed ginseng root tea in a silver teapot. These two species are the descendants of *Panax* populations that inhabited the northern supercontinent before the separation of North America and Eurasia, which explains their common physical and medicinal properties. Once the similarity between the *Panax* species was widely recognized in the mid-1700s, the collection of American ginseng for export to China became a vigorous business for settlers, explorers, and Native Americans in wilderness areas of North America.

Some American medicines also used ginseng as an ingredient, and the American species was listed in the United States Pharmacopoeia from 1840 to 1880. During this time, some Shaker settlements cultivated and marketed American ginseng as "Chinese Seng" for use as a tonic, stimulant, and nerve medicine. Native Americans had similar uses for American ginseng, also regarding it in many cases as a panacea and magical charm. The active ingredients are presumably the ginsenosides, saponin-containing glycosides that occur in *Panax ginseng*, *P. quinquefolius*, and probably also in other species including the dwarf American ginseng (*P. trifolius*).

American pennyroyal (*Hedeoma pulegioides*) and American mandrake (*Podophyllum peltatum*) were also

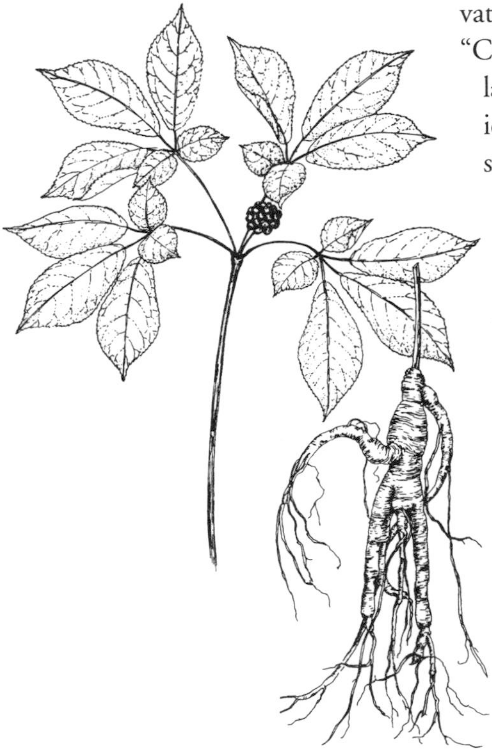

American ginseng (*Panax quinquefolius*) was collected and cultivated as a New World substitute for Asian ginseng (*P. ginseng*).

named for their perceived similarities to known Old World plants. European pennyroyal (*Mentha pulegium*) from Europe was among the mints grown in colonial herb gardens, where it was harvested as a culinary herb, abortifacient, and treatment for congestion, gout, and other ills. Colonists discovered *H. pulegioides*, which had a similar scent and flavor to their familiar pennyroyal; both contain an essential oil composed primarily of the terpene pulegone.

Both pennyroyals can induce abortions if taken in large doses, and even small amounts of their distilled oil can cause irreversible liver damage. American pennyroyal often replaced the familiar European plant, and it is the so-called "squaw mint" described for female troubles in the *Druggists' Handbook of American and Foreign Drugs* (1870). Sarah Orne Jewett described Maine populations of American pennyroyal in *The Country of the Pointed Firs* (1896), in which the wild plants symbolize the barren marriage of her character Mrs. Todd, a widowed doctress who collected pennyroyal on a coastal island.

Settlers seeking European mandrake (*Mandragora officinarum*) in North America instead discovered *Podophyllum peltatum*, a member of the barberry family (Berberidaceae) that also has a large, branched taproot (see chapter 1). It was once commonly known as mandrake, but now the species is more commonly known as May apple, named for its small, apple-like fruits that were sometimes preserved in syrup. May apple was used as a strong emetic and cathartic herb, and it was also used variously to treat jaundice, hepatitis, fevers, and

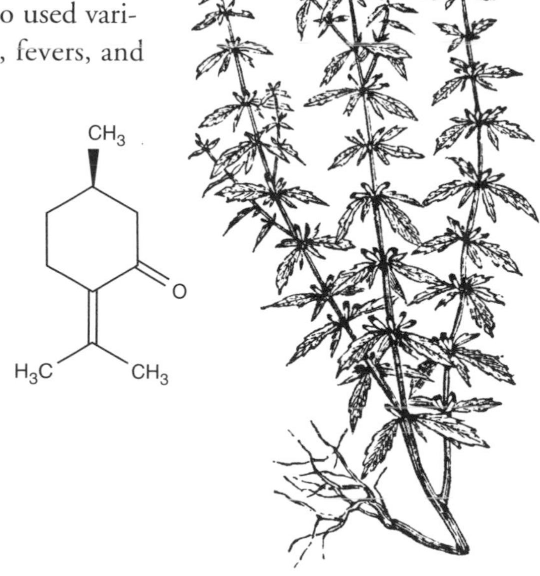

The chemical structure of pulegone, a terpene found in both European pennyroyal (*Mentha pulegium*) and American pennyroyal (*Hedeoma pulegioides*).

American pennyroyal (*Hedeoma pulegioides*) has a scent and flavor that resembles European pennyroyal (*Mentha pulegium*), and both are abortifacients that contain pulegone.

venereal disease. The roots contain a resin composed in part of podophyllotoxin, which is now used to synthesize the anticancer drug etoposide; this may explain the widespread Native American use of May apple roots to treat growths, warts, and possible cancers.

The Doctrine of Signatures was applied to the American flora in other cases as well. The red latex of bloodroot (*Sanguinaria canadensis*) rhizomes suggested its use for blood ailments, but in large doses it causes vomiting, paralysis, and perhaps eyesight distortion; nevertheless, bloodroot became known as a folk cure for dysentery. The alkaloid sanguinarine is antiseptic and shows anticancer activity by affecting cell division (mitosis), but overdoses and toxicity are possible. Nerveroots are lady's slipper orchids (*Cypripedium* spp.), with deeply impressed leaf veins that suggested nerves. These native orchids were known as American valerian, and they were considered a New World equivalent of European valerian, with similar sedative properties useful in treating nervous debility and melancholia; large doses of nerveroots are reportedly hallucinogenic. During his travels in North America, Pehr Kalm noted that *C. reginae* was also used by women in childbirth, and during the nineteenth century, yellow lady's slipper (*C. calceolus* var.

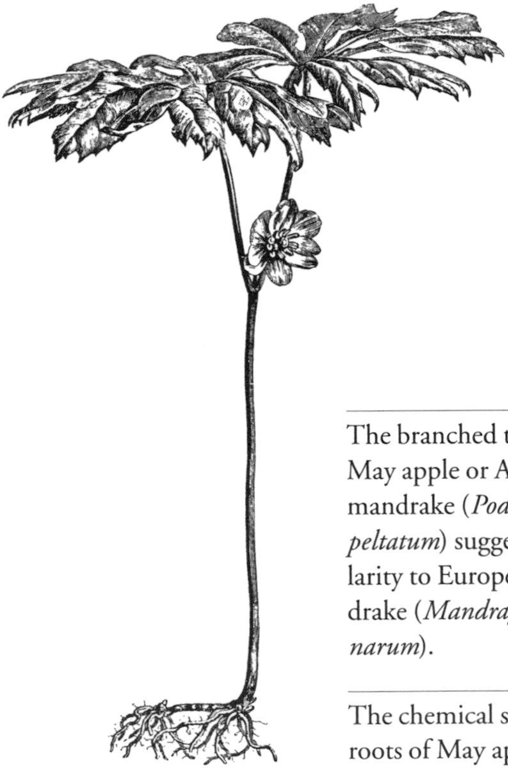

The branched taproot of May apple or American mandrake (*Podophyllum peltatum*) suggested a similarity to European mandrake (*Mandragora officinarum*).

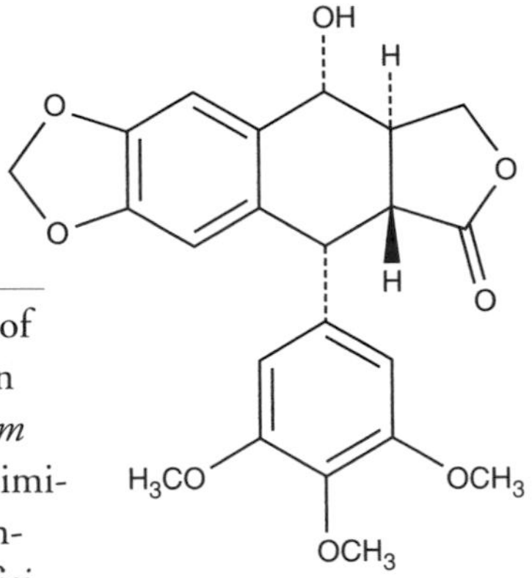

The chemical structure of podophyllotoxin found in the roots of May apple (*Podophyllum peltatum*).

pubescens) orchids were included in patent medicine nerve capsules. Wild *Cypripedium* populations disappeared from over-collection. As a ramification of the Doctrine of Signatures, American folk medicine evolved similar practices even with regard to inanimate objects; for example, some remedies suggested tying red yarn about the neck to stop a nosebleed. These medical traditions evolved during times in which much of the population was illiterate and struggled with a variety of illnesses, conditions, and the need for effective cures.

Benjamin Franklin recognized the need for medical knowledge in the American colonies and published two volumes that addressed topics in practical medicine. One of these was a 1734 reprint of John Tennent's publication *Every Man His Own Doctor, or The Poor Planters Physician.* Tennent, a Virginia landowner who eventually received a medical degree at the University of Edinburgh, promoted the use of Senega snakeroot (*Polygala senega*) as a cure-all; his book was a tract that argued the universal efficacy of Senega snakeroot. He learned about snakeroot from Native Americans who used it to counteract the effects of rattlesnake venom. Considering the symptoms of snakebite similar to pleurisy and peripneumonia, Tennent advocated its use for those illnesses as well as gout, rheumatism, dropsy (edema), and nervous disorders, and he predicted "the highest probability that this root will be of more extensive use than any medicine in the whole Materia Medica." Senega snakeroot became well known in Europe and by 1739 was under cultivation in England. The Shakers later marketed Senega snakeroot for asthma, croup, and other lung conditions, but in America the plant never achieved the status of a panacea promoted by Tennent. The active principles of Senega snakeroot are saponins such as senegenin, an effective expectorant, which may explain its medicinal use for pulmonary ailments.

The second, and more significant, American medicinal botany text published by Franklin was a 1751 edition of Thomas Short's *Medicina Brittanica*, with a new preface, text notes, and appendix on New World medicinal plants. These additions were written by Pennsylvania botanist John Bartram and amounted to an American herbal for practical use. His appendix lists familiar American medicinal plants, with notes on their effects and uses. Some of the names are difficult to interpret because Bartram did not use current botanical nomenclature; his publication predated 1753, the date of *Species Plantarum* by Carolus Linnaeus, which introduced the systematic use of binomial botanical nomenclature. *Medicina Brittanica* was clearly a physicians' reference, but Bartram's list reveals some of the American plants that had been adopted as familiar cures by the mid-eighteenth century. He mentioned the use of sarsaparilla (he called it "Aralia Caule Nudo") as a flavoring for drinks; by this time, interest in it for treating venereal diseases had waned among some physicians. The bark of tulip poplar (*Liriodendron tulipifera*, which he called "Liriodendrum") was steeped in rum to make a medicine for treating malaria or "Fever and Ague," while lizard's tail (*Saururus cernuus*) was used for leaf tea and poultices to treat breast pain. He described at length the Native American use of lobelia for treating venereal disease and the use of butterfly weed ("Apocinum") to cure "hysteric Passion."

Since ancient times and until recently, the practice of medicine required the study of botany. The formal study of medicine in nineteenth century America included lectures and examinations that required detailed knowledge of botany. Students were quizzed on floral and vegetative morphology, plant identification, formulae for the preparation of tinctures from plants, as well as the scientific and common names of medicinal species. The opportunity to see medicinal plants in a garden setting followed the botanical tradition of gardens as sites for cultivating plants for study and research. With that in mind, in 1801 David Hosack established the Elgin Botanical Garden, the first physic garden in the United States, in New York City at the present site of Rockefeller Center. Hosack was a professor of botany and *materia medica* at Columbia University; with his own funds, he purchased twenty acres near the city center and requested that other botanists begin to collect useful specimens to populate the new garden. As a physician with extensive botanical training, Hosack wanted to grow significant medicinal plants as demonstration materials for botanical lectures; his goal was to provide concrete botanical knowledge for aspiring physicians, and he sought American plants that were relatively unknown medicinally, including the snowdrop tree (*Halesia carolina*) and the fringe tree (*Chionanthus virginicus*) as well as the wild sarsaparilla or spikenard (*Aralia spinosa*).

In 1811, Hosack issued *Hortus Elginensis: Or a Catalogue of Plants, Indigenous and Exotic, Cultivated in the Elgin Botanic Garden, in the Vicinity of the City of New York*, which included medicinal information about the plants in the garden. The garden site had various soil types and eastern and southern exposures, allowing for the cultivation of ecologically diverse specimens, including plants grown from seeds obtained from the Lewis and Clark expedition and from French seedsmen who provided seeds to Thomas Jefferson.

The diverse plantings expanded to include North American forest trees, a fruit tree nursery, and a hothouse for the tender medicinal plants that he obtained in trade; unfortunately, the physic garden was short-lived and fell into disuse in the years following 1811. While the garden was not a model for domestic counterparts, it did play a brief role in establishing native North American plants in the American pharmacopoeia. Of course, many of these plants had long been grown and gathered as folk remedies, often as a result of knowledge gleaned from Native Americans. By the end of the nineteenth century, many physicians used tincture of fringe tree (one part bark by weight in five parts alcohol) to wash wounds and treat liver ailments and nephritis; yet, long before Hosack cultivated the fringe tree in Manhattan, Native Americans had used the root bark tea brewed from fringe trees for infections and inflammations.

Medicinal Imports

The early American pharmacopoeia was not limited to temperate plants. Colonial ports received shipments of medicinal plants from abroad, which included jalap from Mexico, cinchona bark from South America, and aloe from the West Indies. Jalap (*Ipomoea purga*) is a strong purgative and member of the morning glory family (Convolvulaceae), while cinchona bark (fever-bark) from various *Cinchona* species contains quinine, a malaria cure. Aloe is the familiar *Aloe vera*, also known as *A. barbadensis* or socatrine aloe, named for the African Island of Socatra where the plants originate.

Aloe was used internally as a cathartic and vermifuge and was also a known abortifacient; externally it was used on wounds, and according to legend, Alexander the Great captured the island of Socatra to have a ready source of aloe. All of these plants were available from colonial apothecaries, at least on an occasional basis; of course, imports were costly, and temperate substitutes for all imported drugs were known and grown. Tulip tree or yellow popular (*Liriodendron tulipifera*), dogwood (*Cornus florida*), water avens (*Geum rivale*), and cinquefoil (*Potentilla* spp.) served as various temperate, inexpensive substitutes for cinchona, and several native plants such as American senna (*Cassia marilandica*) were valued for their cathartic or emetic properties.

Even rural practitioners had access to imported drugs. As a midwife in Hallowell, Maine, Martha Ballard attended hundreds of births between 1785 and 1812 and recorded her daily activities in a diary that she began at age fifty. She purchased myrrh (*Commiphora* spp., probably *C. myrrha* from Africa), aloe, and camphor to augment the temperate herbs such as dock (*Rumex* spp.), plantain (*Plantago* spp.), and annual wormwood (*Artemisia annua*), which constituted her pharmacopoeia. Myrrh is a sweet-smelling bark resin with antibiotic properties; it was also prepared as a tincture and used to treat ear infections. She mentioned the external use of camphor (sometimes combined with turpentine, an essential oil distilled from pine wood) in several diary entries, but it was also used in medicines as a flavoring. She used laudanum to relieve labor pains; New England gardens included opium poppies (*Papaver somniferum*), which were used to prepare weak opiate syrups, but laudanum was likely prepared as a tincture of imported opium. Galbanum (*Ferula galbaniflua*) was imported from Asia and used as an expectorant and aid to digestion, and a diary entry in 1794 recorded that she purchased some from a physician. A year earlier she noted the purchase of dragon's blood (probably *Dracaena draco* from the Canary islands), which was used in preparing plasters. Myrrh and aloe were also combined with saffron (*Crocus sativus*) and rum to prepare "elixir proprietatis," described by Child (1844) as "a useful family medicine when the digestive powers are out of order." Her recipe slightly resembles the elixir of vitriol formulated by Adrian Mynsicht, a seventeenth century German physician whose complex elixir combined spices including cinnamon, ginger, cloves, and nutmeg with sage, mints, aloe, lemon peel, galangal root (*Alpinia officinarum*, an Asian ginger relative), and sugar.

Some home cures involved a thorough dosing with emetics or cathartics. Ipecac is derived from the roots of a South American shrub (*Cephaelis ipecacuanha*), which contain the emetic alkaloids cephaeline and emetine. Small doses functioned as expectorants and were used to treat whooping cough and cases of near drowning, and the many substitutes for ipecac had a similar emetic effect, depending on dose: native plants

Aloe (*Aloe vera*) was an imported herb obtained from apothecaries, used externally on wounds and internally as a cathartic, vermifuge, and abortifacient.

such as Indian physic (*Porteranthus trifoliatus*) served a similar purpose, and a tea made by steeping tobacco (*Nicotiana tabacum*) was administered with similar effects. Plants with cathartic effects included several other exotic species, including senna (*Cassia acutifolia*), manna (*Fraxinus ornus*), and tamarind (*Tamarindus indica*). Old World senna (*C. acutifolia*) originated in the Red Sea area and was cultivated in Africa; the leaves and fruits were soaked in water to make a laxative, which was often administered with manna, a liquid collected from stem cuts in flowering ash trees from southern Europe. American senna (*C. marilandica*) was an effective substitute that was cultivated in home gardens. Tamarinds are also leguminous fruits with laxative properties, and their astringent pulp was a favorite sickroom food that was discussed in chapter 3.

Rhubarb (*Rheum officinale*) from China was also a common cathartic, related to the traditional "pie plant" (*R. raphonticum*, discussed in chapter 3) that was cultivated for its edible petioles. Both species contain toxic calcium oxalate crystals in their leaves, but they have edible, laxative rhizomes, the result of the presence of chrysarobin, chrysophanol, emodin, and related anthraquinone compounds. These compounds occur in senna and aloe as well as rhubarb, which explains their similar physiological action. In aloe, the anthraquinones occur in the form of glycosides (attached to sugars) in a latex known as barbaloin or aloin, which irritates the intestinal lining. Anthraquinones also occur in yellow dock (*Rumex crispum*) and related species, temperate herbs that were used both as cathartics and externally to cure skin afflictions; anthraquinones are now also known to be antibiotic against ringworm and other parasitic fungal skin infections. Castor oil was another household cathartic, recommended for children and sometimes disguised in sweetened milk; it was extracted from the poisonous seeds of castor beans (*Ricinis communis*), a member of the spurge family (Euphorbiaceae) that is probably native to tropical Africa.

Castor oil was a common cathartic drug extracted from the seeds of castor bean (*Ricinis communis*).

Cathartics were also often used to treat the problem of parasitic intestinal worms, an almost universal problem that resulted from poor domestic sanitation and cycles of reinfection. Worm eggs were easily spread in homes, yards, and gardens in which animal and human waste accumulated, although the connection between parasitic infections and hygienic practices was not understood until the mid-nineteenth century. Some believed that worms were caused by consuming green fruits or fish during Lent, and traditional herbs such as tansy or the Native American pinkroot were used as effective vermifuges. However, for many colonists, the vermifuge of choice was wormseed or Jerusalem oak (*Chenopodium ambrosioides* var. *anthelminticum*), a Mexican and South American member of the spinach family (Chenopodiaceae). Wormseed (not to be confused with wormwood, *Artemisia* spp., also an effective vermifuge) was naturalized early in North America, perhaps before the arrival of Europeans, and was used in the form of seeds (whole or powdered) or as an oil pressed from the seeds. Whether colonists learned about its medicinal properties from Native Americans or by trial and error is unknown, although a related European species (*C. botrys*) was used for treating shortness of breath and may have sparked curiosity about the similar plant discovered growing in North America. Josselyn (1672) recorded the use of wormseed (he called it "Oak of Hierusalem") in a complex herbal drink prepared from a combination of native and introduced plants, but Cotton Mather was the first to describe the use of wormseed as a vermifuge in 1722.

Since the wormseed plant was widespread and available for the gathering, it was used by all social classes almost as a panacea; it became especially popular with southern plantation owners who

Wormseed (*Chenopodium ambrosioides* var. *anthelminticum*) was introduced from Mexico and South America and was an effective vermifuge that was easily gathered from naturalized populations.

dosed their families and their slaves seasonally to cure them of roundworms and tapeworms, which may have been more of a health issue in the southern colonies. Wormseed reached Europe, sent by William Byrd to Sir Hans Sloane, patron of the Royal Society and donor of the land for the Chelsea Physic Garden to the Company of Apothecaries. After his North American travels from 1747 to 1751, Pehr Kalm sent seeds to Sweden, with an eye toward cultivating the plant in Europe. He found the plant growing commonly along Pennsylvania roadsides and riverbanks and recorded its use as an effective vermifuge for children. Physicians and apothecaries likely used wormseed in prepared medicines, and it may also have been the unnamed active ingredient in various patent medicines advertised to cure patients of intestinal worms. The active principle is now known to be ascaridole, a terpene peroxide that constitutes up to 90 percent of the essential oil of wormseed. As with several other phytochemicals that were used medicinally, the effective dose of wormseed is close to the dose that is toxic to the patient, especially a child.

Sesame, or *benne* (*Sesamum indicum*), from Africa also became a valued medicine for children. In *The New American Gardener* (1857), horticulturist Thomas Fessenden credited the sesame introduced by slaves with curing cases of the deadly "summer complaint" in children, chronic diarrhea, which was also known as cholera infantum and cholera morbus and was caused by poor hygiene. Children became rapidly dehydrated with summer dysentery, but a few fresh sesame leaves in water produced a thin jelly (presumably from a soluble carbohydrate) that prevented dehydration. Jefferson was interested in cultivating sesame as a source of an oil that could substitute for olive oil, but he also knew about the medicinal properties of sesame water in treating dysentery; he advised drinking five or six pints daily as a cure.

Patent and Proprietary Medicines

Medicines that promised remarkable results and contained herbal ingredients began in eighteenth century England with products such as Dr. Brodum's Botanical Essence. Colonists carried the notion of such medicines to America, which spawned an enormous industry based on self-medication during the nineteenth and early twentieth centuries. Some clarification of terms is useful; in fact, most "patent medicines" were marketed in bottle designs that were patented, but the medicines per se often were not protected by applications made to the United States Patent Office. These products are often more accurately known as proprietary medicines, meaning that they were produced and marketed by an individual manufacturer. Trade names and labels were protected by copyrights; however, the formulations were often known only to the proprietors, were subject to change without notice, and perhaps did not contain some or all of the ingredients described on labels. Patenting medicinal formulae would have required disclosing the ingredients, and seventeen years after patenting, the formula would have become public property.

Many proprietary medicines aggressively advertised their herbal or vegetable contents as an appeal to public skepticism about mineral medicines such as calomel. The popularity of calomel, mercurous chloride ($HgCl_2$), originated with its colonial use by Dr. Benjamin Rush, a Philadelphia physician and signer of the Declaration of Independence. In 1793, he was credited with halting a yellow fever epidemic in Philadelphia by treating the sick with blood-letting and purging, using a mixture of calomel and jalap. The eventual overuse of calomel caused mercury poisoning and suffering; physicians guilty of calomel abuse were satirized in song popularized during the mid-nineteenth century by the Hutchinson Family Singers, a band of siblings who performed extensively throughout the United States and Great Britain, often to audiences of social reformers and progressives. To the familiar hymn melody of "Old Hundred," they sang:

> Physicians of the highest rank
> (To pay their fees, we need a bank)
> Combine all wisdom, art and skill,
> Science and sense, in calomel
>
> Howe'er their patients may complain
> Of head, or heart, or nerve, or vein,
> Of fever high, or parch, or swell,
> The remedy is calomel.
>
> When Mr. A. or B. is sick—
> "Go fetch the doctor, and be quick"—
> The doctor comes, with much good will,
> But ne'er forgets his calomel.

He takes his patient by the hand,
And compliments him as a friend;
He sets awhile his pulse to feel,
And then takes out his calomel

He then turns to the patient's wife,
"Have you clean paper, spoon and knife?
I think your husband might do well
To take a dose of calomel"

The song continued for several more verses, and the Hutchinsons' message was clear: Avoid calomel and the quackery of physicians who prescribe it. Suspicions of calomel and physicians generally spawned interest in self-medication with proprietary herb-based tonics.

The botanical contents of proprietary medicines (at least what was listed on their labels) varied considerably. Some products were based on American medicinal plants, such as Dr. Ayer's Extract of Sarsaparilla, Dr. Hammond's Syrup of Tar and Wild Cherry, and Strother's White Pine Cough Syrup. Other North American plants listed as ingredients included hemlock, wintergreen, witch hazel, gold thread, and blackberries. Various conifers such as white pine (*Pinus strobus*) and eastern hemlock (*Tsuga canadensis*) were used to brew teas that were rich in vitamin C and used in folk medicine; their pungent resins may have recommended them for various tonics and cough syrups, and for external use as plasters and ointments, such as Kennedy's Hemlock Ointment, which was advertised for croup, scurvy, bunions, chilblains, and ringworm. Child (1837) described the "strengthening plasters" prepared as home cures using pine, hemlock, and other conifers, and these plasters were later marketed as patent preparations for external use. Advertisements for Green Mountain Balm of Gilead and Cedar Plaster, a product manufactured in Windsor, Vermont, beginning in 1868, illustrated a coniferous forest and described its use as a counterirritant for relieving "lameness, cramps, side pains, old sores, boils, and corns."

Other proprietary mixtures contained European plants such as thyme, sage, catnip, chamomile, celery, dandelion, and mullein and spices such as cardamom. Additional ingredients included imported medicinal plants such as aloe, camphor, senna, cinchona (feverbark), rhubarb, nux-vomica, damiana, guaiacum, and sabadilla. Nux-vomica was derived from the seeds of an Indian tree (*Strychnos nux-vomica*) and contained strychnine, once used commonly in compounded medicines. Guaiacum, a gum from the dense wood of *Guaiacum officinale*, was also a familiar ingredient, sometimes in the form of guaiacol that was derived from the gum. It was introduced into Europe from tropical America in 1526 and was once credited with several medicinal properties, which accounts for its common name of lignum vitae. A product marketed as Guaiatonic was described as a "palatable preparation of guaiacol and creosote"; distilled from coal tar, creosote is now know to be carcinogenic but was used commonly as a wood preservative, disinfectant, and internally to treat tuberculosis and intestinal worms. Seeds of sabadilla (*Schoenocaulon officinale*) contain the alkaloid veratrine, which serves as an emetic but is potentially toxic; it was also used to make a counterirritant salve for treating rheumatic pains and neuralgia. The leaves of damiana shrubs (*Turnera diffusa*) from Central and South America had laxative and purportedly aphrodisiacal properties, and they were often combined in proprietary tonics with nux-vomica and phosphorus. Medicines containing damiana were sometimes advertised as associated with Mormons, suggesting that the plant was responsible for the sexual vitality that led to their large families.

Of course, identification of botanical ingredients is not always straightforward. Balm of Gilead was a common name applied both to a fir tree (*Abies balsamea*) and a poplar (*Populus balsamifera*), both resinous American plants with medicinal uses. *A. balsamea* had been used for urogenital complaints, and *P. balsamifera* was employed both as a stimulant and poultice; one or both of them may have been included in products such as Green Mountain Boys Balm of Gilead and Cedar Plaster. Foley's Pure Tincture Arnica probably used the European mountain arnica (*Arnica montana*) but might have referred to the dried roots, rhizomes, and flowers of the American species (*A. fulgens*), which was used in tonics and to treat wounds. Pennyroyal referred to either the European or American species both known by the same common name.

Sometimes the common names as they appeared on labels offered no clue as to the botanical identity of the primary ingredient, as in the product marketed as Dr. Kilmer's Swamp Root. Indeed, confusion as to the identity of their ingredients was the intention of some proprietors, including William Robert Prince, the grandson of Robert Prince, who established the first Prince family nursery in Flushing, Long Island, in 1737. Robert was followed by his son, William, who issued the first catalog of fruit trees and other woody plants in 1771, but the family business closed around 1865. William Robert Prince then withdrew from the business of horticulture and devoted himself to compounding medicines using nursery stock and other available medicinal plants, his "innate soul-love" that he described in a sixteen-page pamphlet, "A Treatise on Nature's Sovereign Remedials, Eclectic Fluid Compounds, Extracted from Plants" (1864). He was consistently vague about the specific plants that he used, with occasional references to several species including "Pucconia" (perhaps hoary puccoon, *Lithospermum canescences*), "Leonuri Nervine" (probably motherwort, *Leonurus cardiaca*), buchu (*Barosma crenulata*), cinchona (*Cinchona* sp.), and an unnamed speedwell (*Veronica* sp.). Prince was not hampered by geography in selecting ingredients; hoary puccoon is native to North America, motherwort is native to Eurasia, cinchona was imported from South America, speedwells originate in Europe and North America, and buchu was imported from Africa. A later (1884) advertisement for Helmbold's Fluid Extract of Buchu illustrated Hottentots in southern Africa gathering buchu leaves for export and described its use for urinary problems, dyspepsia, female weaknesses, chronic rheumatism, and "cutaneous affections."

Prince was a nurseryman who converted to the lucrative practice of botanic healing; he closed the family mail-order nursery but continued to sell plant-based medicines by mail and offered to prescribe cures for correspondents who advised him of their symptoms. Prince's new business of medicinal botany commingled with theories about "humanized electricity and magnetism," as noted on the title page of his pamphlet. He may have taken his business cues from the practice of Thomsonian medicine, a movement started by Samuel Thomson and based on botanic medicine rather than the heroic practices of calomel consumption and blood-letting. In 1822, Thomson published his *New Guide to Health* and opened his Botanic Medical College in Ohio; he argued for the use of American medicinal plants such as the emetic Indian tobacco (*Lobelia inflata*), which were distilled into medicines using his methods. His *materia medica* also included Old World plants such as tansy, horseradish, mullein, horehound, cloves, and myrrh. His message was one of suspicion of the medical establishment, and he encouraged self-help by selling his secret botanical formulae to those who wanted to practice his style of medicine.

Thomsonian medicine was followed by phrenology, mesmerism, electropathy, and hydropathy, which were practiced (often quite profitably for their practitioners) throughout the nineteenth century. These coexisted with proprietary medicines in a medical marketplace that encouraged experimentation, self-medication, and a wide range entrepreneurial medical practices. The emphasis on American medicinal plants culminated with an interest in

"Indian medicines," often sold as part of medicine shows such as those made famous by the Kickapoo Indian Medicine Company. These became known as "snake oil" medicines, which is probably based on the use of rattlesnake fat in some liniment preparations, but proprietary products such as Bright's Indian Vegetable Pills and Seminole Cough Balsam were supposedly based on Native American medicines.

In fact, some proprietary medicines were based on legitimate Native American practices. Coneflower (*Echinacea purpurea*, *E. angustifolia*, and *E. pallida*) was used by Plains Native Americans as an almost universal remedy to treat snakebite, wounds, toothaches, coughs, and communicable diseases; the plants were also heated to produce the steam used during sweat lodge ceremonies. Early settlers soon adopted the herb, probably gathering it from wild populations and using it as a home remedy.

By the late 1800s, several proprietary tonics contained coneflower; one of these was Meyer's Blood Purifier, developed by Dr. H. F. C. Meyer, who learned about coneflower in Nebraska about 1870. The label claimed that the coneflower tonic cured several complaints, including malaria, diphtheria, gangrene, and rabies; according to legend, Meyer allowed himself to be bitten repeatedly by rattlesnakes and then was cured of the venom's effects by the coneflower preparation. Coneflower was adopted by eclectic medical practitioners who prescribed both herbal remedies and various therapies, and by 1895 coneflower was known in Europe. As coneflower gained popularity as a European remedy, it faded into obscurity in America, but its use as an immune system stimulant has been recently revived; indeed, more is known about the use and effects of coneflower than many other herbal or ethnobotanical medicines. Recent studies suggest that the immune system is stimulated by coneflower polysaccharides, which signal macrophage cells to produce interferon molecules, which in turn promote the migration of white blood cells known as neutrophils into the blood from bone marrow.

Another plant familiar to many Native Americans was butterfly weed (*Asclepias tuberosa*), also known as pleurisy root, which contains the same cardiac glycosides as common milkweed (*A. syriaca*); these are chemically similar to the compounds in foxglove and lily of the valley. Depending on dose, butterfly weed acted variously as an emetic, diuretic, diaphoretic (causing perspiration), and expectorant, and it was used in treating pleurisy, pneumonia, coughs, and lung inflammations. These latter complaints were common, and butterfly weed was easily gathered from wild populations.

Native American women often consumed the roots after childbirth, which may explain its inclusion in Lydia E. Pinkham's Vegetable Compound, a mixture of herbs macerated in alcohol. The effect of unregulated patent medicine doses of cardiac glycosides in women striving for reproductive health and full-term pregnancies is unknown, but beginning in 1875 the Vegetable Compound was sold to both sexes for various reproductive complaints. It was marketed aggressively with testimonial advertising, and Lydia Pinkham promised a reply to customers' personal queries.

The other medicinal ingredients of the original Pinkham mixture were unicorn root (*Aletris farinosa*), black cohosh (*Cimicifuga racemosa*), and golden ragwort (*Senecio aureus*). Unicorn root was used by Native American women to prevent miscarriages and does in fact contain diosgenin, a molecular precursor to the female hormones estrogen and progesterone. Black cohosh was used for female problems, including use in large doses as an abortifacient, which can also cause the onset of premature labor in pregnant women; in laboratory studies it has shown estrogenic effects. Golden ragwort, also known at squaw weed, was taken to prevent pregnancies, for complications in childbirth, and as an abortifacient. Its active principles are toxic pyrollizidine alkaloids, and the plant was considered a botanical substitute for ergot (see chapter 2). These four species were macerated in alcohol, strained, and then preserved with the addition of more alcohol; although the Pinkhams were temperance followers, they readily sold a medicinal mixture of herbs in alcohol. The Lydia E. Pinkham Vegetable Compound label registered at the United States Patent Office in 1876 specified 17.9 percent alcohol, but the alcohol content of the Pinkham Vegetable Compound changed as the tax laws required; it eventually decreased to 15 percent to avoid taxation. The labels of proprietary medicines often noted that alcohol was used as a preservative for the herbal or vegetable contents, but many products had alcohol concentrations that exceeded the alcohol in table wine.

Pinkham's product was advertised as a remedy for female troubles, a wide category of complaints that included mental conditions (nervous breakdown) and physical problems (exhaustion, debility, and prolapsed uterus). Vegetable Compound customers were often struggling to sustain normal pregnancies and produce healthy infants. Testimonial letters claimed that many did bear infants after dosing themselves with the Vegetable Compound, and advertising that predated the Pure Food and Drug Act (1906) also included the unlikely claim that the Vegetable Compound would "dissolve and expel tumors from the uterus." Lydia Pinkham wrote a frank booklet discussing gynecological topics, and personal medical advice was produced by an all-female staff of letter-writers; the typical recommended dose was four teaspoons daily of the Vegetable Compound.

Nor was alcohol the only potentially addictive ingredient in proprietary medicines. Opium collected from the capsules of opium poppies (*Papaver somniferum*) was the active ingredient in many soothing syrups and medicinal cordials, anodyne mixtures that were unregulated and available legally; opiate medicines contained morphine that was both addictive and potentially lethal if an overdose were administered. Dr. Swayme's Wild Cherry Tonic was an opiate mixture that claimed to cure tuberculosis, although its effects were strictly analgesic. An advertising card for Mrs. Winslow's Soothing Syrup illustrated a young mother holding a medicine vial and dandling a plump infant, with the legend "for children teething." During the 1870s, the company offered to its customers "Mrs. Winslow's Annual Receipt Book," a household almanac. Products such as Fluid Opium sold by James C. Wells Druggists in New York promised that "all the valuable medicinal properties of the Opium, exempt from all the noxious ingredients of the crude drug," while opiates in the form of laudanum (opium dissolved in alcohol to form a tincture, first prepared by Paracelsus) and paregoric (tincture of opium with added camphor) were also commonly available.

In January 1876, *The Druggists' Circular and Chemical Gazette* published a short cautionary article, "Abuse of Anodyne Cordials," which warned that homemade versions of Godfrey's Cordial (a mixture of opium, spices, and sugar) had caused accidental infant deaths in England and called for "legislative interference" in the United States to halt the dangerous practice of administering opiates to young children. Adults were more likely to become addicted to opiates than to suffer overdoses, and addiction to opiate mixtures became known as the "soldiers disease" after the Civil War because of the widespread use of opium as an analgesic for the wounded. A similar situation existed with proprietary medicines that contained cocaine, an alkaloid produced by South American coca trees (*Erythroxylon coca*). Cocaine-based toothache

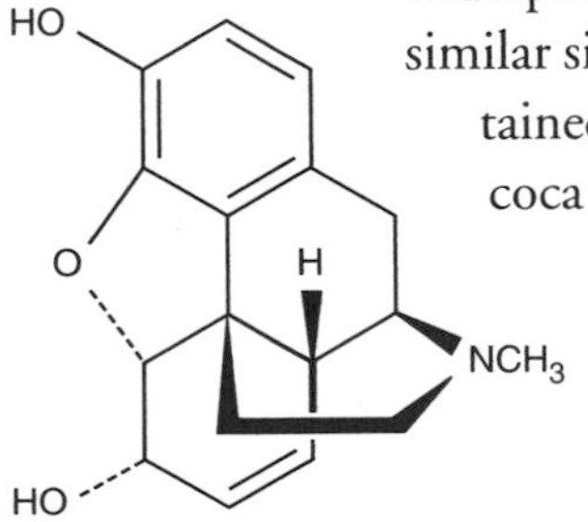

HO
O
H
NCH_3
HO

The chemical structure of morphine, one of twenty-six alkaloids isolated from opium poppies (*Papaver somniferum*).

cures and various coca-based proprietary medicines and medicinal wines contained 2 to 4 percent cocaine, and like opium medicines these caused inadvertent addictions; remedies promising a cure for addiction often also contained opium or cocaine among their ingredients. Coca-Cola originated as a brain and nerve tonic that contained cocaine from the time of its introduction in 1885 until 1904, when the manufacturer removed it at government urging. Coca leaves with the alkaloids removed were still used for flavoring, along with cola leaves (from *Cola acuminata*), which contain the stimulating alkaloids caffeine and theobromine.

The proprietary medicine business evolved in response to the temperance movement, but instead of a decrease in alcohol concentrations, the high-alcohol products known as bitters proliferated. These masqueraded as legitimate medicines in typical bottles, often promoted as curative herbal mixtures, such as Baxter's Mandrake Bitters, Burdock Blood Bitters, Cinchonia Bitters, and Williams Vegetable Jaundice Bitters. These were considered temperance drinks, a pseudo-medicinal means for alcohol consumption, and bitters were often stocked on shelves next to proprietary tonics and elixirs with lower alcohol concentrations. As with other proprietary mixtures, the ingredients in bitters were often unknown or unverified, but the alcohol concentrations sometimes exceeded 40 percent. During the Civil War, Union troops were generously supplied with Hostetter Bitters, which were 44 percent alcohol and delivered by the trainload.

Liniments with alcohol concentrations of 70 percent or more were recommended for both internal and external uses. According to their labels, they contained pungent or resinous plants or their products, including hemlock, myrrh, guaiacum, sassafras, turpentine, and camphor. Such liniment mixtures served as counterirritants for muscle soreness, and, used internally, they were reputed to cure common complaints such as flatulence, coughs, or colic. Indeed, by 1883 the Commissioner of Internal Revenue commented, "To draw the line nicely, and fix definitely where the medicine may end and the alcoholic beverage begin, is a task which has often perplexed and still greatly perplexes revenue officers." Alcohol-free bitters appeared in response to criticism from temperance advocates. One such product was Dr. Walker's California Vinegar Bitters, which promised a "Purely Vegetable Preparation manufactured from the Native Herbs of California." Product names were also sometimes deceptive. For instance, Sulphur Bitters contained no sulfur, but the product did have the distinction of being endorsed by a first lady—the advertisements featured a sepia print of Frances Folsom Cleveland, wife of President Grover Cleveland, and the motto "Use Sulphur Bitters." Out of concern for the purity of ingredients, some pharmacists made their own alcoholic bitters from herbs such as goldenseal (*Hydrastis canadensis*) and gentian root (*Gentiana* spp.), sometimes with the addition of aloe for its laxative properties and cloves for flavor; a typical recommended dose was a spoonful daily, but many no doubt exceeded that amount. Goldenseal contains berberine, an antibiotic alkaloid, and gentian root was a common bitter tonic herb that was used to stimulate digestion. Resourceful Americans also concocted their own alcoholic tonics and bitters by macerating various herbs in wine or brandy; goldenseal, cloves, dogwood and poplar bark, spikenard root, and the berries of prickly ash were among the plants included in homemade bitters.

By the 1890s, the vegetable medicines that promised health and freedom from the dangers of calomel spawned their own problems. Some were ineffective and perhaps served as placebos, while others were toxic, addictive, or otherwise falsely advertised. Proprietary medicines were often untrustworthy products because their contents and efficacy were unknown, although some of the plants listed among their ingredients had a long history of effective medicinal use. Addiction to opium, cocaine, or alcohol was a social problem spawned by the use of patent and proprietary medicines, and some may have practiced self-medication for serious conditions instead of visiting a physician. The boastful claims and labels that misrepresented the medicinal contents of these products were typical business practices, and the vast business of advertising proprietary medicines was lucrative and intentionally deceptive.

In a case of principle over money, *The Ladies Home Journal* under the editorial leadership of Edward William Bok began to reject advertisements for proprietary medicines beginning in 1892. They enlisted the help of the Women's Christian Temperance Union in curtailing advertisements for medicines that were primarily alcohol, and Bok printed a photograph of Lydia Pinkham's gravestone, debunking the notion that women could write to her for a personal reply. *The Ladies Home Journal* was joined by *Collier's Weekly* in a print campaign to inform consumers about products that were possibly ineffective, toxic, or addictive. Magazine articles in these publications and others attacked the questionable business practices of proprietary drug companies and the efficacy of their products, foreshadowing the Pure Food and Drug Act of 1906.

Because nothing was inherently wrong with the notion of ready-made medicines, by the end of the nineteenth century, many families no longer cultivated gardens or retained the practical stillroom knowledge of earlier generations. Physicians, pharmacists, and those needing medicines relied on herb growers and importers to supply the necessary medicinal plants, often purchasing powdered herbs and products offered for sale by American Shaker communities. During the first half of the nineteenth century, Shakers embraced both the food reform movement promoted by Sylvester Graham and the Thomsonian herbal medicine philosophy espoused by Samuel Thomson, and perhaps this was the motivation for their cultivation and marketing of efficacious medicinal herbs. Shakers defined the highest level of American herbalism and were long regarded as the most reliable and meticulous of herb growers and purveyors. Their medicinal plants were often sold directly to physicians and pharmacists, but Shaker herbalists also followed the popular trend and compounded medicines that were sold for self-medication. These were produced with the usual high standards, but they closely resembled many popular products in the marketplace. Vegetable Bilious Pills, Compound Vegetable Cough Balsam, Shaker Tamar Laxative, Syrup of Bitter Bugle, Corbett's Shaker Dyspepsia Cure, and Compound Concentrated Syrup of Sarsaparilla are a few of the many Shaker proprietary medicines produced for mass consumption during the nineteenth century.

Shaker herb growers cultivated both European and American medicinal plants and combined the tradition of European herbalism along with the practical lessons acquired from Native Americans. Their ambitious goal was to practice the most legitimate herbal medicine possible, using medicinal plants that were cultivated, harvested, and dried expertly. In keeping with the times, medicinal wines were also produced; the Shakers of Union Village, Ohio, fermented wine from grapes that were cultivated in vineyards along Lake Erie. Shaker wine was sold by the bottle, with the caveat "Shaker wines are medicinal wines. They are for medicinal purposes. . . . Above all things they are not incentives to drunkenness."

Medicines, Toxins, Food, and Survival

In years past, the categories of food and medicine were less well defined; in sickroom cookery, certain foods and beverages were considered medicinal, and herbs and spices did double duty as culinary and medicinal plants. Medicinal foods, herbs, and spices occupied an essential niche in folk medicine, domestic practices that combined European herbal lore, Native American practices, and experimentation in the face of adversity. Sundry domestic methods reflect daily struggles with physical and mental illness, parasitism, injury, and seasonal malnutrition. Folk practices spread orally and by means of household manuals and almanacs that shared printed information about domestic problems, cures, caveats, and the core of essential knowledge that was needed for survival and comfort.

If a particular remedy "worked," the placebo effect might explain an apparent cure, or perhaps the ailment was a temporary complaint that disappeared regardless of treatment or medicine. In other cases, cures were the result of effective medicinal chemistry. Before pharmaceutical antibiotics were available, wounds and infections were treated with familiar plants in ways that seemed to inhibit bacterial infections and parasitic fungi such as ringworm. Onions and garlic were used in poultices and are now known to contain alliin and allicin, sulfur-containing compounds that are antibiotic against both bacteria and fungi. Some even claimed success in curing viral diseases such as rabies (known as hydrophobia) and warts with topical applications of onion or garlic. The presence of mustard oils and capsaicinoids (see chapter 4) explains why various mustards and peppers were also effective poultices; mustards are both counterirritants and antibiotics, while peppers function as numbing poultices on rheumatic pains. Decomposing apples were another familiar poultice mentioned in various household manuals; the fungus *Penicillium expansum* causes soft rot in apples and produces the fungal toxin (mycotoxin) patulin, which defends its "territory" from bacterial attack by destroying both Gram positive and Gram negative bacteria. Patulin is now also known to be carcinogenic, which explains why rotten apples should not be pressed for cider, but its external use to fight infection was probably often effective.

The scope of home remedies is vast; in *The Family Nurse* (1837) and *The American Frugal Housewife* (1844), Child compiled scores of "cures" that were familiar to her and others, including several that depended on kitchen and pantry ingredients: vinegar (for dysentery, colic, and poultices), sage (for intestinal worms), tea (for dysentery), blueberries (for digestive ailments), horseradish (for colds and sore throats), and parsley (for the edema known as dropsy). Bleeding was often stopped with astringent substances such as salt, sugar, or tea. Sickroom cookery called for simple gruels made of oatmeal, cornmeal, or rye; puddings based on tapioca or blancmange; and stewed fruits. The bark of slippery elm (*Ulmus rubra*) was cooked into a mucilaginous tea favored for children and convalescents; it was used commonly in many homes and was often purchased already dried and powdered. These various foods were considered curative in their own right, but other folk cures relied on wild plants, such as the familiar treatment for mouth cankers that relied on goldthread (*Coptis groenlandica*, also known as canker root), which contains the antibiotic alkaloid berberine.

The practice of medicine was primitive for lack of equipment and diagnostic tools, but its basis in plant secondary compounds was often legitimate and sometimes similar to modern medical practices. We shudder at unsafe and foolish practices of folk medicine, but let us not be guilty of arrogance; future generations may regard twenty-first century pharmacology, surgery, and chemotherapy in the same way that we regard quackery and questionable herbal cures of the nineteenth century. Colonial women were skilled domestic healers, but during the nineteenth century many dubious medicines were marketed for daily consumption by women who self-diagnosed and self-medicated, with the risk of addiction and toxicity. Ultimately the Pure Food and Drug Act of 1906 aimed to prevent "the manufacture, sale, or transportation of adulterated or misbranded or poisonous or deleterious foods, drugs, medicines, or liquors," but not until many had been addicted or deceived by anodyne cordials and dubious tonics. Survival begets ingenuity, and dose often distinguishes efficacy from danger, but some botanical medicines were ineffective or potentially dangerous. Certain home remedies required extremely toxic wild plants, including jimsonweed (*Datura stramonium*, once known commonly as "stramonium"), which was recommended as a cure for both asthma and alcoholism; it contains several hallucinogenic tropane alkaloids and was potentially fatal if ingested. Botanically speaking, the demand and desire for botanical drugs demolished populations of native North American medicinal plants, including ginseng, goldenseal, and nerveroots. The phytochemical irony is that the secondary compounds that evolved in nature as protective molecules rendered the plants medically useful and became a liability for their survival.

CHAPTER EIGHT

Wood, Fibers, and Textiles

HOUSEHOLDS required a diversity of botanical products. Colonial dwellings and furniture were constructed of wood, and wood also served as a fuel for cooking and heating. Textile production required specific plant fibers and dyestuffs, and various plants provided materials for basketry and other types of weaving. Structural tissues from plants were used in most of the household goods and objects that furnished colonial homes; furnishings, brooms, mattresses, candles, bowls, baskets, boxes, and tools were all fashioned from wood or other plant parts. In a practical sense, form dictated function, and New World colonists soon learned the hardness, tensile strength, appearance, and other practical properties of native North American plants. Colonial dwellings were built of the available local timber, which included various oaks, chestnut, red maple, white pine, and hemlock. Native cherry and black walnut woods were favored for furniture because of their beauty, and black walnut was also exported from Virginia to England in the early 1600s. As land was cleared, potash (from "pot ashes," various potassium-containing compounds such as potassium carbonate and potassium hydroxide) was made from the wood that remained; potash was exported or used locally as an alkali in soap-making, as a mordant in dyeing cloth, and in the manufacture of glass.

Wood in particular illustrates the diversity of available materials for craftsmen. Massive timbers were hewn from tree trunks that grew straight and resisted fungal decay, while furniture was made from woods with distinct figures and close grain. Tool handles required close-grained woods that absorbed shock and resisted breakage and splintering, while cooperage relied on woods that tolerated bending. Oak was easily split into building shingles and strips for basketry, while willow branches provided flexible woody stems for osier work. The non-woody stems of various grasses and rushes were also sturdy enough for basketry, seat weaving, and various types of wicker work.

Anatomy dictates all of the properties that make plants useful for various structural functions. Plant cells have cell walls with various shapes, thickness, flexibility, and chemistry, and these have varying properties when the cells are aggregated into tissues. A similar correlation between plant anatomy and practical uses is seen in botanical fibers such as linen, cotton, and hemp, each with unique properties that reflect the evolutionary adaptations of the plants that produce them.

Wood Form and Function

Trees begin life as herbaceous seedlings. Their young stems at first consist of soft tissue with embedded vascular bundles, each comprising an inner layer of water conducting tissue (xylem) and an outer layer of food-conducting tissue (phloem). Within the first year or two of growth, a layer of cambium cells grows horizontally from the center of each vascular bundle, and these merge to form a continuous internal cylinder of dividing cambium cells. Cambium cells are small and boxlike, but through repeated cell divisions they account for all horizontal growth and the increase in circumference in woody plants; cambium cell division accounts for the increase in tree girth from a slender sapling to a massive tree up to several feet in diameter. As cells divide on the inside face of the cambium, they differentiate into the wood (xylem), which makes up most of the volume of a tree trunk. As the cambium layer divides on its outer surface, the new cells develop into phloem, the inner living layer of the bark. Phloem transports the sugars that are made by photosynthesis, and the cambium cells retain the ability to divide indefinitely. Both of these layers and their essential functions are destroyed when bark is injured or removed, which explains the familiar caveat not to peel bark from tree trunks.

In temperate regions, cambium activity is seasonal, which results in the distinct annual rings that distinguish temperate woods. Growth begins at the end of winter with large diameter, thin-walled xylem cells (springwood), and growth ends a few months later with smaller diameter cells with thicker cell walls (summerwood). An annual ring consists of both bands and represents the layer of wood produced by the cambium in a single year. Oak and yellow pine have springwood and summerwood that are markedly different, while the transition from springwood to summerwood in maple, basswood, and white pine is less obvious to the eye. Annual rings can be counted in temperate trees to determine their age, but no such bands develop in trees that grow under relatively uniform tropical conditions.

Close examination of wood reveals the microscopic differences that distinguish conifer woods (softwoods) from woods made by flowering plants (hardwoods). In conifers, the wood consists of tracheids, elongated cells that are dead and hollow at maturity; they function both in water conduction from the roots up to the crown and in support of the tree trunk. The wood of pine and other conifers also has resin canals that contain the pitch long used for making rosin and turpentine, and the fossilized resin was used to make amber jewelry. Conifer resins likely evolved as compounds that discouraged herbivores and protected trees from bacteria and fungi by effectively sealing tree wounds. In contrast, hardwood trees differentiate into two other cell types in addition to tracheids. Vessel elements are hollow cells arranged into vessels, which are vertical, tubelike water conduits; these specialized conducting cells often have wide diameters. In contrast, fibers are flexible, thick-walled, elongated cells that provide support and tensile strength to hardwoods. Softwoods lack wood fibers, which explains why they are generally less dense and more prone to compression and damage from external forces than hardwoods.

The wood of conifers such as pine, hemlock, fir, and spruce consists of closely packed tracheids, and water flows vertically from tracheid to tracheid through thin areas in their cell walls known as pits. In flowering plants, water transported through the vessels and generally follows a direct pathway. The wide diameter of vessel elements weakens wood, but the presence of fibers in hardwoods compensates for the loss of mass and strength. Of course, some woods exhibit unexpected strengths—balsa wood is a hardwood (from a flowering tree, *Ochroma pyramidale*), but it is very soft due to the presence of many thin-walled cells. Douglas fir (*Pseudotsuga taxifolia*) is valued for its hardness, even though as a conifer its wood is composed of only tracheids.

Although wood cells are no longer alive when they function in water conduction, their cellulose walls are strengthened by the presence of lignin, a carbon-based compound comprising six-carbon aromatic alcohols. Lignin usually accounts for 15 to 25 percent of the dry weight of woody plants, and next to cellulose it is the most abundant organic compound in nature. Lignin prevents compression, while cellulose imparts flexibility and resistance to tension. Wood strength is dictated by density, a property that depends on cell wall thickness and the amount of lignin in the wood; wood density can be measured by comparing equal volumes of wood and water to determine specific gravity, the ratio between wood mass and water mass. High specific gravity indicates a high concentration of lignin in a known volume of wood, and the wood with the highest specific gravity is lignum vitae, the exceedingly dense wood of *Guaiacum officinale* and *G. sanctum* trees from Central and South America and the West Indies. In addition to its medicinal properties and use in synthesizing vanillin (see chapters 6 and 7), lignum vitae was used to make the mortars in which spices were pounded with pestles. Its specific gravity is 1.32, indicating that a dried block of lignum vitae is 1.32 times the mass of an equal volume of water and that the wood will not float in water. Temperate woods usually have masses that are less than that of water, resulting in specific gravity numbers that typically range from 0.35 to 0.65. Woods with relatively high specific gravity include sugar maple (*Acer saccharum*, at 0.63), persimmon (*Diospyros virginiana*, at 0.64), and black locust (*Robinia pseudoacacia*, at 0.66). All of these woods were used when hard woods were needed; cutting boards were cut from maple, and both maple and persimmon were used to make shuttles, bobbins, and spools for the weaving industry because their wood grains remain smooth despite friction. Black locust resisted wood fungi and was used to craft fence posts and railway ties (see chapter 9).

Xylem and phloem function in the vertical transport of water and food, but living cells in the bark must also be able to exchange food and water with living cells inside the tree trunk. Horizontal transport in tree trunks occurs in the ribbonlike bands of cells known as vascular rays, which extend across the cambium and allow for communication between the inner and outer areas of a tree trunk. Vascular rays vary in size, from a single cell to several cells in thickness, and their presence affects wood appearance and can increase the tendency of wood to split vertically. Vascular rays are particularly visible when the wood is quartersawn into boards parallel to a radius of the log. Oak and sycamore rays are large and tend to reflect light, which distinguishes their wood figures. Figure is the overall pattern presented by a piece of wood—the sum of the grain (the direction of the conducting cells), the annual rings, and the appearance of the vascular rays.

As a natural product, wood is hygroscopic, meaning that it responds to changes in moisture and humidity; cell walls expand with moisture and contract with drying, and these changes are reflected in the subtle seasonal movements that occur in wooden structures. For instance, as they experience many cycles of expansion and contraction, wood shingles gradually push out the nails that hold them in place. As a remedy to counteract movement, square nails with greater surface area and more friction were used to nail wooden shingles in place. Boards for buildings had to be dried evenly or warping resulted, which could be corrected by dampening the wood and weighting the concave side. Household objects were also affected by the hygroscopic nature of wood. Wooden bowls over time transformed from circular to slightly oval in shape, as the wood cells dried and contracted in diameter, and itinerant artists painted both sides of the boards used for portraits to avoid differential swelling and warping of their paintings.

Dwellings

Seventeenth century colonists sought to duplicate their European houses by constructing half-timbered buildings and thatching them in the traditional manner with bundles of straw and reeds. These proved to be insufficient protection because North American winds and heavy rains penetrated the plaster walls between the timbers and the layers of thatch. Faced with a harsher New World climate, colonists found more resistant building materials in the virgin forests of North America; sturdy colonial houses were soon constructed almost wholly from trees. The framing of post-and-beam dwellings was usually done with white oak (*Quercus alba*), which resembled familiar European oaks, or less frequently with red oak (*Q. rubra*), chestnut, (*Castanea dentata*), hemlock (*Tsuga canadensis*), or white pine (*Pinus strobus*). Trees were selected for their upright growth and straight bark, which usually indicated that the wood grain

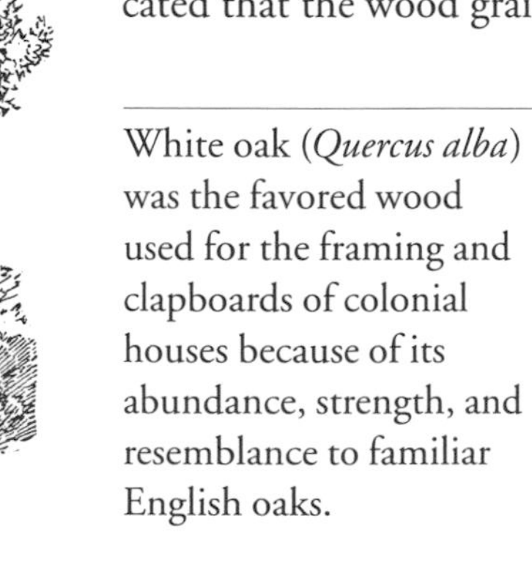

White oak (*Quercus alba*) was the favored wood used for the framing and clapboards of colonial houses because of its abundance, strength, and resemblance to familiar English oaks.

would be straight internally and that the timbers would tend not to warp or twist as they dried.

Logs were cut in the winter and accurately hewn into massive square timbers, which were then stacked and separated with narrow boards for uniform, slow drying. Colonial builders acquired practical knowledge of plant anatomy: for instance, they knew that as wood dries, the outer annual rings dry and contract first; this results in cracks and checks if wood dries too rapidly. These caused possible areas of weakness in timbers that were used to frame buildings and made it difficult to cut the joints accurately. Before the timbers were trimmed and the joints cut, they were inspected for potential problems, including damage from insects or fungi and knots. Knots, the embedded remains of side branches, were considered to be zones of weakness if they occurred on the lower side of a timber that received tensile stress; for this reason, knots were usually removed with a notch cut into the wood, which was considered less weakening than the presence of a knot.

Houses were erected on wooden sills, situated on foundations of stone or locally produced brick and often with a cellar; floor joists were inserted into the sill with dovetail joints. The mortise-and-tenon joints used in framing a house involved making an opening (mortise) in a timber into which the trimmed end (tenon) of another timber was inserted. The tenons were typically cut somewhat longer than necessary, so that they would not be pushed out as the wood of the mortise continued to contract. Once the joint was assembled, it was pegged with trunnels, nails cut from hardwood and pounded firmly into a hole that traversed the mortise-and-tenon joint. The best trunnels were cut as square pegs and then shaved to have an octagonal cross section, which provided added tightness and resistance to movement once it was in place.

The sides of a post-and-beam house were preassembled on the ground in sections known as bents; each bent required the strength of several men to lift it, typically one man for each fifty pounds of timber weight. Needless to say, the framing of houses required a community effort that was rewarded with food and an opportunity for socializing. Once they were raised, the bents were stabilized with knee braces and joined with a beam, a weight-bearing timber with great strength. This was the so-called summer beam, a term probably derived from *sumpter*, an old English word for a mule or beast of burden, and it was often planed to have a decorative appearance since it was visible in the room below it.

The roof was constructed with rafters, and once they were in place, as part of a celebration a small sapling known as the "wetting bush" was nailed in place. Once the wooden skeleton of a house was erected, the spaces between the timbers were filled with nogging, some combination of stone, clay, brick, and mortar, which was reminiscent of seventeenth century half-timbered construction. In place of nogging, New England colonists soon began to sheath their houses with wooden boards applied horizontally; these at first were sawed by hand and used while still rough. Exterior walls were covered by clapboards and shingles, often with clapboards on the front of a house to satisfy the notion that boards presented a more attractive exterior. During the seventeenth century, these were made using logs cut from white oak trees into lengths of four feet, which were then split and cut radially by hand into narrow boards. By the early eighteenth century, clapboards were mass produced by sawmills in various types, and they differed in shape, overlap, and in how they were applied. Nothing was wasted; even the sawdust and wood chips from sawmills were collected to insulate the ice blocks stored in icehouses. White pine eventually became the most commonly used wood for clapboards because it was resistant to harsh weather and was easily worked, but clapboards were also made of basswood (*Tilia americana*), white cedar, and other types of pine.

Wooden shingles typically covered the roof of a post-and-beam house as well as three or four of the exterior walls. These were first split from oak, using sections with straight grain and no knotholes. However, even perfect oak logs required trimming; the hard, central heartwood resisted nails; and the outermost sapwood was not weather resistant. Heartwood in various trees tends to become occluded and more deeply pigmented with the deposition of wastes; in oaks it becomes remarkably dense and hard, valuable traits in timbers but unwelcome in wood for shingles. Sapwood is the outermost, most recently produced xylem that functions in water conduction, but in oaks, this young wood was often attacked by wood fungi. Chestnut and white pine shingles were also produced, but the most desirable shingle wood was white cedar (*Chamaecyparis thyoides*), a conifer that colonizes swamps from Maine to Florida.

White cedar logs required no trimming and the wood was remarkably resistant to water and decay. During the eighteenth century, extensive white cedar swamps in New Jersey provided the sheathing for innumerable shingled houses, and these wetlands were largely deforested by the beginning of the nineteenth century. The surprising discovery in white cedar swamps of layers of waterlogged but sound logs led to the mining of submerged cedar, a practice that continued until the Civil War. Remarkably, these logs had resisted decomposition for hundreds and possibly thousands of years.

The interiors of early colonial houses were finished with wainscoting of pine boards, applied either horizontally or vertically with tongue-and-groove joints. By the early eighteenth century, flat wainscoting gave way to raised panels, often surrounding the fireplace in the best room of the house; the rest of the room was typically finished in horizontal wainscoting, with plaster applied from the chair rail to the ceiling. Plaster, a mixture of calcium hydroxide, sand, and animal hair applied to oak lathe, was whitewashed or wallpapered. Wallpapers were imported or painted by hand (see chapter 10), and in some houses a landscape scene was painted onto a board above the fireplace; woodwork was painted in colors such as Indian red, Prussian blue, and yellow ochre. During the mid-eighteenth century, more ornate wood paneling and cornices became commonplace, but by 1800, wooden interior walls were outmoded and replaced by walls covered in plaster and wallpaper. Doors were often crafted of more valuable woods, such as black cherry (*Prunus serotina*) or imported mahogany (*Swietenia mahogani*), and the inner surface sometimes matched the wainscoting of the room; American craftsmen also incorporated the Christian symbol of a cross into door designs.

Flooring was first made of oak, following the English tradition, but pine soon became the wood of choice for floorboards. Boards were sawed in random lengths and wide widths from stands of virgin white pine (*Pinus strobus*) and yellow pine (*P. australis*). These were left unfinished in many houses until the beginning of the nineteenth century, but floors were kept clean by a weekly scouring with abrasive sand. This was often done on Saturdays, and then a layer of white sand was strewn on the floor and swept into decorative patterns; a thicker layer of

White cedar (*Chamaecyparis thyoides*) provided abundant wood for shingles, some of it collected from submerged logs that had resisted decay for centuries.

sand on kitchen floor boards protected them from grease, candle wax, and other stains.

Furniture

Houses and household furnishings, from Pilgrim chests to elaborate Victorian parlor suites, were made of wood. Furniture woods are often cut to reveal the wood figure, the sum of the various structural components of wood that influence its appearance, including the grain (the direction of the xylem cells), annual rings, vascular rays, pigmentation, and texture. The surface texture of wood is dictated by the diameter of its vessel elements; oak, chestnut, and other woods with particularly large diameter vessel elements have visible longitudinal grooves, each composed of vessel elements arranged end to end. Rays often add luster to wood that is quartersawn, and annual rings may produce bands that differ in their color once the wood is finished. Finished color was also important; tropical woods such as West Indian mahogany and various rosewoods (Brazilian *Dalbergia nigra* and related spp.) were later valued for their deep pigmentation, and yellow, lustrous satinwood (*Zanthoxylum flavum*) from the West Indies provided contrast when used as an inlay or veneer. Various woods and their figures moved in and out of style, depending on taste and fashion. For instance, oak was used not only for building the first colonial houses, but also to craft seventeenth century

The wood of white pine (*Pinus strobus*) was used for clapboards, shingles, wainscoting, and floor boards because it was easily worked but was strong and weather-resistant.

furniture. Pilgrim furniture (1620–1670) was made of oak as well as pine, birch, and maple; these were functional pieces such as storage chests, decorated with carvings and designed for practical use. Jacobean styles (1670–1694) were copied in America and also relied on oak as the primary wood for their heavy designs and turnings. The wood carvers in the vicinity of Hadley, Massachusetts, produced heavy carved chests, and other practical furniture such as stools, trestle tables, Bible boxes, benches, and bedsteads were also crafted of American oak during the seventeenth century.

Eighteenth century cabinetmakers avoided oak in crafting fine furniture, perhaps because of the comparatively coarse surface texture that is caused by the large diameter of its vessel elements. American craftsmen made furniture in the William and Mary style (1694–1710) from pine, maple, and black walnut (*Juglans nigra*), sometimes embellished with wood veneers cut from burls, large growths that may occur on tree trunks. Burls are common on walnut and maple and may originate from dormant buds that begin dividing to make fan-like masses of wood tissue that coalesce. Burl wood is characterized by wood figures with patterns of swirls and distorted annual rings; these made striking patterns when the wood was cut thin and applied as a surface decoration, sometimes in opposing directions with the figure matched in the center. A typical use of burl wood was as veneers on the drawers of tall chests crafted in the William and Mary style.

The trend in American Chippendale (1750–1775) and Federal (1775–1830) furniture styles was also toward replicating English furniture styles using American woods. Philadelphia furniture-makers used hickory (*Carya* spp.) to make the American style of Windsor chairs, modifications of an English design that according to legend was favored by King George I. Hickory was used extensively for fuel and for smoking meat for preservation, but as a furniture

The wood of shagbark hickory (*Carya ovata*) and related species was used for the spindles of Windsor chairs and other furniture parts where tensile strength was important.

wood it was known for its hardness and tensile strength; the spindles of American Windsor chairs were more slender and strong than the spindles of Windsor chairs turned from English oak. Craftsmen turned the spindles and legs from seasoned hickory and pounded them into the seat, which they crafted from green wood. As the seat contracted, it held the spindles and legs fast, resulting in a chair that was both strong and firm. These chairs were typically painted, which hid the coarse hickory grain that closely resembled oak.

Typically, a single piece of furniture was crafted from the wood of various native species, perhaps a table top and legs made from birch (*Betula* spp.), a drawer veneered in maple, and the drawer itself made of pine. The New England furniture-makers used maple, cherry, pine, and birch, as well as the mahogany that arrived as ballast on ships carrying molasses and raw sugar for the manufacture of American rum.

Mahogany logs and boards placed around casks kept them from shifting in the hold of a ship, and mahogany also became part of the slave trade triangle that revolved around slaves, molasses, and rum (see chapter 6). Cabinet-

Colton and Fitch's Modern School Geography (Fitch 1859) illustrated the logging of West Indian mahogany trees (*Swietenia mahogani*) growing in Central America.

makers purchased the mahogany wood inexpensively and used it as a valuable resource in furniture-making, often combined with American woods such as maple or birch veneers for color contrast or with satinwood inlays. The deep color of mahogany was enhanced by finishing the furniture with boiled linseed oil, beeswax, and energetic rubbing. Cabinetmakers also sometimes stained birch to resemble mahogany, although other hardwoods were often finished with copal varnish (derived from the fresh or fossilized resins of various exotic trees) or shellac and smoothed by rubbing with a mixture of pumice and oil.

The wood of sugar maples (*Acer saccharum*) sometimes exhibited wood figures that were particularly valued in fine furniture; bird's-eye and curly maple are wood figures that appear in individual sugar maple trees. Bird's-eye wood figures are apparently caused by cambium anomalies that cause conical depressions in the annual growth rings and the formation of eyelike patterns in the xylem. These markings are visible in boards cut tangentially to the long axis of the tree and occur in a random pattern. Trees in the northern range of the species tend to show the anomaly more frequently, and the wood figure sometimes develops only in some parts of a tree trunk; the cause is unclear but may be the result of climate or mechanical stress. Cabinetmakers sought maple wood with the bird's-eye figure as well as wood with similar markings that also occurred in ash, walnut, and birch. Wood figures in maple are also described as wavy or curly and are seen in boards cut either radially or tangentially, in which the grain of the wood appears to undulate. This wood figure reflects light differentially to produce a pattern of distinct bands, the result of the convex or concave surfaces of the elongated wood fibers. Fiddleback patterns in maple wood had

Sugar maples (*Acer saccharum*) sometimes developed wood figures known as bird's-eye, curly, and fiddleback maple; these were used in making fine furniture veneers.

particularly close wavy bands; the wood was used in making furniture and stringed instruments, and it was particularly common in wood cut near roots and limbs. Fiddleback patterns were discovered by removing some of the bark and examining the wood for evidence of anomalies, and similar wavy growth also occurs in ash, walnut, and birch.

Interesting wood figures were used as thin veneer layers on drawer fronts or table tops, which displayed the patterns in light to their best advantage. *Cabinet Maker and Upholsterer's Guide* published in 1788 by George Hepplewhite encouraged the use of decorative wood veneers and inlays, and furniture in the Hepplewhite style was made with figured northern woods. Bird's-eye and curly maple were found as far west as Ohio and as far south as Pennsylvania, reflected by furniture that was locally crafted from these woods. Walnut also provided another interesting wood pattern for veneers, the featherlike figure of crotch walnut that was obtained from areas of the tree where large branches left the main trunk. Empire style (1810–1840) furniture was sometimes made of pine and covered with veneers of walnut or mahogany.

Walnut and mahogany were also the most common woods in Victorian (1840–1890) furniture, but pine, ash, chestnut, and elm were also used. Some inexpensive pieces such as those made of pine were embellished with painted designs or simulated wood figures painted on with metal combs and dark pigments. Rosewood imported from the West Indies appeared in high-style Rococo revival furniture, such as the parlor pieces made by John Henry Belter of New York City. He used rosewood that was laminated into sheets resembling plywood, and these were used to craft elaborate chair backs that required both carving and bending. The Belter firm also produced remarkably realistic botanical carvings, floral and fruit motifs (often roses and grapes) that were incised deeply into the laminated wood layers of seat backs, bedsteads, and table skirts and legs.

In 1868, English designer Charles Eastlake published *Hints on Household Taste*, in which he recommended oak and walnut for furniture-making and rejected the use of very dark woods. He also advocated straight lines in place of heavy Victorian carvings, principles that were embraced by Americans seeking change after the upheaval of the Civil War. Inexpensive pieces in the Eastlake style were produced by furniture factories in Michigan and in the southern states, and oak again became desirable as a furniture wood for the first time since the end of the seventeenth century. Oak furniture was mass produced in a variety of Eastlake and more ornate styles, often with carvings or pressed designs produced by machinery; pieces such as oak desks and bookcases were used as premiums for soap purchased from the Larkin Company. Pine furniture was also factory-made from native wood, and sometimes the tops and sides of chests and tables were made from a single board. Chairs were made partially from pine, with the spindles and legs turned from hardwood for more tensile strength.

Many families relied on their own ingenuity and a supply of native woods for their furniture, from the time of the earliest colonists through the nineteenth century. Rural furniture was simple and functional, and in the vernacular tradition it was crafted locally from American woods without veneers or elaborate elements of design. Function dictated form; splayed legs prevented highchairs from tipping, and wooden cradles were made large enough to accommodate elderly or convalescing adults. Some rural furniture such as dower chests were painted, which added decoration if the wood figure lacked appeal. Pieces often combined several wood types, depending on what was available and the suitability of various woods for specific uses. Typical choices included maple, pine, chestnut, and "poplar," which was tulip poplar (*Liriodendron tulipifera*), a member of the magnolia family (Magnoliaceae) and a common southern hardwood tree. Sassafras (*Sassafras albidum*) was valued for its spicy scent (see chapter 1) and was used to inlay cradles and make Bible boxes and for carving household spoons and drinking vessels.

While most rural furniture was made for home use, Shaker colonies in America relied on furniture production for part of their livelihood and used woods that were locally available. During the first half of the nineteenth century, the Shaker furniture industry exemplified the ideals of craftsmanship and design tied to function in the rural tradition; in making furniture to use and sell, Shakers eschewed unnecessary ornamentation and favored birch, pine, and the appearance of wood without dark stains. Paint was sometimes used as a finish for straight-backed chairs, and seats were woven from cane, rush, wooden splints, or cotton tape. Most importantly, designs were conceived with the furniture use in mind; armless chairs allowed handwork without interference, and low rocking chairs were useful for such tasks as sorting apples. Shakers also made practical tables, cupboards, chests, candle stands, and storage boxes with simple designs but resilient materials; salesmen spread their wares, arguably the best examples of American rural furniture, as far west as Ohio and throughout the East.

Caning, Basketry, and Brooms

Grasses, rushes, and cattails were used in weaving chair seats, basketry, and various types of straw weavings. Grasses (family Gramineae or Poaceae) are herbaceous plants, but the epidermal layers of their stems and leaves have specialized silica-containing cells, and some also have thick-walled, hardened cells known as sclereids. As a result of their anatomy, grass stems are flexible, tough, and waterproof, which made them ideal structural materials for weaving. The leaves of rushes (*Juncus* spp. in the family Juncaceae) have epidermal layers that may be impregnated with lignin, the hardening compound in wood cells. Rush seats were also woven from plant stems, a weaving technique that was known in ancient Europe and Egypt and used bulrushes (*Scirpus* spp.), a type of sedge. In place of bulrushes, American furniture-makers used cattails (*Typha latifolia* and *T. angustifolia* in the family Typhaceae) to weave seats in Chippendale-style chairs and others.

In his *Travels in North America* (1770), Pehr Kalm called seats woven from cattails "rush seats," and he observed that they were made by weaving together twisted cattail leaves. Seats of rustic, fine, and even mass-produced furniture were also woven using the technique known as caning. Caned seats originated in Asia and were introduced into Europe by the Dutch, who colonized Asia during the seventeenth century; English chair-makers learned to weave the traditional open pattern with octagonal interstices, and the craft arrived in America with colonists. Caned seats were woven from thin rattan canes, stem strips cut from climbing palms (*Calamus* spp. and *Daemonorops* spp.) imported from tropical Asia. This craft was commonly used in making chairs for home use; caned seats were on the

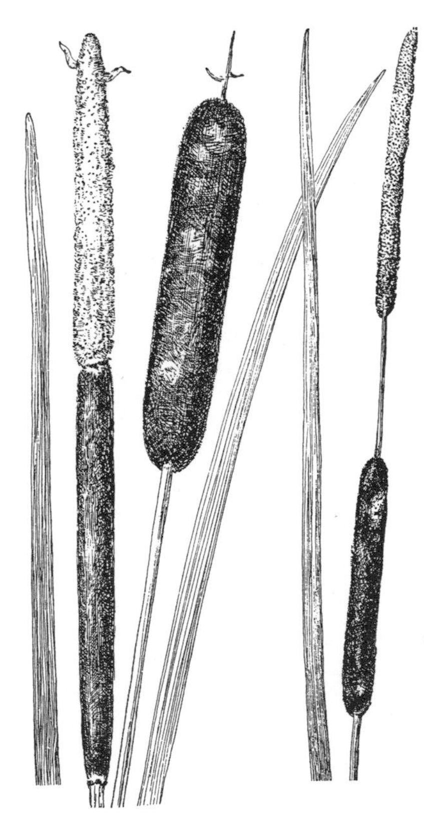

Cattails (*Typha latifolia* and *T. angustifolia*) substituted for Old World bulrushes (*Scirpus* spp.) as a source of material for woven chair seats.

chairs manufactured to supply London households after the Great Fire of 1666, and two hundred years later similar seats were woven on chairs sold to replenish American homesteads after the Civil War. Seat weaving was also done in rural areas with elm (*Ulmus* spp.) bark, which separates into strips when it is soaked in water and provided another tough material that craftsmen wove into seats.

Baskets had a vast number of uses and regional designs, with specific sizes and shapes that were used for gathering and harvesting foods, storage, measuring, winnowing, sieving, and cheese-making. Basketry also used some of the same materials as seat weaving, including imported rattan that arrived in coastal towns aboard whaling ships. Some rattan baskets were woven around hardwood staves inserted into oak bases, the style associated with the Nantucket basket-making industry that started as whaling began to wane in the 1840s. The handles of early Nantucket baskets were crafted from green wood that was easily bent. Wood also was a suitable weaving material, including the thin splints that were separated by pounding the wood of white ash (*Fraxinus americana*) and black ash (*F. nigra*) with a mallet. Ash wood was easily worked because it has large diameter vessel elements in its springwood, which allowed the summerwood to be separated easily into strong splints for both chair seats and baskets. Ash can absorb shock without breaking, which made it a suitable material for both tool handles and weaving, and craftsmen searched for straight trees with few knotholes to make long splints without anatomical defects. Splints were also made from the ring porous wood of white oaks (*Quercus alba*) and basket oak (*Q. michauxii*), materials that were used commonly for basketry in the South. Shakers used thin white oak splints to weave baskets of several sizes and types for various household uses, and these were sometimes lashed with black ash splints and fitted with hickory handles.

Other basketry materials used by settlers in the South included the flexible stems of honeysuckle (*Lonicera* spp.) and willows or osiers (*Salix* spp.), as well as rye straw, corn husks, pine needles, and rushes. German immigrants in Pennsylvania made baskets of rye straw collected from harvested grain crops (see chapter 2). The stems were cut, dried for several months, and gathered into bundles that were coiled and held together with thin oak splints to make round and oval baskets. Shallow rye baskets were made for bread dough, which was wrapped in a cloth and placed in the basket to rise; the thick rye coil kept the basket interior an even temperature for the yeast metabolism.

Beekeepers kept their colonies in domed skeps that were made in the same way, and the rye straw insulated the bee colonies. Skeps required that the entire hive be dismantled to collect the honey, and they were replaced during the mid-nineteenth century by wooden beehives, which allowed removal of just a portion of the honeycomb.

Straw is a broad term for the dried stems of various grasses, including cultivated grains. During the nineteenth century, the braiding of various types of straw was a cottage industry; women and girls made long lengths of plaited straw that were used in making hats and bonnets, and some families also braided imported palm leaves. Straw "India matting" covered many floors, but these were typically imported from China rather than made domestically. In *The American Woman's Home* (1869), Catharine Beecher and Harriet Beecher Stowe recommended straw matting as an alternative to expensive carpets, well-suited to the parlors and sitting rooms of thrifty New Englanders.

Brooms were necessary for household cleanliness, and the tradition of making brooms from twigs or strips of wood dates from ancient times. Colonists made withy brooms from the flexible twigs of hickory, birch, and other trees, tied onto wooden handles with flax cordage or corn husks. These brooms were imperfect tools that shed bits of plant tissue and fell apart with use; indeed, the phrase "flying off the handle" supposedly originated with the tendency of brooms to fall apart with vigorous use. A botanical advance in broom-making came with the introduction of broom corn, a variety of the native African grass sorghum (*Sorghum bicolor*) that was used for hay, forage, and the source of a sweet juice used in making syrup. Sorghum produces flowers in a branched inflorescence known as a panicle, which in the broom corn variety had long straight branches that could be harvested before or after the flowers developed. Benjamin Franklin introduced broom corn to America from Europe by saving a few grains from the straws of a whiskbroom and carrying them back to Philadelphia for planting. This was the beginning of the commercial broom-making industry in America, and by the late 1700s a few farmers were experimenting with broom corn as a crop. Levi Dickinson of Hadley, Massachusetts, cultivated broom corn, and by the 1790s he was crafting excellent brooms from the harvested panicles or "brush." During the nineteenth century, broom corn became an important local crop and the source of raw materials for several broom shops that operated in the Hadley area. Farmers were cautioned not to overcrowd the plants and to harvest the brush while it was still green and the grains were immature if they wanted the product with the best appearance. To make brooms that would tolerate more wear, farmers let the grains mature and the brush harden with age.

These early brooms were of the traditional round design, made by lashing broom corn to pegs inserted into the handle. During the 1820s, Shakers developed the prototype of the modern flat broom that was made by wiring the broom corn to the handle, clamping it in a vise, and stitching it into a flattened layer. By mid-century, factories produced brooms using machinery powered by hands and feet. Broom production on this scale required large quantities of broom corn, but J. H. Walden (1858) cautioned that broom corn "is a hard, exhausting crop for the soil," although its grain provided an excellent food for farm animals. He cautioned that broom corn was best grown by experienced farmers because of the difficulty and labor involved in threshing, but many farmers nevertheless grew successful crops to supply broom shops and factories. Broom corn grew particularly well in many western territories on land that was suited to corn production, and the plants thrived in particular on alluvial soils. Broom production moved west with settlers, and American brooms were exported to South America, Canada, and Europe; only later did England permit the import of American brooms crafted of broom corn, which replaced traditional twig brooms.

Barrels and Boxes

Households required a myriad of wooden containers and tools. Wood was used for boxes and related objects such as firkins, grain measures, and sieves. These included simple square boxes made with basswood or tulip poplar sides and pine bottoms, in which thin layers of basswood or tulip poplar were scored and bent at the corners and tacked to the softwood base. Square boxes were also made with bark sides and slotted bottoms, while round and oval boxes required that the wood be bent by steaming and shaped around a mold. For instance, round containers were bent from thin boards of basswood. Hardwoods lent themselves to bending into circular and oval shapes because the fibers in their xylem provided tensile strength and prevented splintering; the round boxes crafted by Shakers typically had sides molded from thin elm boards and tops crafted from oak.

The arduous process of hollowing a sycamore (*Platanus occidentalis*) log yielded the useful barrels known as gums that were used to make a variety of containers including wash tubs, pails, grain barrels, and even cisterns.

Sycamore wood does not release pigments or odors and resists splitting because its wood cells do not run perfectly parallel with the long axis of the tree trunk; instead, they slope, with the grain periodically reversing to make a pattern known as interlocked grain. These properties meant that sycamore wood was well suited for containers, and later it was also used for making the tubs of patented, hand-agitated washing machines, as well as trunks and stereoscopes.

Barrels crafted of separate boards (known as staves) were eventually preferred for storage, and these were made in two types: slack barrels for dry grains and flours and wet or tight barrels for cider, molasses, and other liquids. As winter work, colonial farmers rived the barrel staves from oak, which they assembled themselves or traded with local coopers for new barrels. Crude staves were also among the earliest goods exported from the colonies. Slack barrels were made from red oaks, including pin oak (*Quercus palustris*) and blackjack oak (*Q. marilandica*), while wet barrels used white oak (*Q. alba*). The mature wood of white oak is watertight, the result of a remarkable microscopic anatomical wood trait; its vessel elements contain tyloses, blockages that form once the wood no longer functions in conduction. Tyloses occur when nearby living cells protrude into the vessel elements and prevent the passage of water through the vessels. White oak was also used for making watertight buckets, and the image of the "old oaken bucket" became a nineteenth century symbol used to promote abstinence and the joy of drinking pure, fresh water. Author Samuel Woodworth described the poetic image of an oak water bucket in 1818, later popularized as the lyrics to "The Old Oaken Bucket," which delivered a subtle temperance message:

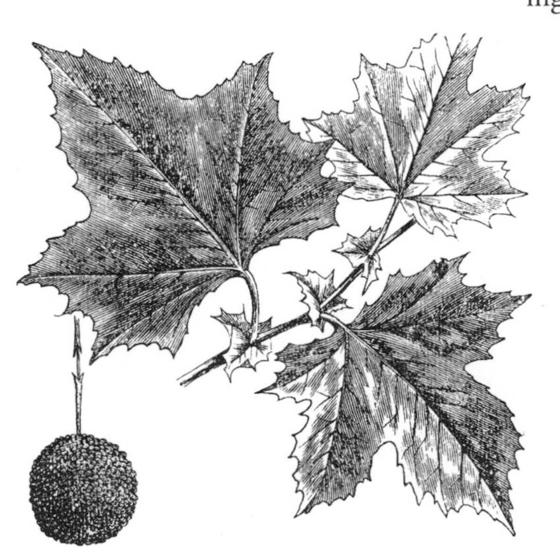

Logs of sycamore trees (*Platanus occidentalis*) were hollowed to use as wash tubs and grain barrels, and sycamore wood was also used for trunks, stereoscopes, and slack barrels.

The moss covered bucket I hailed as a treasure,
For often at noon, when returned from the field,
I found it the source of an exquisite pleasure,
The purest and sweetest that nature can yield.
How ardent I seized it, with hands that were glowing,
And quick to the white pebbled bottom it fell;
Then soon, with the emblem of truth overflowing,
And dripping with coolness, it rose from the well.
The old oaken bucket, the iron bound bucket,
The moss covered bucket that hung in the well.

The reference to the "iron bound" construction of the bucket refers to a late colonial era change in the coopers' craft, suggesting that Woodworth's bucket was not very old when he wrote this verse. Early buckets and barrels were held together at the top and bottom with narrow hoops cut from hickory or chestnut, which were overlapped and held together with notches. By 1800, coopers began to craft wet barrels and water buckets using iron hoops, an advantage because the staves were typically thicker than those made for the slack barrels used for dry storage. Iron hoops were installed while hot, so that as they cooled and contracted, the white oak staves were pulled together more tightly. In addition to red oaks, typical woods for slack barrels included sycamore, with hoops made of birch or cedar rather than iron.

Plant Fibers

Botanical textile fibers are various plant tissues that can be spun into threads for weaving, which makes them quite different from the tough, tensile fibers that occur in the wood of flowering plants. Plant fibers differ widely in their origins; they include cotton (*Gossypium* spp.), which produces long hairs (trichomes) on its seed coats, and flax (*Linum usitatissimum*), which has elongated support fibers associated with its food-conducting tissue (phloem). Textile fibers consist primarily of cellulose, the major substance in plant cell walls, and they function in strengthening and supporting the cell wall. Cotton hairs are spun into cotton threads, and flax fibers (known botanically as bast fibers) are spun into linen. Cellulose is a flexible polysaccharide comprising hundreds of linked simple sugars, and it makes strong threads when plant tissues are spun for weaving cotton and linen cloth.

The properties of plant fibers were discovered during ancient times, and their uses evolved with various cultures. Linen has been used for cloth-making for ten thousand years, beginning in the Mesopotamian region, Egypt, and Europe. Flax thrived in western Europe during the Middle Ages and traditional practices governed its cultivation and use; it was used for clothing and household textiles and was an essential crop for survival. Wild flax is a small perennial shrub, but the cultivated plants were harvested after about three months of growth; these were perfected by artificial selection into unbranched varieties with long bast fibers. Branched varieties of the same species were cultivated for the oil produced by their seeds, the linseed oil that is used in wood finishing, varnishes, and paint, and farm animals ate linseed cakes made of flax seeds with their oil pressed and removed.

Mature flax plants were harvested, dried in the sun, and then drawn through an iron comb that removed the seed capsules; the seeds were saved for the next crop, used as animal feed, or sold. Flax fibers required about twenty laborious steps, beginning with retting (from the Dutch word *roten*, which means "to rot") to remove the bast fibers from the stems. Traditionally this was done by spreading the plants on the ground, where they were dampened by the dew and where bacteria slowly decomposed the thin-walled cells of the stem and left behind the fibers, a process that was sometimes hastened by submerging the plants in a pond or river. The remaining stem tissues were dried, pounded in a flax-brake, and combed (known as hackling or hetcheling) several times. This prepared the long bast fibers for spinning, usually a task done by women or girls on a small, foot-operated spinning wheel, but the prepared linen fibers were a potential fire hazard that required careful storage away from fireplaces and candles. Spinning was messy, noisy work; many women spun their wool first

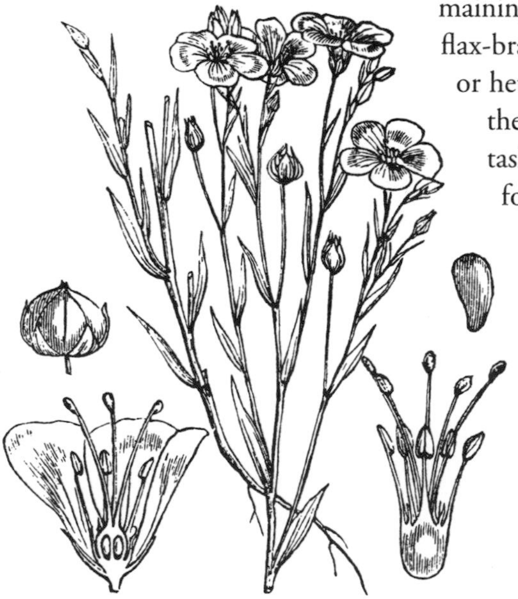

The bast fibers of flax (*Linum usitatissimum*) were harvested and laboriously prepared for spinning and weaving into linen or linsey-woolsey.

and moved on to flax in the late winter, often in an unheated room, although some flax wheels were in kitchens. The large looms were also typically set up in unheated rooms, and the weavers (often men who were professional weavers) converted the spun fibers into cloth for household linens, bedding, and light clothing. Another option was to weave flax with wool to make a heavier cloth such as linsey-woolsey, which was used for winter clothing.

Early colonists often relied on imported textiles, but by the mid-seventeenth century local leaders encouraged the production and spinning of flax in the American colonies. By 1750, Irish settlers who arrived in Londonderry, New Hampshire, were producing linen using methods they introduced from Ireland, and the area became known for its fine fabrics. Children were also taught to spin, and when imported linen was taxed in the years before the Revolution, Americans became self-sufficient in linen production. Like the liberty teas that replaced imported tea, homespun linen was encouraged as another symbol of American independence from England; Susannah Clark's 1773 verse about liberty teas mentioned "homespun garbs" (see chapter 5), and because of the labor involved in the home production of cloth, no scrap of fabric was ever wasted. Clothing was mended and cut down for smaller family members, and worn clothing was converted into bed quilts. Well-washed linen was often used to make infants' clothing, and old sheeting or table linen was cut into diapers that were known as "clouts."

Some homespun cloth was also bartered with shopkeepers for cotton muslin and calico, and shopkeepers advertised for the various textiles that they sought to purchase from industrious local weavers. Nevertheless, textile production in American homes was never a universal enterprise; some colonists purchased their fabric from professional weavers or from importers who supplied linen and cotton cloth, wool, and silk, including fancy goods such as velvets and block-and-plate–printed fabrics. These fabrics had appeal for many over simple homespun cloth, and by the beginning of the nineteenth century, in most areas only a minority of families grew flax and owned spinning wheels and looms. Mid-nineteenth century farm manuals still provided detailed instructions for flax culture and methods for retting, advising that the seeds required broadcasting on rich, friable soil and benefited from the application of calcium in the form of plaster of Paris.

Nevertheless, well into the nineteenth century, linen cloth of various types was essential for clothing and household uses. Linen was also used as the basis for samplers, instructive stitchery that displayed young girls' sewing skills and good training. Both plain and ornamental sewing were valued skills, but ornamental designs advertised a girl's training and qualities as a potential mate; *The Young Lady's Friend* (Farrar 1836) noted, "A woman who does not know how to sew is as deficient in her education as a man who cannot write" and described needlework as "a corrective for many of the little irritations of domestic life." Good needlework demanded self-discipline, and the accuracy of cross-stitched samplers depended on the regular weave of their linen backgrounds; in addition to alphabets, numerals, biblical passages, and moral verses, samplers often included stylized botanical designs.

Natural waxes embedded in the walls of flax bast fiber cells give a luster to linen that persists indefinitely, and samplers were often worked in dyed silk threads that also have a natural sheen. Although silk is not a plant fiber, silk production required a dependable source of mulberry leaves needed to nourish caterpillars of the silkworm moth (*Bombyx mori*). Silk is composed of the structural protein fibroin, which is synthesized in the silk glands of the caterpillars. In four days, a single silk gland cell produces a billion fibroin molecules, a process fueled by the food energy supplied by mulberry trees; each cocoon is a continuous silk filament that can be painstakingly unwound if the cocoon is soaked in warm water. The process is labor-intensive and depends on the success of culturing both the insects and the mulberry trees that they require.

Silk was first imported to Europe from Asia, but in 1622, King James I sent silkworm eggs and the Asian white mulberry (*Morus alba*) to Virginia, where planters were compelled by law to cultivate mulberry trees and were fined if they did not. His intention was to establish a colonial silk industry to supply England, but silk culture failed in Virginia and was supplanted by tobacco.

During the 1830s, American farmers looked to silk production as a means

Seventeenth century Virginia planters were compelled to cultivate white mulberry (*Morus alba*) as part of the royal intention to begin a colonial silk industry in America.

of self-sufficiency, and agricultural periodicals such as the *New England Farmer* published numerous articles on silk culture from 1835 to 1838. Some local success was achieved with silk production in Georgia, Connecticut, Pennsylvania, New York, and Massachusetts. Mechanical devices such as the Brooks's Patent Silk Spinning Machine, advertised in the *New England Farmer* in 1834, promised to unwind the silk from cocoons and convert it efficiently to sewing silk. Such machinery was necessary for silk culture on a massive scale; for instance, in 1832 a typical silk enterprise was established in Northampton, Massachusetts, where Samuel Whitmarsh planted 25 acres of mulberry trees with a cocoonery that accommodated two million caterpillars, spawning a silk thread industry that survived until the 1930s.

During the 1830s, oriental mulberry trees (*Morus multicaulis*) were introduced from China, and these trees were advertised as a replacement for the typical silkworm diet of white mulberry and native American mulberry (*M. rubra*) leaves and sometimes lettuce. Oriental mulberries produced two leaf crops each year, and they were anticipated to be so beneficial to the silk industry that speculation inflated the price of single twigs to five dollars. The plants were not reliably winter hardy and died back during the harsh winter of 1844; many entrepreneurs who had invested heavily in silkworms, mulberries, and silk production equipment suffered. Many cocooneries closed after the failure of oriental mulberry culture, but some American silk was still produced throughout the nineteenth century and used for manufacturing thread, cloth, handkerchiefs, and ribbons.

Like linen, hemp (*Cannabis sativa*) is also a bast fiber, and it was used for cordage, including the bedropes that were woven through a bed frame to hold

Hemp (*Cannabis sativa*) was a source of bast fibers used for various types of cordage, including bedropes and the rigging of ships.

the mattress. Hemp was a common farm crop that involved careful management. It is a dioecious species, with separate staminate (pollen-producing) and pistillate (ovule-producing) plants; following pollination, the staminate plants were uprooted from the field, and the pistillate plants were left to produce seeds. Once the crop was mature, the hemp plants were harvested, threshed, and retted, and the hemp seed was saved for the next growing season. Nineteenth century household manuals recommended using the seed to feed household canaries, which along with its common cultivation for cordage fibers explains the spread of naturalized hemp along roadsides and in disturbed sites.

Hemp also provided cordage for sailing ships, ropes that were manufactured in the long coastal buildings known as ropewalks. Clotheslines were made from imported Manila grass ropes, composed of fibers from the petioles (leaf stalks) of *Musa textilis*, a relative of bananas (*M.* ×*paradisiaca*). Faced with possible shortages of supplies during the Civil War, southerners experimented with other bast fibers, including Indian hemp (*Apocynum cannabinum*) and milkweed (*Asclepias syriaca*), in addition to the traditional use of leaf fibers from yucca (*Yucca filamentosa*) to suspend meat in smokehouses.

Cotton originated as several species of the genus *Gossypium*, members of the mallow family (Malvaceae) that are native to Central and South America, Asia, Africa, and Australia. The fruit is a capsule known as a boll, which splits into five valves and reveals a mass of white cotton fibers that are attached to the seed coats of the mature seeds. About ten seeds usually mature in each capsule, and each seed may have up to twenty thousand hairs on its surface, accounting for the dense mass of fibers inside each boll. The bolls mature two or three months after pollination, and slaves were employed in the labor-intensive process of harvesting the cotton on large plantations. Cotton was grown as a crop in Virginia by 1607 and became a plant of commerce in the southern colonies by the beginning of the 1700s. Even with the opening of American cot-

Flowers of cotton (*Gossypium* spp.) mature into capsules or bolls, which required a vast labor force for harvesting the mass of cotton fibers attached to the seeds.

ton mills in the 1780s, it was not until the nineteenth century that cotton cloth became commonplace, although some women did spin imported cotton into yarn for weaving.

With Eli Whitney's 1792 invention of the cotton gin and the increased demand for cotton exports to Europe, slavery increased during the years prior to the Civil War. Since cotton production depended on slave labor, reform-minded New Englanders used linen exclusively; Bronson Alcott, Charles Lane, and their followers in the utopian "Con-Sociate Family" at Fruitlands farm selected linen as the moral choice for their clothing. Flax cultivation exploited neither slave labor like the cotton industry nor animals like the shearing of wool, and in "Transcendental Wild Oats," Louisa May Alcott ironically noted their use of linen "till we learn to raise our own cotton." This was optimistic, because semitropical *Gossypium* was unlikely to thrive on a Massachusetts hilltop; cotton culture required a warm growing season of fifty to eighty days, careful preparation of the soil and seedbeds, and conscientious fertilizing with manure and wood ashes, lime, or powdered bones.

From agricultural descriptions, it appears that the cotton plants that were grown in the American South were primarily varieties of *Gossypium hirsutum*, the upland cotton that originated in Central America and Mexico. Southern coastal islands had naturalized populations of *G. barbadense*, originally from the Andes, which has particularly long, soft fibers; this is the variety known as sea island cotton that was used to produce excellent cloth. Each cotton fiber consists of the cellulose cell wall of a single seed coat hair, and these hairs vary in length. The longer ones are known as lints or staples and the shorter hairs are known as linters; the two types twist and spin together easily, and the finest thread and cloth were made from the longest cotton fibers, such as those of *G. barbadense*. Cotton fabrics were imported from Europe and India, but colonists also spun and wove the cotton that was grown in the southern colonies to make their own dimity, cheesecloth, flannel, and other domestic textiles. As with linen, homespun cotton fabrics were often bleached in the sun to lighten the natural fiber color.

During the second half of the eighteenth century, Boston children learned to spin in special schools. Parents were encouraged to send their children to work converting cotton and flax into textiles, and the majority of mill workers during the nineteenth century were women and children. Child labor made possible the inexpensive prices of commercial textiles; parents sometimes brought very young children to the mills to help tend the looms and increase their earnings, and mill owners and politicians alike promoted child labor in textile mills as a cure for the societal ills of poverty and indolence. With the opening of large commercial mills, yards of cotton sheeting, shirting, and bright calicos appeared in shops and were an inexpensive, convenient alternative to homespun cloth. Late nineteenth century commercial looms used shuttles crafted of dogwood (*Cornus floridus*), a close-grained wood that became smoother and more resistant to abrasion with use. Although the trees are also native to New England, dogwood was harvested from Virginia woodlands and shipped to mills in both New England and Britain, where it became a popular substitute for traditional boxwood shuttles. Many skilled home weavers were also recruited to work at their home looms weaving factory-spun fibers for the marketplace; some wove rugs from cotton and wool, while others continued to produce homespun fabric to augment their incomes.

In addition to cloth for clothing and household linens, cotton had some miscellaneous domestic uses. Cotton batting, or "cotton wool," was used to fill quilts and was saturated with oil to bandage burns. In *The House Book: A Manual of Domestic Economy* (1843), Eliza Leslie recommended that cotton batting be used in winter to seal drafty window sashes and cracks. Cotton rags were used in weaving rugs, and cotton and linen rags of various types were used in making paper. Textile fibers provided the cellulose that composed colonial paper, which because of the length of the cotton and flax fibers was stronger than later papers made of wood pulp. Cast-off clothing and household linens were a valuable commodity to colonial papermakers, an industry that began with William Rittenhouse, who established the first New World paper mill northwest of Philadelphia in the late 1680s. The current of the Monoshone Creek powered the mill, which converted cloth into pulp using a series of heavy wooden pestles; later mills used "hollanders," drums covered with iron knives, to process the rags to pulp. The slurry of plant fibers was hand-molded into sheets of paper on rectangular sieves of brass wires on an oak frame, and wires formed into letters or simple patterns impressed a watermark onto each sheet.

Papermaking depended on a ready supply of cotton and linen until after the Civil War, when makers began to use wood pulp. Rags were in continual short supply, and newspaper editors and advertisements appealed that rags be saved. Local legislation mandated that the dead be buried in woolen clothing to save valuable cotton and linen, while entrepreneurs experimented with alternative sources of botanical cellulose for papermaking, including straw, corn husks and stalks, various leaves, cattails, sedges, sugar cane, hemp, seaweed, and moss. Perhaps the most unlikely source of recycled fibers for papermaking were Egyptian mummies. Isaiah Deck, a chemist, suggested importing ancient mummies as a source of linen fibers that were in short supply in the mid-1800s; paper mills in New York and Maine soon followed his suggestion. *The Druggists' Circular and Chemical Gazette* (April 1873) described the 1866 visit of a New York businessman to Alexandria, where he purchased and exported to the United States "mummies from the catacombs" to be converted to pulp for papermaking. This was nothing new; according to the *Gazette*, a journal published in Syracuse, New York, had advertised ten years earlier that one of its issues was printed on "ancient habitants of the land of the Pharaohs."

Mummy paper was not ideal for printing because it contained the various oils and other botanical substances used in embalming and preservation, but the *Gazette* noted that a Boston clergyman shocked his congregation by announcing that the paper on which he wrote his sermon on the Israelites might contain the bodies of their Egyptian masters. In fact, some paper mills used only the linen wrapping, while others processed entire mummies. Egypt finally outlawed the sale and export of mummies in 1923, well after Victorian Americans experienced a period of intense fascination with Egyptian artifacts generally. Mummies toured the United States for public viewing, and under the guise of educational entertainment, some were unwrapped from their linen coverings before attentive audiences.

Dyestuffs and Tannins

Yarns and cloth dyed various colors were used for clothing and various household textiles, and dyes were known from botanical, animal, and mineral sources. Like imported spices, dyes were luxuries rather than necessities, and traditional dye plants were introduced from Europe to use in colonial households. Many of these had ancient histories, and more dye plants were discovered in the North American flora. Dye plants were purchased, collected from the wild, or cultivated in home gardens, and eventually some were grown and used commercially. Despite the brilliant natural colors of flowers and fruits, most of the useful pigments were found in vegetative parts of the plants such as leaves and roots; the red, blue, and violet anthocyanin pigments of flowers were not permanent when used to dye various fibers. Indeed, many of the pigments isolated from various plant parts are "fugitive" when they are used to dye various fibers, meaning that they fade with washings or exposure to sunlight.

The processes involved in preparing dyes and dyeing cloth and yarn required a considerable knowledge of practical chemistry to make the pigments adhere to the fibers and keep them from fading over time. Mordants were used to make plant pigments more permanent as dyes, a practice that was discovered by trial and error during ancient times, perhaps with the discovery that certain dyes were more permanent or had different hues if they were prepared in pots of different metals. Mordants were often metal ions, which in colonial homes were supplied coincidentally by the use of iron, tin, or copper pots, but some dyers also added rusty nails or metal-containing compounds such as alum (aluminum potassium sulfate). The metal ions supplied by mordants serve as a "bridge" between fibers and the dye molecules. For instance, wool (either spun or unspun) was usually treated with the mordant for several hours or overnight to prepare it for the dye pot. As the wool soaked in the mordant solution, the positively charged metal ions were attracted to the negative charges on the protein; once the wool was added to the dye bath, the metal ions were attracted to the negative charges on the dye molecules, which held the dye in place on the wool fibers. Mordants rendered a dye more colorfast and also increased the range of hues that a single dye plant could produce, which depended upon the metal that was used.

Dyes that required a mordant were known as adjective pigments, while those that required no mordant were substantive pigments; these designations varied depending on whether animal fibers (wool and silk) or plant fibers (flax and cotton) were being dyed. Another common technique was the use of acids or bases to increase the depth of color from plant pigments. This worked with plant dyes that are weak acids, which release the hydrogen ions ($H+$) that impart acidity and leave behind negatively charged dye molecules; the dye molecules then cling to weak positive charges on the surface of the fibers. With the addition of more acid such as vinegar to the dye bath, more $H+$ ions cling to the fibers, attracting more dye molecules; the addition of a base such as ammonia (typically in the form of urine) resulted in fewer dye molecules attaching to the fibers. The addition of urine or vinegar to the dye bath was a common practice and made it possible to obtain a range of colors from one plant.

Colonial housewives typically had a variety of dyes to use on yarn, cloth, and faded clothing that needed redyeing. Blue was probably the most desirable color and was first obtained from woad (*Isatis tinctoria*), a member of the mustard family (Cruciferae or Brassicaceae) that was cultivated in Europe since the time of the Romans. Woad leaves yield a blue pigment when they are crushed and were used by ancient Britons as war paint, but an involved procedure evolved for their use as a fabric dye. Woad leaves were gathered, crushed, and packed into balls for drying; later the leaves were moistened and "couched" (fermented) and then dried and fermented again before they were used as a dye. Indigotin, an insoluble pigment, was converted to soluble indigo white during the process, and textiles were soaked in this solution; the blue color did not appear until the fibers dried and the indigo white oxidized to form indigotin. Indigo (*Indigofera tinctoria* and related species), an Asian legume (family Leguminsoae or Fabaceae), also produces indigotin (in the form of the glycoside indican) and was known to ancient Europeans, but woad producers halted its import into woad-producing countries until the seventeenth century. They predicted correctly that indigo would replace woad as the most desirable blue dye.

Indigo was eventually imported from India and spread across Europe, and in the late 1730s, Eliza Lucas Pinckney cultivated a successful crop on the family plantation in coastal South Carolina using seed stock from the West Indies. Indigo replaced rice on many plantations, and slave labor was used in both indigo cultivation and processing. Massive vats held water and bundles of indigo plants for fermentation; the water was drained into other vats, where slaves used bamboo sticks to aerate the water until dye particles began to precipitate. The strained particles were indigotin, which was dried and pressed into cakes that became the most commonly used colonial

Indigo (*Indigofera tinctoria*) plants were fermented to release indigotin, which was mixed with urine or wood ashes to produce a soluble form of the most popular colonial blue dye.

dye. An indigo dye bath was prepared by pulverizing a cake and mixing the powder with urine or wood ashes; this mixture was often allowed to ferment for several days, while bacteria converted indigotin to its soluble form. Indigo pots were notoriously bad smelling, but the pigment was nevertheless used to dye vast amounts of cloth for clothing and household uses; indigo-dyed rags were also used in making the blue paper used as wrapping for cones and loaves of refined sugar (see chapter 6) and sewing notions.

Quantities of processed indigo were exported to England until after the Revolution, when cotton replaced indigo as a crop and indigo from India supplanted American exports. In the American colonies, itinerant peddlers sold various dyes, including indigo and madder (from the mature roots of *Rubia tinctorium*), native to Asia but cultivated commonly in Greece for its brilliant red pigment. Madder was imported to the colonies from France and Holland, but it was long associated with British redcoats whose uniforms were dyed with madder. Madder was used for domestic dyeing as well as the commercial block-printing pigments of calicos; its pigments (alizarine and purpurine) produced red, purple, and brown colors on cotton and were used along with pigments from buckthorn fruits (*Rhamnus* spp.), Brazilwood (*Caesalpinia echinata*), and indigo in making various colorful calico patterns. Other imported textile dyes included annatto, a bright yellow pigment derived from the red arils that cover the seeds of *Bixa orellana*, a shrub native to tropical America; annatto was also used for coloring butter and margarine, a butter substitute that was first marketed at the end of the nineteenth century. The heartwood of tropical American logwood trees (*Haemotoxylon campechianum*) contained the deep purple-red dye hematoxylin, which reacted with tannins in the wood to produce a reliable dark black color. Hematoxylin is still used as a stain in preparing slides for microscopy, but it was particularly valued by colonists as the first truly black dye, made by combining logwood with quercitrin, a deep yellow pigment from the bark of the North American black oak (*Quercus velutina*).

Butternut trees (*Juglans cinerea*) yielded gray, tan, and brown, depending on whether the bark or husks were used and perhaps the time of collection; the inner bark of butternut was used to dye

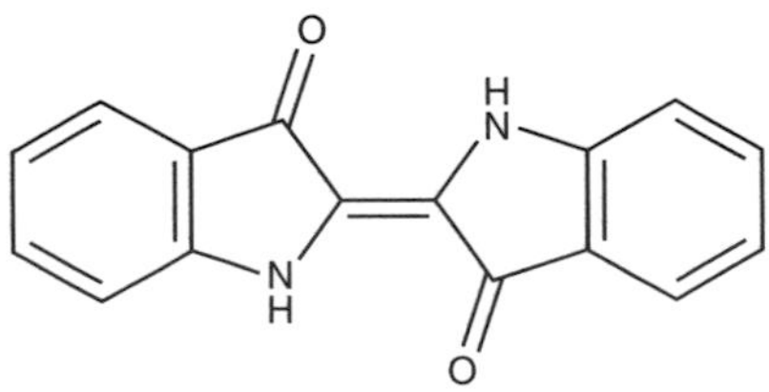

O
H
N
N
H
O

The chemical structure of indigotin, first known from indigo (*Indigofera tinctoria*) and related plants and made synthetically in 1878.

the shell jackets worn by Confederate soldiers during the Civil War. Black walnut (*J. nigra*) husks, twigs, bark, and leaves were used variously for green, brown, and black dyes, and the nuts or bark of various hickories (*Carya* spp.) and oaks (*Quercus* spp.) produced yellow, green, tan, and brown.

Lydia Maria Child (1844) recommended barberry bark (*Berberis* spp.), balm flowers (*Impatiens* spp.), peach leaves (*Prunus persica*), and onion skins (*Allium* spp.) as inexpensive alternatives to more costly dyes, noting that "alum might serve to set the color." Familiar garden herbs were also inexpensive sources of dyes, including yellow pigments from the leaves of tansy (*Tanacetum vulgare*), weld (*Reseda luteola*), safflower (*Carthamus tinctorius*), and lily of the valley (*Convallaria majalis*). Plant parts of one species sometimes yielded different colors; for instance, bedstraw (*Galium verum*) roots yielded a red pigment, while the flowering shoots were used to make a yellow dye. The pigments of some plants also varied with the season or the type of fiber that was being dyed. Marjoram shoots (*Origanum vulgare*) imparted a red-brown color to linen and a purple hue to wool.

Some dyes required frequent renewing, even if mordants were used. Combinations of pigments were sometimes made, either as mixed dyes or by overdyeing to get a desired color. For instance, indigo combined with the juice of goldenrod flowers (*Solidago* spp.) produced a pleasing green. Considerable experience and knowledge of practical chemistry were needed to get the anticipated or even acceptable results, although some domestic dyeing was no doubt by trial and error. Plant dyes were eventually largely replaced by inexpensive synthetic aniline dyes; William Henry Perkin accidentally discovered the first aniline dye in 1856, when he was attempting to synthesize the medicinal alkaloid quinine, which occurs naturally in feverbark trees (*Cinchona* spp.) and was used to treat malaria. He called his new dye mauve, and it replaced indigo as a dyestuff. Indigotin was also made

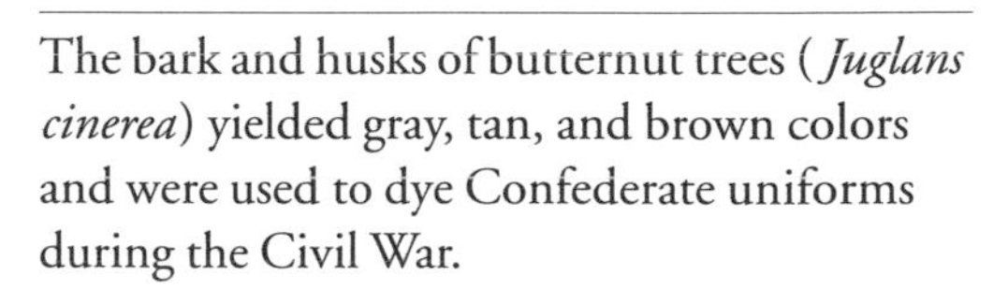

The bark and husks of butternut trees (*Juglans cinerea*) yielded gray, tan, and brown colors and were used to dye Confederate uniforms during the Civil War.

synthetically several years later in 1878 and was eventually less expensive to produce than the natural product.

In addition to plant dyes, the colonial fullers' industry used two European plants that were introduced by colonists arriving from England. Bouncing bet (*Saponaria officinalis*), sometimes known as soapwort or fullers' herb, is a member of the pink family (Caryophyllaceae) and contains saponins in its leaves and roots. Saponins are glycosides that foam when they are agitated in water; since the Middle Ages they were traditional substitutes for soap and were lathered in water to remove grease and dirt from wool or fleeces. Some wool cloth was woven, cleaned, and finished at home, but the industry perfected the use of mills to clean wool mechanically using absorbent clay (known as fullers' earth) and the beating motion of wooden thumpers or pestles. Fullers used the mature heads of teasel (*Dipsacus fullonum*, sometimes known as fullers' teasel) to raise the nap of the clean, dry cloth. Teasel bears numerous small flowers in a tight cluster, and each flower is surrounded by an involucre of spiny bracts. Several of these mature heads were attached to a wooden handle to make an implement for finishing woolen textiles. Woolen mills were located near running water, which may account for naturalized teasel plants in wool-producing areas; Henry David Thoreau (1986) recorded in journal notes that seed-containing teasel heads floated downstream from wool factories or washed up on the shores of the Concord River.

Some of the dyes derived from various tree barks are tannins, complex molecules that are water soluble and yellow or brown in color. These bitter, astringent compounds were among the earliest secondary compounds that evolved to deter herbivorous animals; dinosaurs were probably discouraged from eating cycads and other Mesozoic vegetation because of their high tannin content, and high doses of tannins interfere with digestion by binding to digestive enzymes and dietary proteins. Tannins are also produced in high concentrations in ferns, conifers, and some flowering plants (including tea and some of the tea substitutes discussed in chapter 5), and high tannin concentrations often occur in tree bark. The tanning process used the astringent properties of tannins to produce the leather needed for harnesses, saddles, breeches, aprons, shoes, boots, and carriage tops.

Some attempted tanning at home, but local tanners did much of the work in exchange for a portion of the leather that they produced from raw animal hides and skins using tree bark as a tannin source known as tanbark. Tanning was an essential colonial industry, which started when colonists arrived who knew of the use of English oaks in tanning leather and organized the first American tannery in Virginia in 1630. New England tanneries thrived because of the abundance of hemlock in northern forests, while southern tanneries used various oaks, such as chestnut oak (*Quercus montana*), and chestnut (*Castanea dentata*). Black oak (*Q. velutina*, also known as tanbark oak) and eastern hemlock (*Tsuga canadensis*) yielded particularly high amounts of tannin, which was harvested by felling trees and stripping them of their bark. Farmers often used the waste tanbark in place of manure to line their hot frames, where bacterial and fungal breakdown of the bark released sufficient heat to warm lettuce and other early crops during the winter months.

Black oak wood was then sometimes used in making slack barrels, but chestnut oak and hemlock trees were often left to decompose on the forest floor after they were stripped of thick bark on the lower trunk. The harvested bark was ground into particles the size of wheat grains using a tanbark mill, in which an ox dragged a corrugated stone wheel in a circular trough. The first such mill was constructed in 1633 by Peter Minuit in New Netherland (now New York), and Francois Michaux observed tanners and tanbark mills in every town during his travels through New Carolina with his father, botanist Andre Michaux in the late 1700s. Skins or hides were layered in large vats with one-inch layers of bark, covered with water, and kept submerged in the tannin solution for up to a year. Tannins prevented decay by binding with proteins in the tissues while also darkening and dehydrating the skins and hides; tanners also beat the newly formed leather with clubs to increase its density and toughness even more.

Writing ink also depended on tannins. Galls form in trees and shrubs in response to attack by parasites (often wasps), and oak galls have a particularly high tannin concentration that was used in making early commercial and homemade inks. Tannins were extracted in water from the crushed galls, which in turn reacted with iron

Black oak (*Quercus velutina*) was also known as tanbark oak, the source of tannins for leather preparation and wood for slack barrels.

sulfate, the common mordant also known as copperas or vitriol; this reaction produced ferrotannate, a dark pigment that appeared as the ink was exposed to oxygen. Dyestuffs such as logwood, Brazilwood, and indigo were sometimes added to impart color until the reaction occurred between the tannin and the iron sulfate. In addition, ink mixtures and watercolor paints also often contained gum arabic (a thickening agent from *Acacia* trees that was discussed in chapter 6), which helped the pigment to flow more smoothly from pen to paper and prevented it from seeping too deeply into the cellulose fibers of the paper. Simple inks were concocted at home from pulverized galls, acorn tops, or bark soaked in water with rusting iron nails; iron was required to form the ferrotannate, and the result was a dark suspension that was concentrated to use as ink or to dye cloth.

Household Miscellany

Housekeeping depended on a variety of plant materials that were gathered from both the wild and cultivated fields, such as straw and corn husks used to stuff mattresses. Horsetails known as scouring rushes (*Equisetum hyemale* and some related spp.) have a rough texture, which made them useful for scouring pots and pewter. Their epidermis is covered with silica in the form of opal, a mineral that strengthens the stems and probably protects them from herbivores and desiccation. Like ferns and clubmosses, horsetails were among the earliest vascular land plants, with evolutionary origins in the Devonian period; both the extinct, arboreal forms and living species (often known as "living fossils") have ribbed stems with distinct joints and whorls of extremely reduced leaves, all protected by the characteristic silica layer. Soap-making depended on lye leached from wood ashes, and lampblack, paints, and turpentine were household products made from pine wood. Floors were often covered with floorcloths or "oylcloths" that were made of flax or cotton canvas; these were covered with coats of oil-based paint and then decorated with stencils or block prints or painted a solid color, such as the green floorcloths that Thomas Jefferson requested for Monticello. These were first made in England at the beginning of the eighteenth century, and they were imported to America or made by local craftsmen until their popularity waned with Frederick Walton's linoleum patent in 1860. Of course, linoleum is also a botanical product, composed of cork, rosin, and linseed oil pressed on to a canvas or burlap backing that was probably made of flax or hemp.

Within their natural ranges, woods were selected for tool handles and kitchen implements based on their structure, density, and the ease with which they were worked. Woods with high specific gravity were used when strength and resistance to shock were important. Ironwood, or hop hornbeam, (*Ostrya virginiana*) has exceedingly tough, dense wood, which was used as levers or ax handles, and tool handles were also made from ash, hickory, sassafras, holly, and other hardwoods that were reasonably dense and resisted splintering or shearing. Kitchen tools were fashioned out of local woods that were easily carved, resisted splintering, and did not release tastes or odors to foods. Buckeye (*Aesculus octandra*) and beech (*Fagus grandifolia*) were easily carved into kitchen utensils, and beech and yellow birch (*Betula lutea*) made clothespins that did not split. Bowls carved of maple burls did not warp and shrink because the wood grain in burls occur in swirls rather than in a straight grain.

Nineteenth century agriculturalists wisely encouraged the management of woodlots as if they were valuable cash crops, perhaps because vast areas of the Northeast had been deforested by farming. Although in some areas the remnants of virgin forests remained, woodlots were typically populated by second- or third-growth trees. Nevertheless, rural families benefited by preserving and managing forested land that provided them with household raw materials, including wood for buildings, furnishings, barrels, fences, fuel, edible nuts, and even bark. Anecdotal reports record that in lean times, such as the extremely cold summer of 1816, some farmers collected and milled birch bark to augment their meager grain stores.

Even small households required enormous amounts of firewood for cooking and baking, heating water for bathing and laundry, and heating rooms. All woods provided combustible fuel for fireplaces and stoves, but hardwoods were desirable because softwoods such as white pine burned too quickly and resulted in flammable creosote deposits in chimneys. Pine was used as kindling, but hardwoods such as hickory, white oak, maple, ash, and birch provided long-burning, steady fires that were suitable for baking and roasting. Hickory was also used in smoking meats as a method of food preservation (see chapter 5), and chestnut and other oak species were also burned but yielded more moderate heat. Wood required splitting and drying, so woodcutting was done several months in advance, usually in the winter because snow allowed the transport of the cut logs by sled. Woodlots were often clear-cut rather than selectively harvested; the tree roots contained enough stored food that they soon sprouted into saplings that grew rapidly, and within fourteen years firewood could again be cut from the site. In addition to heating with fireplaces and eventually stoves, houses were insulated from winter drafts by layers of cut vegetation banked against the foundation and first floor, a method practiced in the mid-eighteenth century and perhaps earlier. Layers of branches, leaves, sawdust, bark, corn stalks, and other botanical refuse prevented the contents of root cellars from freezing and drafts from penetrating the house interior. These piles of vegetation may also have generated some warmth as bacteria and fungi decomposed the dead organic material and their metabolic reactions released energy, in the same way that compost and sawdust piles heat as decomposition occurs.

Although pine was a poor fireplace fuel and its knots were undesirable in construction timbers, pine knots (usually those of pitch pine, *Pinus rigida*) did provide a thrifty source of illumination that replaced candles in many homes. Puritan colonists probably learned about the use of pine for illumination from Native Americans, and Governor John Winthrop described candlewood to the Royal Society in 1662. Knots contain a concentrated quantity of combustible pitch, which provided sufficient bright light for reading and writing when the knots were burned in a corner of the fireplace or hearth. Small pine sticks were also sometimes burned like candles, and the widespread use of candlewood continued well into the nineteenth century by those who could not afford candles. In lean times, pioneer families sometimes burned rags or wicks saturated in lard, and the stems of rushes were stripped to their piths, dipped in tallow, and used as a candle substitute.

Candles were usually made by dipping cotton wicks in melted beeswax or tallow that was boiled and skimmed; sometimes saltpeter (potassium or sodium nitrate), alum, or camphor (distilled from the bark of *Cinnamomum camphora*) were added to make the wicks burn more slowly. Wicks were also made from linen or hemp fibers or spun from the hairs on common milkweed seeds (*Asclepias syriaca*). Candles as well as soap were made of bayberry wax when it was locally available, harvested by boiling the drupes (stone fruits) of coastal bayberry shrubs (*Myrica pensylvanica*) and related species. These aromatic candles were less likely to melt in hot weather than those made of tallow. In *Home Life in Colonial Days*, Alice Morse Earle (1898) noted that in 1687 the Long Island village of Brookhaven instituted a fine of fifteen shillings on anyone harvesting the valuable fruit before mid-September in an effort to conserve local bayberry populations.

Self-sufficiency required ingenuity and the ability to adapt Old World domestic traditions to New World environments and raw materials; colonists first used American species such as white oak that resembled familiar European trees, and they soon learned about other useful plants through experimentation and observing Native American practices. Trees provided much of the raw material for survival, and colonial exports to Europe were also often from American forests, including timber, cabinet woods, pine pitch, turpentine, and charcoal. Wood was converted into charcoal by piling several cords into a charcoal pit, covering the logs with layers of saplings and leaves, and slowly burning the pile for several days to obtain bushels of charcoal. Charcoal was used as a domestic fuel and also in smelting the iron used in making pots, hinges, and tools. Other traded commodities included hemp and tobacco, but many settlers were largely self-sufficient. Using a variety of Old and New World plants, many settlers built their houses and furnishings, produced their own cloth, raised their own grain and other crops, and cared for the ill. Americans who staked claims on the American prairies constructed modest buildings from sod "bricks" cut from soil layers held together with dense plant roots; in the absence of trees, they took advantage of the heavy root growth in prairie habitats and used this in place of wood for dwelling construction.

By the early nineteenth century, most forests in southern New England had been cleared for farming, fuel, or timber, and wood became a costly commodity. Some coastal cities and towns received shiploads of wood from Maine and the Maritime Provinces of Canada, but by 1833, anthracite coal was mined in Pennsylvania and shipped north for use in cooking and heating. Many converted to coal because of the scarcity of wood, which had once been the most extensive New World resource. Of course, coal is also a botanical product, mined from vast mineral deposits remaining from fossilized, extinct plants that inhabited Carboniferous swamps from 345 to 280 million years ago.

CHAPTER NINE

Domestic Landscapes

The earliest homesteads were rigorously practical. Settlers built sturdy houses and often situated outbuildings to the northwest, which shielded their dwellings from harsh winter winds. In this little enclave, the dooryard between the house and the outbuildings was the site of such utilitarian activities as wood-chopping, laundry, and soap-making. The gardens were often located behind the house, where they were planted and enlarged as needed, while fruit trees were planted in the dooryard, often with a southern exposure that hastened fruit production. Gardens supplied a variety of fruits, vegetables, and herbs, and were laid out in blocks or raised beds surrounded by hardened dirt or gravel paths. While some families may have valued a tidy landscape, dooryards and kitchen gardens evolved as places for work, with little regard for aesthetic appeal. Floriculture was not yet an avocation, but some women may have cultivated small flower beds near their homes' entrances; these were probably planted with weedy species that flowered profusely, required little care, and were perennials or self-sowing annuals. Grass prevented mud and erosion in the dooryard, and it was scythed occasionally to keep it trimmed.

Eighteenth century homesteads included orchards and trees planted for shade, decoration, and as markers for roadsides and boundary lines. Nurserymen supplied seeds and saplings to those with formal gardens and a particular interest in horticulture, but rural families often planted trees from seed they collected or transplanted seedlings from wooded areas. Trees were often planted to mark significant events, such as births or marriages, and homesteads were typically flanked by a pair of shade trees that were planted when the house was built. Parlor gardens were often cultivated on the south side of the house in beds with a variety of ornamental plants including roses, lilacs, perennials, and annuals. The composition of these gardens varied with the women who kept them, depending on their taste and needs, and the layout varied as well; some had a central path or were bordered by paths, and they may have included a plot of grass edged by perennials or annuals. An appreciable overlap existed between plants cultivated for pleasure and those cultivated as medicinal and domestic herbs, and various useful plants were also cultivated for the appeal of their flowers, foliage, and scent. Seeds and cuttings of herbaceous plants were shared, and when spontaneous mutations appeared, such as doubled flowers, new petal colors, or variegated foliage, these were often preserved, propagated, and shared. Eighteenth century gardens evolved from Old World tradition and New World resources, resulting in European and American plants cultivated in a common ground. The agricultural and horticultural literature encouraged the planting of new varieties and recently discovered species, and landowners with the means had access to a variety of plant materials. Ironically, perhaps, more diversity was often found in the gardens of merchants and gentlemen who lived in towns than in the dooryards and kitchen gardens of rural farming families.

The extent of botanical diversity in eighteenth century gardens reflected curiosity and ingenuity, particularly among those who pursued horticulture as an avocation. The Annapolis garden planted during the 1760s by William Paca, a signer of the Declaration of Independence who became governor of Maryland, reflected his interest in plants of all sorts. Paca was an urban resident and certainly not dependent on agriculture for subsistence, but his two acres included intensively cultivated kitchen and fruit gardens, as well as formal beds and parterres, European shrubbery, seasonal displays, a landscaped pond, and various North American wildflowers and trees. Admirers described it as the finest garden in Annapolis, and although it clearly was not typical of most eighteenth century homesteads, his garden did illustrate what was available for cultivation in a relatively small property, including plants native to North America. In contrast, John Custus of Williamsburg was "a great admirer of all the tribe striped gilded and variegated plants" and cultivated a garden in town of imported specimen plants during the early eighteenth century. Custus corresponded with Peter Collinson, an influential Quaker gardener who shipped rare species and variegated cultivars of European plants from England in return for the native North American species sent to him by Custus; each gardener desired what was rare in his country. The Custus garden became a living museum of plants that were difficult to obtain in colonial America; horse chestnut trees, variegated conifers, white currants, and double tulips survived the voyage and flourished in Williamsburg. At the same time,

Collinson cultivated North American umbrella magnolias, fringe trees, and wildflowers that Custus readily obtained and shipped to London.

The English landscape gardening movement influenced some eighteenth century American gardens, which were designed to include naturalistic plantings of trees and shrubs. Although their gardens were never as large as the English prototypes, George Washington and Thomas Jefferson planted mixed groves of ornamental tree species at Mount Vernon and Monticello, but landscape gardening on this scale was far beyond the reach of most colonists. Woody plants were nevertheless the primary interest among eighteenth century American seedsmen and nurserymen, and in 1790 the Prince Nursery on Long Island listed almost exclusively trees and shrubs; the only herbaceous plants noted specifically were tuberoses (*Polianthes tuberosa*), popular tender Mexican perennials with intensely fragrant flowers. A similar situation existed with eighteenth century Boston nurserymen, who rarely listed herbaceous plants in their advertising, with only occasional newspaper mentions of ornamental flowers such as marigolds or nasturtiums.

Traditional practices continued in many gardens, such as John Hancock's Beacon Hill garden in which the beds were edged in the English manner with a shrubby variety of European boxwood (*Buxus sempervirens* var. *suffruticosa*); however, the evergreen shrubs often suffered in climate extremes, and many abandoned the use of boxwood hedges to frame garden beds. Despite the frustrations and labor of trimmed boxwood, Alice Morse Earle (1898) lamented, "And may I not enter here a plea for the preservation of the box-edgings of our old garden borders? I know that they are almost obsolete—have been winter-killed and sunburned." English yew (*Taxus baccata*) and privet (*Ligustrum vulgare*) proved more resistant and practical, and American gardeners adopted them for garden hedges. While colonists learned by trial and error which European species would reliably survive North American winters, horticultural interest existed among botanists and seedsmen in discovering new North American species to adopt and cultivate as ornamental plants.

Bernard McMahon arrived from Ireland in 1796 and established himself as a seedsman and nurseryman in Philadelphia. He issued a broadside in 1802 or 1803 that listed seven hundred twenty species and cultivars, and his later catalog (1804) included mostly plants native to North America. These are now regarded as the first American seed catalogs and reflected his interest in propagating both European and American ornamental plants. McMahon also influenced landscape design through *The American Gardener's Calendar*, which he first published in 1806, with several later editions; it was a horticultural manual arranged by months, and his essay on "Ornamental Designs and Plantings" argued the case for naturalistic garden design. McMahon recommended that Americans abandon formal pathways and geometric beds and instead cultivate trees, shrubs, and flowers in natural thickets connected by winding paths. Although he relied heavily on English horticultural practices and described European species that were difficult to obtain, McMahon also encouraged American gardeners to cultivate native plants and not to shun American wildflowers because they are common or indigenous. He asked, "What can be more beautiful than our Lobelias, Asclepias, Orchis, and Asters?" and went on to recommend American species of several other genera, including *Monarda*, *Ipomoea*, *Phlox*, *Spigelia*, *Cassia*, and *Rudbeckia* for garden cultivation. He argued reasonably, "In Europe plants are not rejected because they are indigenous; on the contrary, they are cultivated with due care; and yet here, we cultivate many foreign trifles, and neglect the profusion of beauties so bountifully bestowed upon us by the hand of nature."

With the return of the Lewis and Clark expedition in 1806, McMahon assumed responsibility for the seeds and plants collected by Meriwether Lewis, which Jefferson insisted be cultivated in a systematic way because they were the property of the expedition. Through the efforts of McMahon and others, several species collected by Lewis soon became familiar in domestic landscapes—perhaps most significantly, Osage orange (*Maclura pomifera*), which was native to the Mississippi valley and named for the Osage tribe of native Americans in Missouri. These trees were the evolutionary remnants of an ancient species, which had a wider geographic distribution between periods of glaciation over the last several million years. Their remarkable green fruits resemble pungent, rough-surfaced softballs, and evolutionary biologists now hypothesize that they were once perhaps dispersed by the mastodons and other extinct large Cenozoic mammals. Seeds of Osage orange were collected by Lewis in the St. Louis garden of Pierre Choteau, and McMahon grew the shrubs to maturity. With pruning, they formed dense, thorny shrubs that produced additional branches in the form of root suckers. These proved useful as hedge plants and were known commonly as "bodark" hedges, derived from *bois d'arc*, French explorers' name for the wood that was often used to make bows. By the mid-nineteenth century, thousands of miles of Osage orange hedges were planted annually, although many were later replaced by barbed wire. Other familiar garden plants collected by the Lewis and Clark expedition included prairie flax (*Linum lewisii*), Missouri currant (*Ribes odoratum*), and snowberry bush (*Symphoricarpos albus*), and several of these American species were propagated and sold by McMahon in his nursery. Ironically, some American species fared better in the English climate than in their native habitats; clarkia (*Clarkia pulchella*) thrived in English gardens as a favorite annual, and British breeders perfected numerous red, pink, purple, and white cultivars with single and double flowers.

In *A Treatise on the Theory and Practice of Landscape Gardening* (1849), Andrew Jackson Downing followed the lessons taught by McMahon and described naturalistic landscapes that provided growing areas for trees, flowering shrubs, lawns, and a variety of annuals and perennials. Mindful of the smaller scale of American homesteads, he interpreted the style of gardening that was first articulated in England by Sir Humphrey Repton. At the end of the eighteenth century, Repton had encouraged gardeners to cultivate a wide variety of plants, including flower gardens near houses and informal tree groves in the background. Like Repton, Downing's notions of a cultivated landscape required a vast diversity of woody and herbaceous plants, including woody groves comprising "spirited" trees (those with the pointed form found in many conifers) interplanted with hardwoods. He specified shrub thickets to screen utilitarian areas such as vegetable gardens and the drying lawn for laundry. When he suggested that flower beds be surrounded by lawn rather than gravel and that the beds contain single varieties of low annual plants, Downing proposed the modern concept of flower beds. Some flowers were also planted outside the house windows, reminiscent of early parlor gardens. Downing later complained in *Rural Essays* (1854) that American gardens included few native herbaceous and woody species, and that these plants were overlooked in favor of the "verbenas and fuchsias of South America" and "trees of Europe and Northern Asia." Of course, this was the stylish mode for those with estates of at least an acre, but New Englanders often resisted change. Their mid-nineteenth century gardens were often cultivated as they were in the eighteenth century, perhaps with the addition of new ornamental plants as they became available and familiar. Some gardeners sustained a scientific curiosity about American plants, epitomized by Jefferson, but they were certainly not typical; most gardeners were probably more intrigued by tropical herbaceous plants with long flowering seasons. In contrast, Jefferson cultivated many species that others would have likely ignored or rejected as being common, insufficiently decorative, or otherwise unworthy of garden space. One such plant was twinleaf (*Jeffersonia diphylla*), a modest American wildflower with deeply cut leaves, named for Jefferson by American botanist Benjamin Barton in 1792.

The vast nineteenth century literature of household manuals and agricultural texts and periodicals encouraged the ideal of a well-maintained domestic landscape. The title page of *The American Woman's Home* (Beecher and Stowe 1869) illustrated a landscaped homestead above a vignette of three generations of women at work in their garden; gardening was part of a woman's sphere of influence and responsibility, and an untidy yard and dwelling implied slovenly domestic standards and intemperance. Gardening and landscape design were part of the ambitious scheme of nineteenth century social reform, promoted by authors such as Lewis Allen, who described in *Rural Architecture* (1852) the attractive flower borders "so appropriately the care of the good matron of the household and her comely daughters" and argued for "a decided plan of arrangement for all the plantations and grounds . . . nothing left to accident, chance, or after-thought." In *The Flower Garden* (1851), Joseph Breck noted the moral benefits of gardening and believed that women in particular were drawn to flowers, and in *Plain and Pleasant Talks about Fruits, Flowers, and Farming* (1859), Henry Ward Beecher recommended gardening as a good activity for women. Lydia Pinkham (or more likely, her staff correspondents) encouraged women to spend time outdoors, walking, riding, and digging with a trowel. Despite reformers' efforts, not all dooryards were the idealized "front yard garden which seemed to us a very paradise in childhood" described by Sarah Orne Jewett in *Country By-ways* (1881), and sometimes local pressure was applied to a neighbor to improve a slovenly yard.

During the second half of the nineteenth century, gardeners evolved a taste for botanical extremes; plants with atypical growth forms, large or variegated leaves, peculiar flowers, or other morphological idiosyncrasies were sought and acquired for Victorian landscapes. Tropical and semitropical herbaceous plants were valued for their bright flower pigments and continuous flowering, and these were planted in elaborate beds cut into lawns. Mass production popularized garden furniture—such as cast iron benches with designs resembling rustic branches, ferns, or grapevines. Gardening became a leisure pursuit for those with the time and means to devote to an avocation, while the yards of nineteenth century working-class families remained devoid of ornamental features, with the exception of possible flower beds. Shade trees and shrubs were planted to mark property lines, but woody plants were otherwise not particularly important except as elements of decoration. In *The Art of Beautifying Suburban Home Grounds of Small Extent* (1870), Frank J. Scott advocated trimming and training woody plants into "hedges, screens, verdant arches, arbors, dwarfed trees, and all sorts of topiary work." He illustrated hemlocks (*Tsuga* spp.) judiciously pruned and trained to create an architectural tower over a garden gate, and he suggested closely planted sassafras (*Sassafras albidum*) for cultivating shady bowers.

Later in the century some of the ornamental features found in estate gardens began to appear in more modest properties, such as circular beds filled with exotic annuals and woody cultivars with weeping or dwarf growth forms. Those who could not afford cast iron had the option of purchasing or making inexpensive garden furniture crafted from branches and twigs, part of the picturesque garden style that included verdant rockeries and ferneries. Trailing growth was appealing in rockeries and ferneries, and plants with pendant fronds or twining stems were frequently cultivated for their growth form. Kenilworth ivy (*Cymbalaria muralis*) was introduced from western Europe and cultivated for its unique property of colonizing rock walls; the stems (pedicels) that bear the fruits elongate and "plant" the seeds in crevices, and the trailing stems form adventitious roots at their nodes. The kidney-shaped leaves and violet flowers draped on rock faces were valued by gardeners seeking a natural effect.

The horticultural tension between botanical diversity and the verdant simplicity of a well-tended lawn heightened with the invention and availability of lawnmowers. Some preferred a vast variety of cultivars and plantings, while others viewed turf as a "smooth natural carpet over the swelling outline of the smiling earth," as described by Downing in *Rural Essays*. Without a doubt, botanical diversity in household gardens suffered with the planting of lawns; land once occupied by various domestic cultivars was used instead for the monoculture of grass.

Trees, Shrubs, and Woody Vines

The first colonists discovered a forested continent with a vast diversity of woody plants, but the earliest colonial gardens emphasized herbaceous plants with edible, medicinal, and household uses. Perhaps it seemed unnecessary to cultivate trees in a wooded land, or perhaps the struggle to clear sites for buildings and farms rendered tree planting and cultivation an unlikely endeavor. Native trees were natural resources for food, medicine, timber, and dye, and by the eighteenth century, trees were also valued as elements of the landscape. The appointed outdoor meeting places for American patriots were known as "liberty trees," a tradition that began with a grove of elms in Boston; sheltering trees were used in other towns as places to gather and discuss rebellion against England. Trees also appeared on Revolution-era flags and American shilling currency, particularly designs of stylized conifers that symbolized the northeastern environment and colonies.

Woody plants were selected for cultivation based on their growth form, evergreen or deciduous habit, flowers, and hardiness; scent from essential oils was also a consideration, given the number of early cultivated trees and shrubs with pungent foliage or fragrant flowers. Cultivated woody plants often originated from the native trees and shrubs that already grew on the land, selected and transplanted as seedlings or grown from seed. Conifers were selected for their growth forms and evergreen foliage and included several species native to North America. The Annapolis garden of Paca included white cedar (*Chamaecyparis thyoides*), trees that inhabit swamps in nature but that also thrive when planted in moist but well-drained soils. In 1790, the Prince Nursery listed several other American conifers for sale, including Weymouth's pine (probably white pine, *Pinus strobus*), pitch pine (probably *P. rigida*), and Virginia pine (perhaps *P. pungens*). The Prince catalog listed only vernacular names, so in some cases plant identities are uncertain; Latin binomial names were first used consistently by Linnaeus in *Species Plantarum* (1753) and were not yet used by all plantsmen at the end of the century (see chapter 1). "Hemlock spruce fir" was probably the eastern hemlock (*Tsuga canadensis*), and the balm of Gilead listed as an evergreen is the northern balsam fir (*Abies balsamea*). It should not be confused with the American balm of Gilead poplar (*Populus balsamifera* or the cultivated hybrid *P. candicans*), a familiar problem when one vernacular name can be used for two or more quite different plants.

The interest in conifers had medicinal underpinnings, which seemed to increase during the nineteenth century with concern about tuberculosis; conifer-scented air was considered good for consumptives, and many agreed with Lydia Child (1837) that "it was even deemed healthy to have these trees near dwellings." *The Druggists' Circular and Chemical Gazette* (March 1876) noted that "the air of pine forests always appears grateful to the lungs, and has been considered wholesome, although of its absolute curative influence there is little evidence. . . . The idea of pine forests exercising a balmy influence on the lungs is a very ancient one." Nineteenth century pioneers migrating west carried seeds and seedlings of white pines and various eastern hardwoods to their new prairie homes, but they often discovered that trees from the East did not survive in new habitats. Western soils often lacked the specific soil fungi necessary for tree growth—mycorrhizal fungi that grow symbiotically with plant roots and promote nutrient uptake. Ectomycorrhizal fungi (ECM) include various mushrooms, and their underground filaments (hyphae) invade the roots of certain genera such as oaks (*Quercus* spp.) and hickories (*Carya* spp.). Maples (*Acer* spp.) and ash (*Fraxinus* spp.) require vesicular-arbuscular mycorrhizal fungi (VAM), small soil fungi that can colonize a variety of plants. These two distinct types of mycorrhizal fungi also explain the difference in the vegetation that colonizes abandoned homesteads and farms; early stages of succession involve trees with less specific VAM, while the last trees to establish themselves successfully are those with specific relationships with ECM.

Jefferson made the acquisition of new garden specimens a family effort, and he particularly favored trees. His correspondence revealed his ongoing passion for new plants, and his family gathered for their installation; trees were planted for both appeal and practical purposes such as shade and fencing. His plantings at Monticello far exceeded the scale of most domestic gardens, but his zeal reflected his passion for the American landscape that others replicated on a smaller scale. In a 1793 letter to his daughter Martha Jefferson Randolph, he described the landscape around their home:

> I never knew before the full value of trees. My house is entirely embosomed in high plane trees, with good grass below, and under them I breakfast, dine, write, read, and receive my company. What would I not give that the trees planted nearest round the house at Monticello were full grown.

The plane trees (*Platanus occidentalis*) Jefferson knew were sycamores, North American relatives of oriental plane trees (*P. orientalis*), and they were among the first North American trees that were cultivated in England. London plane trees (*P.* ×*acerifolia*) resulted from hybridizing the American sycamore with oriental plane trees. Sycamores and plane trees were also known as buttonwoods (because of the spherical heads of mature fruits) and as "Virginia maples" because of the superficial similarity in their leaves; indeed, the epithet *acerifolia* means "maple leaf."

Jefferson grew other trees that interested him, including related American and Old World species such as the European horse chestnut (*Aesculus hippocastanum*) and the closely related American buckeyes (*A. octandra* and *A. pavia*). He also grew American and European elms (*Ulmus americana* and *U. procera*). American elms were widely planted as street trees during the nineteenth century, and they were valued for their stress tolerance and vase-shaped silhouette. Henry David Thoreau (quoted by Campenella 2001) described in his journals the silhouettes of American elms as markers of "rural and domestic life," a way to identify villages at a distance; he particularly valued their golden fall color, which he imagined as "yellow canopies or parasols over our heads and houses . . . making the village all one and compact—an *ulmarium* which is at the same time a nursery of men!" Most elms were later killed by Dutch elm disease (caused by the fungi *Ophiostoma ulmi* and *O. nova-ulmi*), which was introduced by infected elm logs imported for wood veneers and first observed in 1930. Jefferson's pride in North American trees was also reflected in the cultivation of several other native species including pawpaw (*Asimina triloba*), persimmon (*Diospyros virginiana*), and hazelnut (*Corylus americana*). He also planted Asian curiosities such as ginkgo (*Ginkgo biloba*), mimosa (*Albizia julibrissin*), and golden rain tree (*Koelreuteria paniculata*), as well as the European gold chain tree (*Laburnum anagryoides*). Exotic and American trees alike were prized, but as Minister to France from 1784 to 1789, Jefferson took pride in distributing American tree seeds to his European friends.

Finding new specimens to plant was an ongoing enterprise. Jefferson obtained some plants from McMahon and other botanical friends, but most of the woody plants that he cultivated at Monticello were available from commercial nurserymen. Both John Bartram's botanical garden near Philadelphia (first planted in 1725) and the Prince Nursery on Long Island began as private gardens and later entered the nursery trade; Bartram and Prince both published detailed lists of the plants that they supplied, which reveal that native and non-native woody plants were commingled in eighteenth century gardens. Like Jefferson, these nurserymen chose plants to propagate based on their attributes and little regard for whether the species originated in North America, Europe, or Asia. Cultivated and nursery-propagated North American trees included several with bright fall leaf pigments such as sassafras (*Sassafras albidum*) and sugar and red maples (*Acer saccharum* and *A. rubrum*). Magnolias and other genera with showy, insect-pollinated flowers also attracted attention because most temperate trees are wind-pollinated and have small, unisexual flowers. Cultivated North American magnolias included cucumber tree (*Magnolia acuminata*, named for the appearance of its flowers once the petals dropped and the fruit matured) and umbrella magnolia (*M. tripetala*), characterized by its terminal clusters of large leaves. Southern magnolias (*M. grandiflora*, sometimes known as Florida laurel) and swamp magnolias (*M. glauca*, sometimes known as rose laurel) were also brought into cultivation, and all four species were available from Bartram's botanical garden. Andrew Jackson planted a southern magnolia near the White House (visible in the engraving on a modern twenty dollar bill) in memory of his wife, Rachel Jackson, who died in 1828, and magnolias also framed the dooryards of many southern homes.

Tulip trees (*Liriodendron tulipifera*) bear showy yellow, orange, and green flowers and are also in the magnolia family (Magnoliaceae), although some regarded the trees as poplars because of their broadly columnar shape. Jefferson considered tulip trees among the most remarkable American trees and grew them at Monticello. He included their winged fruits in a box of seeds and dried fruits that he sent to Lafayette's aunt, Madame de Tesse, in France; he also included several American oaks in his shipment, noting that "of the oaks I have selected the *alba* [*Quercus alba*, white oak], because it is the finest of the whole family, it is the only tree which disputes for pre-eminence with the *Liriodendron*." Jefferson also sent the southern catalpa (*Catalpa bignonioides*), which botanical collector Mark Catesby had discovered growing in the mountainous, inland regions of North Carolina. Catesby noted in *The Natural History of Carolina, Florida, and the Bahama Islands* (1731):

> This tree was unknown to the inhabited parts of Carolina until I brought the seeds from the remote parts of the country. And though the inhabitants are little curious in gardening, yet the uncommon beauty of this tree had induced them to propagate it and 'tis become an ornament to many of their gardens and probably will be the same to ours in England.

Catalpas bear showy white flowers in upright panicles, and their fruit is a slender capsule that contains winged seeds. Southern catalpa leaves are a food source for sphinx moth caterpillars, which some settlers used as fishing bait in place of earthworms, and some catalpas were planted for this purpose.

The northern catalpa (*Catalpa specioides*) is more winter hardy and also increased its range through human activity; its wood resists rot, and railroad companies planted groves of northern catalpas near western tracks as a ready supply of wood for replacing ties. By the end of the nineteenth century, nurserymen also offered for sale Japanese hybrid catalpa trees, advertised as a remarkably fast-growing, hardy species suitable for both "ornamental plantations" and timber; these were probably hybrids of the native *C. bignonioides* and *C. ovata*, a Chinese catalpa that was frequently cultivated in Japan.

Silverbells (*Halesia carolina*, *H. monticola*, and *H. diptera*) are southeastern trees that attracted the attention of many explorers and travelers with their numerous, pendulous white flowers. These were probably the "snow-drop trees" listed in the Prince catalog, but the same common name was also used for North American fringe trees (*Chionanthus virginicus*). Fringe trees are flowering shrubs in the olive family (Oleaceae), and their narrow petals suggest white fringe; they were first introduced by Bartram and soon cultivated by Paca and others during the late eighteenth century. Lilacs (*Syringa vulgaris* and *S. persica*) and forsythia (*Forsythia*) are also shrubby olive relatives, but neither originated in North America. Lilacs arrived in England from Persia during the sixteenth century; both purple and white varieties were valued for their fragrance and hardiness and commonly planted in dooryards and parlor gardens. *Forsythia* was a later arrival, a nineteenth century introduction from his China travels by Robert Fortune, who called it "golden bells." Upright and weeping species (*F. viridissima* and *F. suspensa*) were popular in Victorian gardens and yielded several hybrids and cultivars. In other woody species, weeping growth sometimes appeared as a spontaneous mutation, which appealed to nineteenth century gardeners who favored trees and shrubs with distinctive growth forms. Cultivars with this atypical growth included weeping ash (*Fraxinus americana* var. *pendula*) and weeping beech (*Fagus sylvatica* var. *pendula*). Weeping willows (*Salix babylonica*) that had normally pendulous

Southern catalpa (*Catalpa bignonioides*) was native to the mountains of North Carolina but was less winter hardy than the northern catalpa (*C. specioides*).

branches were also popular; the species is from northern China, but Linnaeus believed that it originated in Babylon and named it accordingly. Poplars are deciduous trees that were valued for their columnar growth, and both American and European species were cultivated in American gardens. Washington transplanted wild poplars to line the serpentine roads at Mount Vernon, and the Prince Nursery sold Lombardy or Italian poplars, male plants selected from wild populations of *Populus nigra* (now known as *P. nigra* var. *italica*) found growing along the Po River in Lombardy. These were prized for their exceedingly narrow (fastigiate) growth form and were so widely planted that many mid-nineteenth century gardeners and landscapers tired of their unique silhouette and fragility. The roots of Lombardy poplar damaged pavement and invaded sewer pipes, and as early as 1826, American elms were planted to replace the Lombardy poplars that lined some Boston streets.

Other colonial garden shrubs and small trees included American hollies. English holly (*Ilex aquifolium*) was common in the southern colonies (Jefferson called it "Virginian holly"), but the plants did not survive northern winters. Washington also favored hollies and experimented with growing native hollies from seed and transplanting the seedlings, and cabinetmakers sometimes used close-grained holly wood for inlaid designs because it resisted splintering. Gardeners also began to cultivate the more hardy American holly (*I. opaca*) and the deciduous holly known as black alder or winterberry (*I. verticillata*), which lacks the characteristic evergreen foliage but is valued for its bright red berries. The epidermis of American holly leaves lack the shiny surface cuticle of English holly, but the leaves and berries of the native trees were a reasonable facsimile of traditional Christmas holly; by the end of the nineteenth century, early conservationists were concerned about the destruction of native holly trees for Christmas decorations. Only the pistillate trees were plundered because they produce the characteristic berries, but over-collection likely affected the potential of some wild populations to reproduce from seed.

Mountain laurel (*Kalmia latifolia*) is also an evergreen shrub, a genus named by Linnaeus to honor Pehr Kalm, who botanized extensively in North America (see chapter 1). Despite their superficially similar evergreen leaves, *Kalmia* shrubs are not true laurels like bay laurel (*Laurus nobilis* in the laurel family, Lauraceae) but instead are members of the heath family (Ericaceae). The old common name "calico bush" referred to the markings on the pale pink petals that resembled a printed design, and the shrubs were useful only for their visual appeal in gardens and hedges. Mountain laurel and related species are potentially lethal if their leaves are ingested, including lambkill (*K. angustifolia*, which was also known commonly as sheepkill and calfkill); these small shrubs were cleared from pastures and meadows where animals grazed, and honey was potentially poisonous if bees gathered large amounts of nectar from *Kalmia* flowers, the result of the toxic compounds andromedotoxin and arbutin. City-dwellers were not troubled by mountain laurel toxicity, and the nineteenth century horticulturist Joseph Breck contended "There is no shrub, foreign or native, that will exceed this in splendor, when well grown." Related American rhododendrons were also cultivated, listed in the Prince Nursery catalog broadly as "Rhododendrun." These may have included the impressive rose bay rhododendron (*Rhododendron maximum*), which interested plant collectors who botanized in the Great Smoky Mountains; this species was probably the "Great Mountain Kalmia or Rhododendron" that merchant Collinson grew in his London garden as early as 1736. Azaleas were also cultivated, and all are now classified as American rhododendrons; these included rhodora (*R. canadense*), the flame azalea (*R. calendulaceum*), and the Pinxter azalea (*R. nudiflorum*). Many variants appeared among the cultivated plants, and in *The Art of Beautifying Suburban Home Grounds of Small Extent* (1870), Frank J. Scott described Pinxter azaleas as the "Upright American Honeysuckle. Flowers pink, white, striped, red, and purple. Superseded in cultivation by new varieties." Nineteenth century gardeners continued to cultivate mountain laurel as well, and Scott noted that varieties were available with flowers ranging from white to red.

The Prince Nursery listed several aromatic plants in 1790, including spicebush or Benjamin bush (*Lindera benzoin*), a North American species in the laurel family (Lauraceae) that was used to flavor rum (see chapter 6). The common name *spicebush* is also used for *Calycanthus floridus*, which was known to eighteenth century nurserymen as Carolina allspice, strawberry bush, or sweet shrub, and was listed by the Prince Nursery as "the sweet scented shrub from Carolina"; Jefferson knew it as "bubby flower shrub." In both of these spicebushes, the crushed leaves release essential oils with a strong, spicy scent. Sweet gale (*Myrica gale*) and wax myrtle or candleberry (*M. cerifera*) were also cultivated as aromatic shrubs similar to bayberry (*M. pensylvanica*); their fruits were used in some colonial households as a source of wax for candle and soap manufacture, and the shrubs were considered aromatic curiosities. In *The History and Present State of Virginia* (1705), Robert Beverly described in detail,

> a Berry, of which they make a hard brittle Wax, of a curious green Colour, which by refining becomes almost transparent. Of this they make candles, which are never greasie to the Touch, nor melt with lieing in the hottest weather . . . it yields a pleasant Fragrancy to all who are in the Room.

Bartram grew *Myrica* species and advertised them for sale; he sent their seeds to his English patrons each year to be grown as pleasant garden plants. Another strongly aromatic garden shrub was dittany, or gas plant (*Dictamnus alba*), a species in the citrus family (Rutaceae) that was an ingredient in several compounded medicines and long valued as an expectorant and diuretic. As a garden plant, it had a flammable property; on hot, still days, the essential oils evaporated into the atmosphere around the shrubs, creating a gas that ignited with a spark. These oils also caused photodermatitis, similar to the irritation and blisters caused by rue (*Ruta graveolens*) if the foliage were handled in bright sunlight. Despite these hazards, dittany was cultivated for its showy white or pale pink flowers and lemon-scented leaves, as well as for its medicinal uses. The late eighteenth century garden notes kept by Jean Skipwith at Prestwould Plantation in Virginia recorded that white dittany was valued as an ornamental shrub and easily raised from seeds sown in the fall.

During the nineteenth century, roses were cultivated both as ornamental plants and for practical uses, including those cultivated from stock selected from native American shrubs as well as garden roses with ancient origins in Europe and Asia. Eighteenth century European roses included the French rose (*Rosa gallica*), cabbage rose (*R. centifolia*), cinnamon rose (*R. cinnamomea*), damask rose (*R. damascena*), and white rose (*R.* ×*alba*). Jefferson grew these, as well as the Cherokee rose (*R. laevigata*), which he obtained as seed from Governor John Milledge of Georgia; despite

The atmosphere surrounding a gas plant or dittany shrub (*Dictamnus alba*) became saturated with flammable essential oils on still summer days.

its name, the species originated in China, arrived in the American colonies from England, and is now commonly naturalized in the South. Rose water was made by steeping or distilling rose petals in water, usually from the flowers of *R. gallica*, which was known as the apothecary rose because of its astringent properties. Shakers also grew cabbage, damask, and white roses for use in treating hemorrhages and bowel complaints, and rose petals were combined with the pith of sassafras branches in an infusion for treating eye inflammation. Rose hips (see chapter 4) were also valuable sources of vitamin C and were used traditionally in the preparation of jellies, jams, teas, and syrups, and some were fermented into wine and vinegar; each is the mature hypanthium of a rose flower that contains small, single-seeded fruits (achenes).

Some North American roses were also cultivated, such as *Rosa palustris* and *R. virginiana*, and by the early 1800s rose breeders experimented in earnest with selection and hybridization to develop new ornamental cultivars with desirable traits such as deep petal colors, hardiness, disease resistance, and longer flowering times. Everblooming China roses (*R. chinensis* and its hybrids) were introduced in the early 1800s, including the green rose (*R. chinensis* var. *viridiflora*) that was planted as a curiosity in many southern gardens and is still grown as an heirloom cultivar. The hardy Japanese species *R. rugosa* was favored in coastal areas because it tolerated salt, and *R. multiflora* was propagated for root stocks and hedges, but by the end of the nineteenth century both had naturalized in many areas and became notoriously invasive species. In the *American Flower Garden Directory* (1839), Robert Buist recommended two hundred fifty rose varieties from at least two thousand rose cultivars, and many small gardens included several different roses. Single plants were often planted for display in front yards and in lawns, but nineteenth century gardeners often cultivated larger rose gardens behind their houses because of the short flowering season. By mid-century, hybrids began to replace species roses, and rose cultivars proliferated rampantly; in 1867, tea roses

Rosa gallica, the apothecary rose, was grown both for the ornamental and astringent properties of its petals, which were distilled to make rose water.

(*R. odorata*) from China were hybridized into a vast number of hybrid tea roses, and Victorian gardeners could select from roses with petals of every color but blue. Roses were sometimes planted in elaborate rosariums, rose-filled parterres in complex, symmetrical designs; climbing roses were trained over trellises or archways, while shrub roses were cultivated in the beds.

Ornamental shrubs were also selected and propagated from the genera *Hydrangea* and *Viburnum*, which in some species have flowers in their large inflorescences that are showy but sterile; these lack functional stamens and pistils but have enlarged, petallike sepals. In *How Plants Grow* (1858), botanist Asa Gray called these "neutral flowers" that were "good for nothing except for show," which explains the sole function of sterile flowers in attracting pollinators to the inflorescences. Gray noted that even in wild hydrangeas and viburnums, the inflorescences bear some sterile flowers, a trait increased by selection in horticultural varieties. North American hydrangeas included *H. quercifolia*, with flowers in long panicles and numerous sterile flowers that transform from white to purple.

These were cultivated as hardy native shrubs, but the Asian *Hydrangea macrophylla* attracted great popular attention when it was first sent from China by Sir Joseph Banks in 1790. This species became the familiar pink and blue garden hydrangeas with massive inflorescences of sterile flowers, in which the sepal pigmentation in some varieties depends on soil pH, based on the chemistry of the anthocyanin pigment known as delphinidin 3-monoglucoside. Anthocyanins are water soluble and occur in the large central vacuoles of the sepal cells, where their chemistry is affected by the uptake of soil aluminum that may be dissolved in the groundwater. Acidic soils (with a pH of 5.5 or lower) have free aluminum ions that bind with the anthocyanin molecules and a copigment to produce blue colors, while more basic soils (pH of 6.0 or higher) yield pink pigments because the aluminum ions are bound to other soil chemicals.

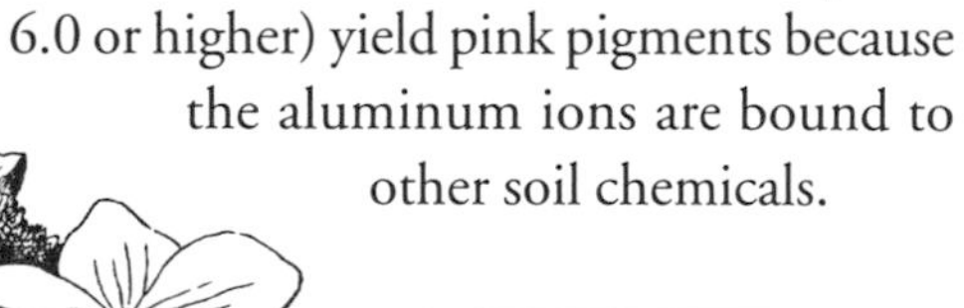

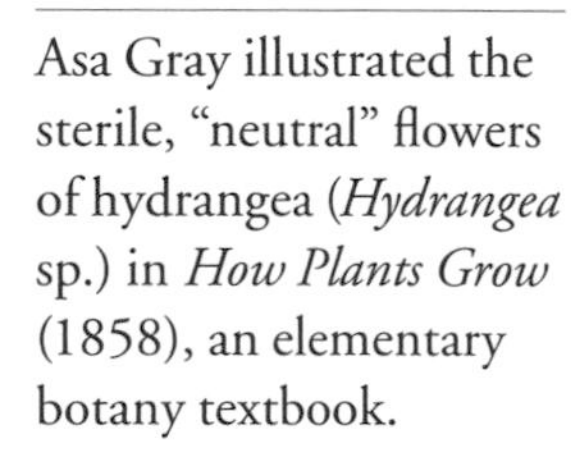

Asa Gray illustrated the sterile, "neutral" flowers of hydrangea (*Hydrangea* sp.) in *How Plants Grow* (1858), an elementary botany textbook.

In contrast, the sterile flowers of viburnum shrubs are typically white or off-white, which explains the common name of snowball bush for the large inflorescences of sterile flowers in cultivated varieties of *Viburnum opulus*, the European Guelder rose or cranberry bush. Varieties with fertile flowers produce edible red fruits, but these are unrelated to the cranberries (*Vaccinium* spp.) discussed in chapter 4. Snowball bushes (*V. opulus* var. *sterile*) were described by seedsman Joseph Breck as having "Large white bunches of flowers like those of the Hydrangea." Shrubs of *V. macrocephalum* imported from China were even more showy, described by landscape gardener Downing as "a new and splendid species. . . . M. Van Houtte describes it as found growing in the gardens about Chusan, China." Both cultivars were developed as garden plants through selection that eliminated fertile flowers from their inflorescences; viburnum became known in nineteenth century gardens as familiar ornamental shrubs, and knowledgeable gardeners learned to propagate the shrubs asexually by layering or root cuttings.

Woody vines cultivated in eighteenth century gardens included several North American plants, including "poison oak" (*Toxicodendron radicans*, now known commonly as poison ivy), with its shiny compound leaves and vigorous growth. It was cultivated at Monticello, even though severe dermatitis results from the resinous oil that contains urushiol, a mixture of related allergenic compounds. In *Wild Flowers and Where They Grow* (1882), Amanda Harris warned amateur botanists about its "acrid oil" and characteristic "hairy woody stem, three leaflets, and greenish white berries," but she agreed that it was a handsome vine "which rivals the five-leaved Virginia Creeper [*Parthenocissus quinquefolia*] in being one of the two most truly decorative vines in New England." Smoke tree (*Cotinus coggygria*) was a commonly cultivated European and Asian member of the same family (Anacardiaceae), which sometimes caused similar allergic dermatitis.

Jefferson also cultivated red-flowered trumpet vine (*Campsis radicans*) and coral honeysuckle (*Lonicera sempervirens*), which presaged the popularity of many red-flowered, hummingbird-pollinated cultivars during the nineteenth century. European flowers lack bright red petal pigments because hummingbirds do not occur in Europe. Pure red flowers were a novelty that attracted attention as they were discovered and introduced, and a remarkable number were brought into cultivation, including flowering vines, bedding plants, and North American wildflowers. In *Travels in North America* (1770), Pehr Kalm observed several red-flowering plants, including phlox (*Phlox glaberrima* and

P. maculata), cardinal flower (*Lobelia cardinalis*), and bee balm (*Monarda didyma*), and he noted "we must own the land is undoubtedly adorned with the finest red imaginable."

In 1790, the Prince Nursery listed North American wisteria (*Wisteria frutescens*), a stout woody vine with pinnately compound leaves known historically as the Carolina kidney bean tree. It was grown as a climbing plant over arbors and walls, but it had smaller inflorescences than the Asian wisterias introduced during the nineteenth century, including *W. sinensis* from China and *W. floribunda* from Japan. These grew to immense proportions on arbors and porches, with intertwining, woody stems and large numbers of purple or white flowers in pendulous clusters.

Both native and exotic species of clematis were also cultivated, beginning with North American virgin's bower (*Clematis virginiana*), which produces white flowers and fruits in a cluster with long, plumose styles. Leather flower (*C. viorna*) had purplish flowers and was probably also brought into cultivation in its native range from Pennsylvania to Mississippi, but by the nineteenth century, gardeners favored the Japanese *C. florida* for its large flowers and the European *C. flammula* for its sweet fragrance. Nineteenth century gardeners cultivated balloon vine or heartseed (*Cardiospermum halicacabum*), a widespread tropical vine that produces membranous capsules containing black seeds, each with a small, white heart-shaped mark. Native passionflowers (*Passiflora incarnata*) were also cultivated, both for their curious flowers and their large edible berries, known as maypops or maracocks. Linnaeus named the genus for its legendary, coincidental resemblance to aspects of Christ's crucifixion; for instance, the fringed corona (filaments growing from the base of the petals) in many *Passiflora* species suggested the crown of thorns, and the five stamens were thought to symbolize the five wounds.

The Chinese wisteria (*Wisteria sinensis*) produced massive woody vines with pendulous inflorescences of purple flowers.

Herbaceous Garden Plants

Early flower gardens originated with herbs that were cultivated for practical uses, and by the eighteenth century many domestic plants cultivated for enjoyment were also perennial herbs with dual uses as medicines, dyestuffs, or household plants. Heartsease (*Viola tricolor*), carnations and pinks (*Dianthus* spp.), lady's mantle (*Alchemilla vulgare*), foxglove (*Digitalis* spp.), hollyhock (*Althaea rosea*), and lily of the valley (*Convallaria majalis*) had visual appeal as well as useful chemistry, as did many mints and aromatic composites. Opium poppies (*Papaver somniferum*) were cultivated for preparing narcotic syrups, but variations in petal colors and shapes made them attractive as garden flowers; feathered, carnation, curled, and fringed opium poppy varieties were cultivated in borders. Garden plants were cultivated from seeds or vegetatively by means of divisions or cuttings, and desirable spontaneous mutations such as new petal colors or markings, double petals, or leaf variegations were preserved and propagated. Foliage mutations such as variegated leaves were also selected as horticultural oddities for herbaceous gardens. Variegated leaves lack chlorophyll in some parts of their leaf blade, a disadvantage in nature because they are often less hardy and vigorous than fully pigmented plants of the same species, but the variegated form of European goutweed (*Aegopodium podagraria*) proved to be an exception. Goutweed reproduces both by seeds and stolons (horizontal stems) that easily sprout new shoots if they are severed. Its old medicinal use was for treating gout, but the variegated form was probably cultivated as an ornamental plant rather than medicinal herb; it escaped cultivation as a herbaceous perennial and easily naturalized in the Northeast during the 1860s and 1870s. In goutweed and some other species, mutations produced variegations that bred true when the plants were started from seed, making it possible to grow from seed the variegated cultivars of grasses and tuberoses advertised in nineteenth century seed catalogs. In other cases, the variegated pattern began as a localized mutation (known as a chimera) at the growing point (meristem) of the stem, and the cells descending from that part of the meristem inherited the distinctive pattern. These mutations appeared as sports on normally green plants, which were propagated asexually to produce new horticultural varieties.

Herbaceous plants were commonly known by their vernacular names, which often resulted in multiple names being applied to one plant or one vernacular name being used for two or more distinct species. Vernacular flower names had folk origins and were often descriptive in some way; for instance, the common name *gillyflower* was used for July-flowering plants and was applied to two sorts of European plants grown in colonial gardens, clove gillyflowers (*Dianthus caryophyllus* and *D. deltoides*) and stock gillyflowers (*Matthiola incana*). Clove gillyflowers are wild carnations (family Caryophyllaceae) that, according to Gerard's *Herball* (1633), originated in Poland, while stock gillyflowers are unrelated coastal mustards (family Cruciferae or Brassicaceae) with a similar clovelike scent. Clove gillyflowers were used as a folk substitute for the scent and flavor of imported cloves (see chapter 6), while stock gillyflowers were valued for their diverse petal colors and similar scent; seventeenth century housewives also prepared a flavored cordial of clove gillyflowers, which explains its early common cultivation as the herb known as sops-in-wine. According to legend, European villages raised just one variety of stock gillyflowers to keep the seed lines and petal colors distinct, and these European varieties were eventually introduced to American gardens. The traditional gardening practice was to water the seeds with brine and mix salt in with the soil, a practice that was thought to improve the growth of coastal plants. By the end of the nineteen century, American seed catalogs offered both dwarf and double-flowered stock varieties, including mixtures that produced flowers in twenty different shades of white, red, pink, yellow, and violet.

Native North American plants were also brought into cultivation for their color and fragrance, as well as for medicinal uses (see chapter 7). Bee balm or Oswego tea (*Monarda didyma*) is a North American mint that attracted attention for its red, hummingbird-pollinated flowers, and it was grown both as a garden plant and for medicinal teas. Gardeners also cultivated and exchanged several other native species that they transplanted or grew from seed, including columbine (*Aquilegia canadensis*), dog tooth violet (*Erythronium americanum*), bloodroot (*Sanguinaria canadensis*), liverleaf or "liverwort" (*Hepatica acutiloba* and *H. americana*), shooting star or American cowslip (*Dodecatheon meadia*), trillium (*Trillium erectum* and *T. sessile*), Solomon's seal (*Polygonatum biflorum*), and black-eyed Susan (*Rudbeckia hirta*). Several of these had Native American medicinal uses or were interpreted according to the Doctrine of Signatures to be useful in treating specific ailments, including liverleaf and bloodroot (see chapter 7).

During the nineteenth century, more native herbaceous plants also became known as suitable for gardens, including butterfly weed (*Asclepias tuberosa*), described by Buist in *The American Flower Garden Directory* (1839) as the

"Finest genus of all. Orange, dry situations"; it was also a well-known medicinal plant (see chapter 7). Annual and perennial phloxes were cultivated from wild populations across North America, including the annual *Phlox drummondii* that Thomas Drummond first collected in Texas and sent to England in 1835. The plants were cultivated in English borders, and garden varieties were exported to North America later in the century. Wild *P. drummondii* plants were up to eighteen inches in height and had rose-red petals, and some horticultural writers noted that frustrated gardeners used hairpins to stake the long stems to the ground. By 1888, *Burpee's Farm Annual* advertised a dwarf variety "excellent for edging beds of the Grandiflora varieties." These large-flowered phlox had flowers in white, purple, scarlet, rose, blue, and salmon colors, some distinguished by a white "eye" or stripes.

Several other phlox were cultivated during the eighteenth century, including *Phlox divaricata*, often known as wild sweet William (not to be confused with *Dianthus barbatus*, known as sweet William) that Jean Skipwith cultivated in Virginia after the Revolutionary War. She kept notes and records of the native and exotic perennials that grew in her Virginia garden, including "flowering roots," the old term for species that are planted and propagated asexually by bulbs. Hyacinths (*Hyacinthus* spp.), grape hyacinths (*Muscari botryoides*), narcissus (*Narcissus* spp.), and various tulips (*Tulipa* spp.) all produce bulbs, overwintering structures that consist of a short stem surrounded by overlapping, fleshy leaf bases known as bulb scales. Bulbs store carbohydrates that fuel spring growth, and they also reproduce asexually by producing small bulblets in the axils of their bulb scales, two valuable survival strategies. "Tulipomania" had seized Holland as a speculative horticultural fad in the 1630s, although it is unlikely that Pilgrims arriving in Plimoth from Leiden carried *Tulipa* bulbs on the *Mayflower*. Dormant bulbs were easily imported from Europe, and colonial newspapers advertised their sale in port cities. For gardeners who could afford them, imported bulbs provided new horticultural stock, and several of them were highly selected florists' flowers. Bulbs were often expensive rarities; for instance, Jefferson's garden notes record that although he grew various flowering bulbs at Monticello, they were not cultivated in large numbers. In 1807 he planted twenty tulips and twenty-seven hyacinths, a modest display by modern standards, and in 1812 he received small shipments such as "two Roots Parrot Tulips . . . red, green, and yellow mixed" from McMahon. Imported bulbs were not easily replaced, and offsets were nurtured to maturity and flowering. Jefferson also grew some native bulb-producing plants such as the North American Turk's cap lily (*Lilium superbum*) and "Columbia lily" (probably *Fritillaria pudica* collected during the Lewis and Clark expedition), and perhaps the European white lily (*L. candidum*), which later became a Victorian favorite.

Like seeds and cuttings, bulb-producing plants were often shared among gardeners, and both old varieties and new introductions were included in perennial beds and propagated by both seeds and bulblets. *Narcissus* species easily naturalized in wooded sites and persisted in old homesteads and farms, including the species known as jonquils (*N. jonquilla*). These differ from daffodils in having shorter tubular corollas, and they were named for the similarity of their long leaves to rushes (*Juncus* spp.). Another long-lived bulb plant was also considered the most grand, the crown imperial lily (*Fritillaria imperialis*), which has showy tiers of pendant orange or yellow flowers. The plants were easily propagated, originally introduced from Persia by way of Constantinople, and they were popular garden plants in England when Gerard (1633) noted that "This plant likewise hath been brought from Constantinople amongst other bulbous roots, and made denizons [sic] of our London gardens, whereof I have great plenty." Crown imperial lilies were atypical among early English garden flowers in having no domestic uses other than ornament—as noted by Gerard, "The vertue of this admirable plant is not yet knowne." They were nevertheless desired for their form and flowers; after several requests to McMahon, Jefferson obtained and cultivated these lilies at Monticello, later sending propagated bulblets to his retreat at Poplar Forest.

Scent was another attraction, and herbaceous plants with particularly strong floral odors were often cultivated. Tuberoses, mignonette, and heliotrope were all valued in domestic landscapes where floral scents were a pleasant diversion and helped to mask unpleasant odors. Of course, floral scents evolved as chemical rewards for visiting pollinators; many of the most fragrant flowers are pollinated by nocturnal moths who locate their nectar sources using scent, but beetle-pollinated flowers are often strongly scented, such as various wild roses and the water lilies (*Nymphaea odorata*) that some gardeners cultivated in ponds or barrels. Heliotrope (*Heliotropium arborescens*) varieties from South America have scents that some thought resembled cherries or vanilla, and the small flowers of Egyptian mignonette (*Reseda odorata*) were commonly grown in window boxes. Indeed, several fragrant flowers are among the species named in a mid-nineteenth century parlor song performed by the Hutchinson Family Singers; beginning in the 1830s, they toured the United States with musical performances that promoted social causes such as temperance and emancipation and parodied social norms, such as the lyrics of "Calomel" (see chapter 7). Judson Hutchinson became known for performing "The Horticultural Wife," a genteel 1850 parody of the nineteenth century obsession with gardening. Historically speaking, the lyrics provide a list of the most popular eighteenth and early nineteenth century domestic flower garden species and presage later nineteenth century horticulture, particularly the popular trend away from borders of traditional perennials and wildflowers to exotic bedding plants.

The Horticultural Wife
She's my myrtle, my geranium,
My sunflow'r, my sweet marjoram,
My honeysuckle, my tulip, my violet,
My hollyhock, my dahlia, my mignonette.

Ho! ho! she's a fickle wild rose,
A damask, a cabbage, a china rose.
Ho! ho! she's a fickle wild rose,
A damask, a cabbage, that everybody knows.
She's my snowdrop, my ranunculus,
My hyacinth, my gilly flow'r, my polyanthus;
My heart's ease, my pink, my water lily,
My butter-cup, my daisy, my daffy-down-dilly.

Ho! ho! she's a fickle wild rose,
A damask, a cabbage, a china rose.
Ho! ho! she's a fickle wild rose,
A damask, a cabbage, that everybody knows.
I am like a scarlet runner, that has lost its stick,
Or a cherry, that is left for the dickey birds to pick;
Like a watering pot, I'll weep, like a pavion, I'll sigh,
Like a mushroom, I'll wither, like a cucumber, I'll die.
Ho! ho! she's a fickle wild rose,
A damask, a cabbage, a china rose.
Ho! ho! she's a fickle wild rose,
A damask, a cabbage, that everybody knows.
I am like a bumble-bee that don't know where to settle,
And she is a dandelion, and a stinging nettle;
My heart's like a beet-root, choked with chick-weed,
My head's like a pumpkin, running off to seed.

Ho! ho! she's a fickle wild rose,
A damask, a cabbage, a china rose.
Ho! ho! she's a fickle wild rose,
A damask, a cabbage, that everybody knows.
I've a great mind to make myself a felo-de-se,
And finish all my woes on the branch of a tree;
But I won't! for I know that at my kicking you'd roar,
And honor my death with a double encore.

Ho! ho! who would suppose,
I'd suffer so much by that fickle wild rose.
Ho! ho! who would suppose,
I'd suffer so much by that fickle wild rose.

The verses conjure images of eighteenth century borders and beds, including perennials (violets, hollyhocks, heartsease, daisies, myrtle, and pinks), bulbs (tulips, hyacinths, snow-drops, and daffy-down-dilly, an old English name for daffodils), and fragrant plants (garden roses, mignonette, water lily, marjoram, and "polyanthus"—probably tuberose, which is *Polianthes tuberosa*). Before they were used as edible legumes, scarlet runner beans (*Phaseolus coccineus*) from the uplands of Central America were grown as herbaceous ornamental vines, and geraniums and dahlias were typical of tropical bedding plants with bright petal pigments and long flowering seasons. Gardeners typically grew these plants as annuals because they were not winter hardy, but in their native habitats their growth is often perennial, and their stems become woody with cambium formation and the onset of secondary growth.

Many popular bedding plants had temperate relatives, but many gardeners preferred densely planted beds of the exotic species for the bright colors they provided over most of the growth season. For instance, garden geraniums (*Pelargonium*) are in the same family (Geraniaceae) as the north temperate wild geraniums or cranesbills (*Geranium maculatum* and related spp.), which were grown as garden perennials. *Pelargonium* is native to South Africa and includes the zonal geraniums that were derived from crosses between *P. zonale* and *P. inquinans*, short, shrubby hybrid plants with petal pigments ranging from deep red and pink to white; the horseshoe-shaped pigment markings on *P. zonale* leaves persisted in the hybrid cultivars. The plants were introduced in England in the early 1700s and were widely known as "Geranium Africanum," but it was not until the nineteenth century that they were used to plant low flower beds of single colors surrounded by turf, in the style described by Downing in *A Treatise on the Theory and Practice of Landscape Gardening* (1849). Their flowers, growth form, and scents were all appealing; geranium plants are covered with multicellular hairs containing geranium oils, essential oils composed of terpenes including geraniol. In *Old-Time Gardens* (1901), Earle praised its "clean color, in head and blossom, its clean beauty, its healthy growth" while maligning the overuse of other nineteenth century bedding plants such as coleus (*Coleus blumei*) from Asia and lobelias (*Lobelia erinus*) from the Cape of Good Hope. Rose-scented geraniums such as *P. graveolens* were cultivated in America as pot plants valued for their fragrant leaves; they were also grown in Europe for their essential oils, which were distilled as an inexpensive substitute for the attar of roses used in perfumery and originally obtained from *Rosa damascena*.

Other favorite nineteenth century annuals included petunias and pansies, both with flowers that coevolved with specific pollinators resulting in distinctive floral characters. *Petunia axillaris* from Brazil has fragrant white, moth-pollinated flowers that open at night. It was the first petunia introduced from Brazil to Europe, and it was hybridized with bee-pollinated species (*P. violacea* and *P. integrifolia*) to produce a vast range of red, pink, violet, and white flowers with a variety of markings and strong floral scents in some. Unlike most bedding plants, pansies had a temperate rather than tropical origin, beginning with the historic cultivation of heartsease (*Viola tricolor*), a bee-pollinated herb used to treat heart ailments because of the Doctrine of Signatures' interpretation of its heart-shaped petals. The linear petal markings are visual guides directing pollinators to the nectar deposits, a reward for floral visitors that collect and carry pollen. These nectar guides are marked with xanthophyll pigments that reflect ultraviolet light that is visible to bees, and they occur in both cultivated and wild plants. *Viola* flowers also have bilateral symmetry, known as zygomorphy, in which one petal is adapted as a landing platform for pollinating bees.

Garden pansies originated with crosses between heartsease and related European species such as *Viola lutea* and *V. cornuta*, and their facelike flowers were Victorian favorites. Cultivars were distinguished by a number of petals that bore deep-pigmented spots, and seed catalogs advertised petal colors ranging from golden yellow to blue, violet, and mahogany. Unlike tropical annuals, pansies were best grown in northern exposures because the plants thrived in shaded, cool sites, and gardeners sowed seeds outdoors in the early fall for flowers of maximum size in the following spring. Pansy culture was practiced with care, and some enthusiasts belonged to pansy societies. Circular gardens devoted only to pansies became popular during the 1860s, with usually one variety planted in a bed, and breeders produced hybrids devoid of the characteristic linear nectar guides. The flowers also lent their name to *The Pansy*, a youth magazine published during the 1880s that promoted children's gardening and encouraged diligence, morality, and benevolence.

In addition to pansies, bedding plants with bilateral floral symmetry included the calceolarias (*Calceolaria crenatiflora*), a Chilean species in the figwort family (Scrophulariaceae) in which the lower petals fuse into a pouch-like enclosure that may briefly trap a pollinating bee. Through hybridization, curious calceolaria varieties were developed that appealed to gardeners for their peculiar floral forms; catalogs advertised "Calceolaria hybrida," seed mixtures that promised hundreds of showy floral types with various pigments and markings. Other favored bedding plants with bilateral flowers included butterfly-flower (*Schizanthus pinnatus*) from Chile and torenia (*Torenia fournieri*) from Asia. While bilateral (zygomorphic) flowers are typically bee pollinated, flowers with radial symmetry (actinomorphy) often attract butterflies. Garden verbenas (*Verbena*×*hybrida*) were pink, white, and violet hybrids of various butterfly-pollinated South American verbena species. Their petals are fused to make a floral tube containing the nectar that is easily reached by the long mouth parts of butterflies; various *Phlox* species have petals with similar architecture and are also often butterfly pollinated.

Composites (family Compositae or Asteraceae) have small flowers organized into inflorescences that each resemble a single radially symmetrical flower, and these attract a variety of pollinators. In a botanical sense, the inflorescence functions as a pseudanthium (false flower) comprising dozens of small flowers (florets); the ray florets that often surround a pseudanthium are the "petals." The composites are the largest family of flowering plants and include the dahlias (*Dahlia* spp.) from Mexico, mentioned in "The Horticultural Wife," but ageratum (*Ageratum houstonianum*) from Mexico and African marigolds (*Tagetes erecta*) were also favorite tropical composites in flower beds. Several temperate garden plants that are also composites, including pot marigolds (*Calendula officinalis*), were introduced into England from southern Europe and grown in early American gardens as a medicinal plant, dyestuff, and potherb; nineteenth century cultivars were also used as bedding plants, including some with pale ray florets striped or edged in orange or deeper yellow. The common ox-eye daisy (*Chrysanthemum leucanthemum*), now a naturalized weed, was also introduced from Europe, while sunflowers (*Helianthus annuus*), black-eyed Susans (*Rudbeckia* spp.), and blanket flowers (*Gaillardia* spp.) are North American composites cultivated in Europe before they were commonly cultivated in American gardens.

As with woody plants, nineteenth century gardeners were intrigued by peculiar growth forms in herbaceous plants; seed catalogs often described botanical curiosities, such as a single dahlia with petals that *Burpee's Farm Annual* (1889) noted as "striped, flaked, mottled, and dotted in a grotesque and most charming manner." Floral shapes also attracted attention, such as bleeding heart (*Dicentra spectabilis*), which was introduced from China in 1846 by Fortune and within a few decades was a favorite garden plant in both England and America. *Dicentra* flowers have four petals—two inner white petals enclosed by two outer deep pink petals, each with a prominent nectar spur that forms the heart-shaped lobes of the pendant flowers. A more bizarre floral curiosity was the cruel plant (*Araujia sericofera*) from South America, a milkweed relative (family Asclepiadaceae) known for having white, fragrant flowers that frequently trapped and held nocturnal pollinating moths.

Garden favorites also included the bold inflorescences of various tropical amaranths (family Amaranthaceae) that occurred in several colorful and grossly mutated forms. The globe amaranth (*Gomphrena globosa*) has bracts surrounding the small flowers that remain brightly pigmented when dry, and it was used as an everlasting flower. The inflorescences of celosia (*Celosia argentea*) are also surrounded by pigmented bracts and have a feathery form, in contrast to some monstrous garden varieties such as *C. argentea* var. *cristata*, in which the flowers merge into a massive velvety "cockscomb," often with deep reddish pigmentation in the bracts. This cultivated variety was probably the "velvet amaranth" recorded by Jefferson in his garden book. Love-lies-bleeding (*Amaranthus caudatus*) has pendant red

Cockscombs (*Celosia argentea* var. *cristata*) are amaranths with massive clusters of flowers, each surrounded by pigmented bracts.

or purple inflorescences that were used in making bowers and wreaths. As tropical plants that required warmth for their growth, these and many other annuals were often started in hotbeds along with warm-weather crops such as melons, eggplant, peppers, and tomatoes (see chapter 4).

Distinctive foliage was also valued. The compound leaves of the South American sensitive plant (*Mimosa pudica*) fold in response to touch, an adaptive mechanism known botanically as thigmonastic movement, which probably discourages herbivores. The rapid motion of the leaflets is caused by sudden loss of turgor pressure in the cells at the base of each leaflet, similar to the slower nightly "sleep movements" exhibited by many other legumes. Sensitive plant grows as a perennial shrub in its native South America, but it was grown from seed as a curious herbaceous annual in many North American gardens, including those at Monticello. The flowers of the sensitive plant have reduced petals and showy pink stamens that attract pollinators, and it was considered particularly suitable for children's gardens. Brightly pigmented, variegated leaves such as those of coleus (*Coleus blumei*) and caladium (*Caladium bicolor*) were tropical alternatives to green foliage.

Exotic grasses such as zebra grass (a variegated variety of *Miscanthus sinensis*) were cultivated for their height and growth form, as were various species of yucca, particularly *Yucca filamentosa*, a hardy native species once commonly cultivated for its fibers (see chapter 8); nineteenth century gardeners also grew yucca as an ornamental plant for its massive rosettes of sharp foliage and tall inflorescences of moth-pollinated flowers. Castor bean plants (*Ricinis communis*) were originally cultivated medicinally as a purgative and laxative, but garden varieties were selected with variegated leaves or foliage in shades of bronze and purple. Castor bean plants were grown from seed that is highly toxic due to the presence of the nerve toxin ricin, and seed catalogs boasted that some gar-

Varieties of *Caladium bicolor* from South America were the "fancy-leaved caladiums" grown for their variously patterned, pigmented leaf blades.

den plants grew fourteen feet in height. Canna (hybrids of wild *Canna* spp.) also had large leaves with bronze or purple pigmentation, and their bright red, orange, and yellow flowers and height suited them for planting in the center of beds or as lawn ornamentals.

Indoor Gardens

The late eighteenth century garden notes of Jean Skipwith of Virginia suggest that she grew some plants indoors, including citrus trees (*Citrus* spp.), oleander (*Nerium oleander*), dwarf myrtle (*Myrtis communis*), "Prickly Lantana" (*Lantana camara*), and alstroemeria (perhaps *Alstroemeria pulchella*), but for most Americans gardening was an outdoor activity during the eighteenth century. During the early nineteenth century, indoor plants became more common in middle-class homes; jugs filled with flowers and evergreen branches were placed in fireplaces during the summer months, and asparagus ferns (*Asparagus* spp.) were used to attract flies and keep them from spotting clean walls. By the 1860s, middle-class houses were often outfitted with one or two small conservatories where both tender and temperate plants were raised and displayed all year. Seed catalogs listed plants specifically for indoor culture, including geraniums, calceolaria, coleus, and some cacti and other succulents that could be placed in the garden during the summer months.

Other tropical plants for conservatory planters included abutilon (*Abutilon* spp.), passionflower (*Passiflora caerulea*, a tender South America species), fuchsias (*Fuchsia* spp.), and various ferns, which were sometimes cultivated exclusively in conservatories known as ferneries. Plants with trailing growth or pendant branches were

South American passionflowers (*Passiflora caerulea*) were tender conservatory vines, perhaps cultivated by some for the coincidental religious symbolism of its flowers.

also particularly favored, including spiderwort (*Tradescantia* spp.) and nasturtium (*Tropaeolum majus*), also known as a source of edible seeds that were pickled and used like capers (see chapter 5). Trellises were sometimes used to support vines such as ivy (*Hedera helix*), black-eyed Susan vine (*Thunbergia alata*), and Boston smilax (*Asparagus asparagoides*). This South African relative of edible asparagus (*A. officinalis*) was known for its broad, shiny cladophylls (flattened leaflike stems) that strongly resemble the leaves of twining greenbriers (*Smilax* spp.).

An attached conservatory was considered by many to be an important part of nineteenth century domesticity; in *The American Woman's Home* (1869), Beecher and Stowe described conservatory design and use in the chapter titled "A Christian House." They considered horticulture a family enterprise but observed that with regard to conservatories, "few understand their value in the training of the young," and "children through the winter months can be starting seeds and plants for their gardens and raising valuable, tender plants." A modest conservatory was often nothing more than a glass-enclosed niche with a tiled floor, in contrast to the free-standing glasshouses of large estates that were heated by cast iron pipes. Depending on the temperature that was maintained, these were used as cool houses to shelter large plants such as camellias or to cultivate tropical species that required warmth.

Wardian cases are enclosed glass boxes that originated as collection tools for botanical explorers, but they were soon adapted for home use as small "conservatories" in their own right. In some cases they took the place of a conservatory and served as the centerpiece of a cultivated window display. Wardian case invention resulted serendipitously from a botanical discovery made by Nathaniel Bagshaw Ward, a London physician who pursued his avocation of naturalist in the countryside in the early 1800s. He incubated an insect chrysalis in a container of soil placed under a glass jar, but he soon became more interested in the small plants that germinated and thrived in the humid atmosphere. Ward's observations suggested that some plants might be better grown or shipped in glass cases; by 1834 he had constructed such a case and used it to ship fragile ferns safely to New South Wales in Australia. In 1842, he published "On the Growth of Plants in Closely Glazed Cases," a booklet in which he described the methods that led to the widespread use of Wardian cases. Robert Fortune was the first to employ Wardian cases on a large scale to carry botanical specimens from China to England, and the invention is credited with Fortune's successful introduction of tea from China to India for the East India Company. Other collectors used Wardian cases to introduce banana plants from China to other countries and to carry feverbark trees, the source of quinine, from the New World to the Old World tropics.

By the 1860s, Wardian cases had evolved into large terraria for culturing ferns and other plants that require high moisture and protection from airborne pollutants; various decorative models were sold for use in American homes. Luxury Wardian cases for the wealthy were built on walnut stands and in various shapes and sizes, but household manuals offered directions for assembling glass cases from inexpensive materials, perhaps using an old table as a base. Wardian cases were filled with greenhouse ferns or native mosses, ferns, and wildflowers, which thrived during the spring and summer in dark house interiors. Beecher and Stowe (1869) suggested landscaping the inside of a Wardian case with rocks, shells, and bits of mirror to create the effect of a grotto. They recommended using a homemade Wardian case to grow native plants such as trailing arbutus (*Epigaea repens*), partridge berry (*Mitchella repens*), and ground pine (*Lycopodium* spp.). Wardian cases were built with doors to allow for periodic airing and gas exchange, but they rarely required watering because of the trapped humidity; they were considered easy to maintain and were recommended for women who were busy rearing young children.

As late as the 1870s, physicians debated the merits of indoor plants, concerned that greenery might deplete the oxygen in bedrooms and replace it with harmful "carbonic acid gas" or carbon dioxide. J. H. Walden (1858) was typical of his time in arguing that "House-plants, and bouquets in sickrooms, are injurious; their influence on the atmosphere of the rooms is unhealthy." On the other hand, proponents of Wardian cases suggested them for the interest and delight of invalids, perhaps because the plants grew in an enclosed space so they might pose no danger in a sickroom. An article in *The Druggists' Circular and Chemical Gazette* (July 1873) sought

Wardian cases evolved from collecting cases used by botanical explorers to stationary domestic terraria for cultivating plants that thrived in high humidity.

to resolve the issue of indoor plant dangers, arguing, "As to the injury from vegetation, those of us who have had to sleep at various times in the woods, with but green branches for a pillow . . . know well . . . it is the most health-giving of all luxuries."

Gain and Loss

Human activity changes the landscape in both predictable and unexpected ways. From the moment of their arrival in North America, European settlers cleared agricultural land and introduced European species both intentionally and unintentionally; weed seeds arrived in straw bedding, with garden seeds, and in the soil used as ballast on sailing ships. Some garden plants also escaped from cultivation and began to self-sow and naturalize, becoming familiar weeds or invasive plants in natural habitats. Seventeenth, eighteenth, and nineteenth century gardeners planted many horticultural species that escaped from gardens, but neither colonists nor nurserymen could predict the environmental impact of various imported plants once they reached New World soil.

In *New-Englands Rarities* (1672) John Josselyn listed "Of such Plants as have sprung up since the English Planted and kept Cattle in New-England," and he included several European medicinal plants such as comfrey (*Symphytum officinale*), mullein (*Verbascum thapsus*), and shepherd's purse (*Capsella bursa-pastoris*) that were already self-seeding and growing wild in New England. Several other early medicinal plants naturalized rapidly, including opium poppies (*Papaver somniferum*), tansy (*Tanacetum vulgare*), valerian (*Valeriana officinalis*), and coltsfoot (*Tussilago farfara*). Ecologically speaking, these species are considered weeds—they are widespread, fast-growing plants that often grow where they are not wanted. Some are easily extirpated, while others have underground stems (rhizomes) and roots that are difficult to eradicate. A woody plant can also exhibit weedy growth, such as the common buckthorn trees (*Rhamnus cathartica*) that were introduced from Europe and cultivated for use as a purgative and a dyestuff. The berries yield yellow, green, and purple pigments depending on their maturity, and they are easily dispersed by birds. Buckthorn colonizes both dry and wet habitats, and like many other weeds, its roots resprout if the aerial parts of the tree are damaged.

Depending on the habitat and the species, weedy plants are sometimes invasive, meaning that, like buckthorn, they can replace native plants in natural environments because they grow aggressively. The characteristics of invasive plants are predictable; typically they are naturalized, exotic (non-native) species with rapid growth, high seed production, and efficient dispersal mechanisms for their fruits or seeds. Many have attractive flowers or fruit, unique growth forms, useful natural products, or medicinal properties that encouraged their cultivation. The colonial medicinal plants valerian and coltsfoot have become invasive in some habitats, as has the purple loosestrife (*Lythrum salicaria*) that was cultivated as a medicinal herb, nectar flower for bees, and ornamental garden plant. During the 1800s, purple loosestrife began to colonize North American wetlands as a troublesome invasive weed, a problem arising from its early domestic use as well as its possible introduction by way of the soil ballast in ships. It thrives in sunny wetland sites, in contrast to garlic mustard (*Alliaria petiolata*), another escaped medicinal herb that germinates and grows even in deep shade. Garlic mustard plants each produce hundreds or even thousands of seeds annually and blanket disturbed sites such as roadsides and floodplain forests. Sometimes merely displacing a plant slightly will result in its invasive growth; black locust (*Robinia pseudoacacia*) was originally native to the South, but the trees thrive under a variety of growth conditions. Black locust trees became invasive plants when settlers planted them in other regions as a source of resistant wood for fence posts.

Many invasive species were widely advertised by eighteenth and nineteenth century seedsmen and nurserymen as desirable garden plants; several were originally native to Asia, including Japanese honeysuckle

Black locust (*Robinia pseudoacacia*) was native to the South, but the trees are invasive in areas where they were planted to harvest for fence posts.

(*Lonicera japonica*), Japanese knotweed (*Polygonum cuspidatum*), oriental bittersweet (*Celastrus orbiculatus*), and Japanese barberry (*Berberis thunbergii*). Birds spread the berries of Japanese honeysuckle, which grows rapidly especially in southern states. The variety known as Hall's honeysuckle (*L. japonica* var. *halliana*) undergoes a change in petal color from white to yellow as the flowers mature, which probably made it an interesting plant for cultivation; the naturalized populations originated as garden plants. They were introduced by Dr. George Rogers Hall, who used a Wardian case to transport plants from his garden in Yokohama, Japan; unfortunately this Japanese honeysuckle has invaded southern woodlands where it competes with native vegetation. The appeal of Japanese knotweed likely originated with its superficial resemblance to bamboo, with herbaceous hollow stems that can reach ten feet in height. The hollow stems occur when the pith cells do not divide rapidly enough to keep pace with stem elongation, an indication of rapid growth similar to dandelions. The Royal Botanic Gardens, Kew, distributed Japanese knotweed as a unique garden ornamental during the mid-nineteenth century, when it was also introduced to American gardens. Japanese knotweed had already naturalized near Philadelphia by the 1890s. The plants thrive under environmental extremes such as heat, dry soils, and salinity; they have small white flowers and reproduce by seed, but the robust rhizomes can extend sixty-five feet from an established plant, creating dense thickets of genetic clones that are difficult to extirpate.

Several invasive Asian plants are dispersed by birds, including oriental bittersweet, which is closely related to the native North American bittersweet (*Celastrus scandens*); both produce small capsules, in which each seed is embedded in a pulplike red aril, an outgrowth from the base of the seed. Birds devour the arils and embedded seeds and coincidentally spread them into disturbed areas and waste sites, where the woody vines grow rapidly over shrubbery and up trees. The use of oriental bittersweet in domestic seasonal decorations led to its cultivation and widespread introduction, and the plants grow more aggressively than the North American species. *Euonymus* is a bird-dispersed member of the same family (Celastraceae) that also produces seeds embedded in a fleshy aril. Native *E. americana* shrubs were often cultivated during the nineteen century, but later gardeners often planted Asian *Euonymus*, including burning bush (*E. alatus*), which was carried widely by birds and now also grows invasively in many open spaces. Bittersweet was not the only decorative plant to become a troublesome weed. Dried flowers and wreaths used as decorations in Victorian homes included non-native grasses, such as the Eurasian jointed goat grass (*Aegilops cylindrica*), one of several agricultural weeds introduced by seeds shed inadvertently from imported decorations. Japanese barberry was planted to replace the common European barberry (*Berberis vulgaris*), first cultivated in American gardens during the seventeenth century for its fruits, as a plant for hedges, and as the source of a yellow dyestuff from its wood. Common barberry was often destroyed once the plants were discovered to be the alternate hosts for wheat rust (*Puccinia graminis*). Japanese barberry grows more aggressively and invaded forested areas; birds efficiently disperse its bright berries, and herbivores ignore its prickly branches. Another armed invasive shrub from Asia is *Rosa multiflora*, which was used for flowering hedges and as a rootstock for grafted roses. Birds eat the red hypanthium (hip) that develops from each flower, and the small seeds remain viable in soil for up to twenty years.

Eighteenth century horticulturists were initially pleased with the foliage and growth form of tree of heaven (*Ailanthus altissima*), a Chinese import that was cultivated as a fast-growing specimen tree after its introduction in Philadelphia in 1784. They were sold by nurseries during the 1840s, but in *The Art of Beautifying Suburban Home Grounds* (1870), Scott noted the trees' bad odor and recommended that the once popular species be banned from street plantings. Tree of heaven has pinnately-compound leaves that resemble walnut and sumac, but it differs in having wind-borne samaras (winged fruits) that disperse in large numbers, germinate, and survive even in marginal urban habitats. The trees are a familiar part of the naturalized vegetation in disturbed and abandoned sites, but tree of heaven also colonizes fields and meadows by means of root suckers; allelopathic chemicals from the trees may kill nearby vegetation. Paulownia (*Paulownia tomentosa*), another Asian species, was imported in 1844 for its attractive foliage and showy violet flowers. Scott recorded "a fine specimen in New York Central Park," but the species became invasive in the South; vast numbers of winged seeds disperse from each tree, and the species soon naturalized extensively in dry, infertile soils and disturbed sites.

Pride of India (also known as pride of China, chinaberry, and mahogany) is *Melia azedarach*, a member of the mahogany family (Meliaceae) and a relative of the tropical trees used in furniture-making. The species arrived in South Carolina in the late 1700s, and Washington, Jefferson, and Skipwith all cultivated the trees, which were known for their purple flowers and bipinnately-compound leaves. In *Uncle Tom's Cabin* in 1852, Harriet Beecher

Stowe described a "a noble avenue of China trees, whose graceful forms and ever-springing foliage seemed to be the only things that neglect could not daunt—or alter—like noble spirits, so deeply rooted in goodness, as to flourish and grow stronger amid discouragement and decay." The mature trees were often surrounded by seedlings that thrived under poor conditions; the abundant berries were strung as beads, and seedlings were shared among neighbors. As might be expected, chinaberry soon became invasive in the South. Norway maple (*Acer platanoides*) also became invasive in a similar way, with trees originally introduced from Europe for street plantings and shade. Bartram sold Norway maple in his Philadelphia nursery during the 1760s, and the dried samaras were imported for planting. The trees typically grew more rapidly than native maples; Scott observed, "More vigorous growth than the sugar and a similar formality of contour," and it proved an ideal, resistant shade tree in urban settings. Large numbers of Norway maple samaras disperse widely, and the seeds germinate reliably, often naturalizing in disturbed sites and becoming invasive in woodlots and along forest edges. The European white poplar (*Populus alba*) was also introduced from Europe in the late eighteenth century for landscape plantings, and like tree of heaven it forms clonal clusters from root suckers, especially after the death of a mature tree. It invades natural areas and prevents normal ecological succession.

Gardeners also cultivated aquatic plants that became ecologically invasive, including yellow flag (*Iris pseudoacorus*), an attractive yellow European iris that escaped from cultivation at the end of the nineteenth century and is now naturalized in wetlands. However, the most notorious cultivated aquatic plant is water hyacinth (*Eichhornia crassipes*), which was sold by nursery catalogs in the late nineteenth century for up to one dollar a plant, a large sum at the time; its lush growth, peculiar air bladders that adapt the plant for flotation, and showy violet flowers made it an attractive curiosity. Water hyacinths are native to South America, where their growth is held in check by natural controls, but escaped plants grew so aggressively in Florida that by 1890s, water hyacinth plants impeded boat travel on rivers and waterways. Their dense vegetative growth also provides breeding areas for mosquitoes, which are vectors of malaria and other diseases. Congress appropriated funds to study water hyacinth control in 1896, an environmental challenge because the plants replicate hundreds of times annually and can regenerate rapidly from fragments.

Invasive plants colonize natural habitats and compete with native plants for soil, light, and water, threatening biodiversity in nature, but cultivated diversity should also be considered with the evolution of domestic horticulture. In *Rural Essays* (1854), Downing extolled the virtues of grass "not grown into tall meadows, or wild bog tussocks, but softened and refined by the frequent touches of the patient mower, till at last it becomes a perfect wonder of tufted freshness and verdure." With changes in landscaping fashion and household needs, lawn blanketed much of the domestic garden space that had earlier been devoted to a variety of herbs and shrubs; these were often the unique garden plants that were discovered as rare sports and shared among gardeners. Domestic gardens evolved from eighteenth century dooryards planted with plots and borders of diverse, useful plants to nineteenth century lawns with occasional, ornamental flower beds cut into the turf; these beds were often monocultures, planted with single species. In contrast, in *Old Time Gardens* (1901), Earle described colonial front dooryards as "wholly of flowering plants . . . little enclosures hard won from the forest." She attributed these early gardens to the English forecourts described in *The English Housewife* by Gervase Markham, a household manual first published in 1615 and used frequently in American homes as a domestic reference (see chapter 7). Many of the plants were derived from cultivated varieties of European medicinal and ornamental species, which for the most part disappeared from American gardens as a result of changes in garden styles and household needs during the nineteenth century. In *Country By-ways* (1881), Jewett recorded the loss of American gardens in which "the yard was kept with care, and was different from the rest of the land altogether. . . . It is like writing down family secrets for anyone to read; it is like having everybody call you by your first name, or sit in any pew in church." She knew that colonial gardening was an early necessity for survival rather than an avocation for cultivating showy exotics and manicured lawns; the women who tended dooryards knew their plants well and cultivated them with care. Nevertheless, some old cultivars vanished long before the current interest in heirloom plants. Botanically speaking, these varieties resulted from the careful selection of plants that were often in use since ancient times; historically speaking, their disappearance is both a cultural and genetic loss.

Asian Origins

A surprising number of garden plants were selected from species native to Asia, as were many invasive species with garden origins, such as Japanese knotweed, oriental bittersweet, and the multiflora rose. About one hundred twenty genera have natural distributions and closely related species in both eastern North America and eastern Asia, including cultivated woody plants such as maple (*Acer*), dogwood (*Cornus*), catalpa (*Catalpa*), tulip tree (*Liriodendron*), and barberry (*Berberis*), as well as *Rhododendron*, *Magnolia*, *Hydrangea*, and *Viburnum*. Knotweed (*Polygonum*), bittersweet (*Celastrus*), and roses (*Rosa*) have similar distributions, as do some herbaceous plants such as bleeding heart (*Dicentra*), May apple (*Podophyllum*), and mayflower (*Epigaea*). Some genera, such as maples and barberries, are also native to Europe, but often the Asian species became the favored plants horticulturally. Bleeding heart from China (*D. spectabilis*) was more showy than the native North American species, also known as bleeding heart (*D. eximia*), Dutchman's breeches (*D. cucullaria*), and squirrel corn (*D. canadensis*). Asian and European barberries were grown as hedges rather than the native barberry (*B. canadensis*), and the North American *Hydrangea quercifolia* was supplanted by the showy Asian *H. macrophylla*. Europeans in contrast often preferred American plants; in *Rural Essays* (1854) Downing described the cultivation of North American tulip trees in Germany, magnolias in France, and rhododendrons and mountain laurels in England.

Horticultural preferences aside, the similar geographic distribution of these genera was of botanical interest to Linnaeus (see chapter 1) and later botanists. In his letters to Charles Darwin, nineteenth century American botanist Asa Gray discussed the floristic similarities between eastern North America and eastern Asia, and he demonstrated statistically that the flora of New England more closely resembled that of Asia than Europe or western North America. Many of these genera show disjunctions, gaps in their distributions that occurred during the Pleistocene when many European species became extinct. Similar species and forests occur in New England and the southern Appalachians and in Japan, China, and Korea, but often not in Europe; as Gray wrote in 1859, "It would be almost impossible to avoid the conclusion that there has been a peculiar intermingling of the eastern American and eastern Asian floras, which demands explanation." Pragmatically speaking, this explains why Asian ginseng (*Panax ginseng*) was easily replaced medicinally by the closely related North American ginseng (*P. quinquefolius*), and horticultural plants were easily shared between Asia and America. Similarities in ethnobotany also evolved; species of the herbaceous wild ginger (*Asarum*) persists in Europe as well as in North America and Asia, and the species had uses both as a medicine and flavoring across its range (see chapter 6).

Gray postulated that plants dispersed and migrated across land bridges between Siberia and Alaska, but a more likely explanation is plate tectonic theory, which explains continental drift. North America and Eurasia formed the northern supercontinent known as Laurasia, which began to break apart more than two-hundred million years ago. A common flora persisted on the continental land masses as the Atlantic Ocean slowly increased in size, but many plants disappeared from Europe and western North America as a result of climate change. Pleistocene glaciers blanketed much of eastern North America, but native species survived in the southern Appalachians and recolonized northeastern North America. The result has been the remarkable botanical similarities between eastern Asia and eastern North America. Woody plants that survived North American glaciations included the mountain laurels and tulip trees that colonists and explorers discovered, cultivated, and soon exported to Europe, as well as the herbaceous May apples that were adopted as the colonial American equivalent to mandrake. Native American and colonial uses of May apple resembled the Hindu uses of the Asian May apple (*Podophyllum emodi*) as a purgative and treatment for skin ailments; Jefferson grew May apple as a botanical curiosity rather than as a household medicinal herb, as one of several North American wildflowers that he cultivated along with native trees and shrubs in the Monticello woodlands. He took pride in the remarkable flora of North America, but he selected plants for Monticello based on their individual botanical merits, regardless of their continent of origin. In this regard, he was probably typical of most gardeners who assembled domestic landscapes from what was practical and pleasing, with cultivated diversity that reflected both household needs and horticultural desires.

CHAPTER TEN

Botanical Lives

WHEN he published *Florula Bostoniensis* in 1814, Dr. Jacob Bigelow envisioned his readers botanizing in the Boston area and learning the local flora; he perceived "a great deficiency of books relating to American plants" and provided locations for various species. However, Bigelow addressed an audience for whom botany was particularly significant or professionally useful, including plantsmen and medical doctors. Children fared quite differently in early nineteenth century New England, where their formal education depended on modest district schools. These provided instruction in basic disciplines: arithmetic, reading, grammar, spelling, history, geography, and penmanship. The teacher's task was difficult; students ranging in age from four years through late adolescence worked at various levels with texts and copybooks supplied by their parents. Chalkboards were not invented until the mid-nineteenth century, so lessons consisted of group work and recitations. Geography instruction provided rudimentary awareness of imported foods, exotic timber forests, and natural habitats, but formal study of the natural world was unknown. Emma Willard's *Geography for Beginners* (1826) illustrated a well-stocked mercantile with coffee, sugar, and molasses from the West Indies; oranges and lemons from Alabama; cinnamon from Ceylon; and clove and nutmegs from the Spice Islands—but this was economic geography, quite different from plant study or natural history.

In 1829, Almira Hart Lincoln Phelps was the first to suggest botany as a topic for young scholars in *Familiar Lectures on Botany*. She was vice principal of the Troy Female Seminary in New York and probably authored this small text to use there, perhaps encouraged by seminary headmistress Willard, who had already published several texts on geography and history. *Familiar Lectures on Botany* became a standard text, commonly known as "Mrs. Lincoln's lectures," which provided "Practical and Elementary Botany, with Generic and Specific Descriptions of the Most Common and Native Plants."

The revised edition (1852) reflected botanical exploration and westward expansion and included plants of the western territories, medicinal plants, and new species of such notable genera as *Magnolia.* Perceiving a need for a more fundamental botany text, Phelps also published *Botany for Beginners: An Introduction to Mrs. Lincoln's Lectures On Botany* (1833) for "the use of common schools and the younger pupils of higher schools and academies." Her object was to direct teachers on "the best methods of teaching the natural sciences" (most probably knew nothing about teaching science), while also providing instruction and motivation for young students to undertake a detailed study of botany, especially flower structure. Teachers were encouraged to use botany to break the monotony of grammar and arithmetic recitation with the explanation, "You have been so long engaged upon a certain set of studies, that I perceive they have become tiresome; I think of introducing a new study into school; tomorrow I shall give a lecture on Botany; you may bring with you all the wild lilies (or violets, or any kind of common flower)." Phelps implored young botanists to make herbarium collections of pressed specimens and provided instructions for pressing, mounting, and labeling the specimens.

At a time when some schools were beginning to teach botany, *Colton and Fitch's Modern School Geography* (1859) illustrated the native flora of Mexico.

In *Botany for Beginners*, Phelps discussed "Physiological Botany," her term for the study of plant parts and their functions, but most of her text was devoted to plant classification. During the early eighteenth century, Linnaeus had developed his sexual system, which classified flowering plants into twenty-four classes and numerous orders depending on the number of stamens and pistils in their flowers (see chapter 1). *Botany for Beginners* taught the Linnaean system. For example, the class Decandria included all flowers with ten stamens; a species in the Decandria was then assigned to an order (Monogynia, Digynia, Trigynia, and so on) depending on its number of pistils. A mountain laurel flower (*Kalmia latifolia*) with ten stamens and one pistil was classified by Linnaeus as Decandria Monogynia, while a milkweed (*Asclepias* sp.) with five stamens and two pistils was classified as Pentandria Digynia. These were arguably the first "hands-on" school science laboratories, in which students were encouraged to collect, dissect, and classify whenever possible. Of course, during the early nineteenth century, botanists were still learning basic botanical facts, and some plant life cycles were not yet understood. Plants assigned by Linnaeus to the Class Cryptogamia (from Greek words *kruptos*, for "hidden," and *gamos*, for "marriage") included bryophytes, ferns, clubmosses and groundpines (*Lycopodium* and its relatives), and horsetails or scouring rushes (*Equisetum*), as well as algae and fungi (now considered as two kingdoms distinct from plants). Mystified by the absence of flowers with obvious stamens, Linnaeus assumed that these plants had "hidden marriages," implying that they had invisible flower parts. It was not yet understood that moss and fern life cycles are genuinely nonflowering, with two distinct generations involving spore production and microscopic eggs and sperm. Phelps also assumed that these plants produced flowers with "stamens invisible, wanting, or very caducous," which means that the stamens fell off almost immediately.

In the interest of respectability, Phelps discussed Linnaean classification at length without referring to it as the "sexual system," as it was widely known among professional botanists and physicians. Of course, this was a nineteenth century modification of terms. A century earlier, New York botanist Cadwallader Colden introduced his young daughter to the Linnaean system, apparently without detrimental effects; Jane Colden Farquahar published an illustrated New York flora in 1749 and became widely known in colonial America for her botanical drawings and her skill in Linnaean classification. She corresponded and exchanged seeds with European botanists and plantsmen until her death during childbirth in 1766. Nevertheless, Phelps was no doubt aware of the controversy caused by Linnaean classification in England, so she defined and illustrated pistils and stamens without mentioning their male and female roles and described the function of pollen as "perfecting the fruit." Despite this error of omission, she provided instruction on flower structure and plant classification using Linnaean principles, as well as the detailed study of leaves, stems, roots, buds, fruits, and seed germination. Her writings reveal scientific observation; the account of seed germination in *Botany for Beginners* accurately described root gravitropism and the requirements for warmth, light, oxygen, and water. Indeed, a young student who worked carefully through her chapters learned both plant morphology and floral diversity, topics that are still discussed in modern botany texts. As might be expected, Phelps credited botanical adaptations and variations to "the Author of nature and the immutability of His law," and she encouraged young people to take up flower study because "Flowers are presents which our heavenly father gives us. . . . We can imitate flowers in wax and various other ways, but who can give them life?" She wrote her texts several years before Darwin proposed natural selection as a hypothesis for evolutionary change in *Origin of Species* (1859), at a time when divine creation was the sole explanation for biodiversity.

Phelps's botany texts were preceded by "Wakefield's Botany," first published in England in 1796 by Priscilla Wakefield as *An Introduction to Botany, in a Series of Familiar Letters, with Illustrative Engravings*. The author, a prolific Quaker writer of natural history and travel books for children, believed that botany was an improving pastime for young girls, superior to "the trifling, not to say pernicious, objects, that too frequently occupy the leisure of young ladies of fashionable manners." The book was organized as a series of natural history lessons designed for individual study; these were relayed by way of fictitious letters written from one sister to another, based on the detailed botany taught by a knowledgeable, but fictitious governess. Wakefield addressed an audience of young girls who undertook the study of botany with the assistance of a governess or parent or perhaps on their own. She also skirted controversy by omitting the reproductive nature of the Linnaean system, and she avoided the extensive use of botanical Latin terminology and descriptions; her goal was to make botany accessible to young women, which she did admirably. An American edition of *An Introduction to Botany, in a Series of Familiar Letters, with Illustrative Engravings* was published in Philadelphia in 1811, and in all, eleven editions were released over forty years. Like Phelps, Wakefield provided a religious rationale for learning botany: "to cultivate a taste in young persons for the study of nature, which is the most familiar means of introducing suitable ideas of the attributes of the Divine Being, by exemplifying them in the order and harmony of the visible creation."

How young women were spending their time was also of concern to Wakefield, who wanted to encourage healthful outdoor activities; in Wakefield's book, sister Felicia described in her letters the benefits of air and exercise during her botanizing trips. *An Introduction to Botany* was generally accurate, although a few interpretations were slightly amiss, such as an explanation that the Venus flytrap (*Dionaea muscipula*) kills harmful insects as a line of defense. Felicia described the process dramatically: "The leaves are armed with long teeth . . . they are so irritable that when an insect creeps upon them, they fold up, and crush or pierce it to death." We now know that insectivorous plants obtain nitrogen compounds by trapping and digesting insects, an adaptation for survival in their nutrient-poor bog habitats, but we cannot fault Wakefield for her interpretation of insectivory as self-protection; at least she considered function and adaptation and not form alone. Judging from the number of editions that were published, *An Introduction to Botany* was selected over competing texts such as *Peter Parley's Illustrations of the Vegetable Kingdom* (Goodrich 1840), which included the rudiments of ecology such as the effects of altitude, latitude, and soils on plant growth. The Peter Parley books and magazines were the work of the prolific Samuel Griswold Goodrich, who wrote a variety of moral tales and natural histories that were published in both England and America. The title page of Parley's botany illustrated an outdoor scene of children engaged in botanical study, and the text proclaimed that botany was "a most agreeable exercise and amusement" for ladies and children. Nevertheless, Goodrich described "The Linnaean, or Sexual System" with a frank discussion of botanical "male and female organs" explained in terms of analogies to men and women, which at the time was probably appalling to many readers.

Social reform movements in the early nineteenth century encouraged young people to cultivate gardens and learn the horticultural lessons of stewardship and appreciation for nature. In the same year that *Botany for Beginners* appeared (1833), seedsman Joseph Breck published *The Young Florist*, which described in detail a plan for a children's garden of concentric floral rings surrounding a rustic arbor constructed of birch branches. Gardening had a educational function in teaching botany, and plants with particular remarkable traits provided "improving" lessons for children. As superintendent of a large garden in Lancaster, Massachusetts, Breck knew the benefits of gardening. In *The Young Florist* he described the recreational and spiritual side of spending time outdoors tending a garden plot and observing birds,

PETER PARLEY'S

ILLUSTRATIONS OF

THE VEGETABLE KINGDOM;

TREES, PLANTS AND SHRUBS.

BOSTON:

B. B. MUSSEY, 19 CORNHILL.

1840.

The illustration on the title page of *Peter Parley's Illustrations of the Vegetable Kingdom* belied the frank discussion of the Linnaean sexual system and reproductive roles of flower parts provided by author Samuel Griswold Goodrich.

insects, and plants; Breck included flowers pollinated by bees, butterflies, moths, and hummingbirds in his children's garden, perhaps intentionally. Various blue and yellow annuals attracted bees, while butterflies visited composites and tubular flowers with deeply hidden nectar. Evening primrose (*Oenothera* spp.) flowers had pale pigments and opened late in the day for pollination by nocturnal moths. Bright red flowers such as scarlet morning glories (*Ipomoea tricolor*) and red zinnias (*Zinnia* spp.) attracted hummingbirds. His plant selection also revealed his intention that children be exposed to interesting growth forms, and his plant selections included such botanical curiosities as sensitive plant (*Mimosa pudica*), with leaflets that fold close to the stem in response to touch; four o'clocks (*Mirabilis jalapa*), which open in the late afternoon; and ice plant (*Mesembryanthemum crystallinum*), a succulent with pellucid epidermal dots that resemble a frosty coating—some even claimed that ice plants were cool to the touch on hot days. Scarlet runner beans (*Phaseolus coccineus*) were trained up poles and over arbors, and exotic annuals such as nasturtiums (*Tropaeolum majus*) and red opium poppies (*Papaver somniferum*) were grown for their bright petals.

Nineteenth century boards of education encouraged teachers and children to plant and cultivate schoolyard flower gardens, and by the end of the nineteenth century, children's gardens both at home and at school were promoted by social reformers working with immigrants in American cities. Some kindergartens eventually included gardening in their curriculum, but the "garden of children" educational concept stems from the philosophy of Friedrich Froebel, who opened the first kindergarten in Blankenburg, Germany, in 1837. Following Froebel's tenets, in 1869 Elizabeth Peabody founded the first English-speaking kindergarten in Cambridge, Massachusetts. According to Peabody's article in *Johnson's Universal Cyclopedia* (1885–86), Froebel believed that both children and plants responded favorably to nurturing and cultivation; he believed that teachers must learn to cultivate their students, in the same way that a gardener learns about his plants and their growth requirements. Like his contemporary Phelps, Froebel rejected the idea of continual lesson drill and believed that children would benefit from activity and investigation, a revolutionary notion.

Harvard botanist Asa Gray published the first edition of his *Manual of the Botany of the Northern United States* in 1848, a comprehensive scientific flora for use by botanists engaged in fieldwork, collecting, herbarium work, and identification. Gray was also interested in teaching botany to young students, and among his papers are manuscript notes from botanical lectures that as a young man he provided for the daughters of John Torrey, his friend and botanical mentor. Years earlier, Gray had written a popular textbook, *Elements of Botany* (1836), and two decades later he followed with *First Lessons in Botany and Vegetable Physiology* (1857), written to serve as an introduction to the rigorous botanical terminology and descriptions in his *Manual. First Lessons in Botany and Vegetable Physiology* was the most authoritative botany textbook available at the time; the first twenty-seven chapters covered morphology, anatomy, form, and function of vegetative and reproductive plant parts, and the last few chapters were devoted to classification.

Along with the usual Linnaean classification, Gray included a natural system, with the explanation that it "is intended to express, as well as we are able, the various degrees of relationships among plants, as presented in nature." Gray had long corresponded with Darwin, so perhaps it is surprising that Gray did not explain plant relationships in terms of natural selection and evolutionary change over time from a common ancestor. Gray's botanical studies contributed to Darwin's theory, and he wrote a favorable review of *Origin of Species* in the *American Journal of Science* (1860), but Gray was typical of his time in weaving moral and biblical lessons into the narrative of children's instruction in natural history. The first chapter of *How Plants Grow* (1858) for use in common schools began with "Consider the lilies of the field" (Matthew 6:28) and included the explanation that "wisdom and goodness of the Creator are plainly written in the Vegetable Kingdom." However, Gray described groups of related species, which was essentially an evolutionary concept; in *How Plants Grow* he explained that "plants are classified according to their relationships . . .

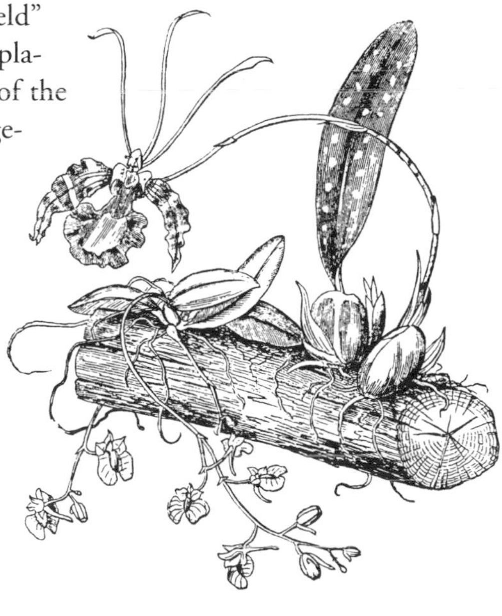

In *How Plants Grow* (1858), Asa Gray illustrated epiphytic orchids and described them as air-plants, capable of living without having their roots in the soil.

according to their resemblances in all effects," but he then noted that classification served as an illustration "of the plan of the Creator in the vegetable kingdom." He was clearly ambiguous about using natural selection to explain diversity to youth. Nevertheless, he included several of Darwin's botanical examples in *How Plants Behave: How They Move, Climb, Employ Insects to Work for Them, etc.* (1872); this popular children's book described insectivorous plants, cross-pollination strategies, vine growth, and other botanical adaptations that clearly support and illuminate Darwinian natural selection.

Gray influenced generations of American botanists to explore and classify the flora of New England; his *Manual* and methodical approach helped to mold Kate Furbish, a botanical artist who in 1880 discovered the rare Maine plant later named in her honor. The Furbish lousewort (*Pedicularis furbishiae*) is partially parasitic on the roots of alders that colonize the banks of the Saint John River in Aroostook County and is a relative of snapdragons and other members of the family Scrophulariaceae. The circle was completed during the 1880s when Furbish met Merritt Lyndon Fernald, a six-year-old from Maine with a precocious interest in botany. Furbish was a frequent visitor to the Fernald home, and young Merritt spent hours watching her paint botanical illustrations. Fernald later became a botanist at Harvard's Gray Herbarium, where he thoroughly revised Gray's *Manual of Botany* to produce the centennial edition (1950), the eighth and most current edition. As a scholar and steward of the New England flora, Fernald was also among the first to recognize the problem of invasive species that originated as ornamental garden plants (see chapter 9); in a paper presented in 1938 at the Franklin Institute in Philadelphia, he discussed weedy species including purple and Japanese honeysuckle, two invasive early garden introductions.

Parlor Botany

Flower study, herbarium-making, and various sorts of floral artwork and needlework were popular nineteenth century avocations for women and girls; each used various media and related in some way to botany or horticulture. Gentility, culture, and flowers were interlinked, as in the "juvenile floral concerts" conceived and directed by Thomas Moses, a nineteenth century artist and musician in Portsmouth, New Hampshire. These parlor entertainments were held in public places, executed on a grand scale for local audiences, and featured flowers, music, and youthful talent. Moses advertised the entertainment by transporting dozens of young female singers through town in a flower-decorated wagon, and his newspaper announcements promised celestial music in a setting of spectacular floral displays. As an entrepreneur, Moses organized these concerts as a way to improve his income as an artist, and he worked diligently to provide a genteel occasion for concert-goers by providing youthful anthems, verdant greenery, and even caged birds.

The parlors of American homes also featured decorative botany and music in a setting of civility. Botanical imagery had its roots in early American homes; the samplers stitched by young women illustrated stylized flowers, vines, and trees, along with alphabets, numerals, and Bible verses. Some needlework flowers were recognizable as roses, pinks, violets, or tulips, while others were generic or fanciful; most trees and vines were also stylized and structurally generic. Colonial needleworkers also embroidered tree-of-life designs that were often incorporated into genealogical samplers, while crewelwork included vines sprouting flowers that resembled roses, tulips, and peonies. Depictions of pineapples and other fruits appeared in the stitchery in samplers, crewelwork, Marseilles quilting, and other forms of needlework (see chapter 4).

Other media lent themselves to greater botanical precision than needlework and were used to simulate nature in ways that exhibited a woman's skill, dexterity, and patience, as well as the extent of her leisure time. In quillwork, an early folk art, small coils of colored paper were used to assemble complex botanical designs or pictures or to decorate household objects, such as sconces. Victorians in particular excelled in the number and variety of their botanical decorations, which included artificial flowers made of wax, hair, paper, shells, or feathers; these were often displayed under glass domes or in shadow boxes that preserved the fragile handicrafts and kept them free from dust. Real flowers were preserved using a variety of methods, such as drying in sand or coating with wax, and some gardeners cultivated everlasting flowers such as globe amaranth (*Gomphrena globosa*) that retained their pigments when they were dried. "Drying and Coloring Natural Flowers," an article in *The Druggists' Circular and Chemical Gazette* (March 1873), encouraged experimentation by exposing flowers to ammonia and nitric acid, which presumably caused permanent changes in petal color by affecting the anthocyanin pigments in their flowers; the flowers were then dried for household use by suspending them in an airy room. Leaves were skeletonized by soaking them in a caustic solution and then gently rubbing away the soft tissue, leaving behind only the complex network of hard-walled xylem cells that make up the veins; these were also arranged and displayed under glass domes. Wild materials such as moss, twigs, and dried grasses were collected and arranged into naturalistic designs or small living displays or placed with the contents of Wardian cases (see chapter 9) along with pots of palms and ferns that tolerated low light conditions. Photographs of parlor interiors suggest that some of the

Nineteenth century domestic manuals suggested cultivating twining and hanging plants in window planters.

"palms" were actually cycads, such as *Cycas revoluta*, non-flowering "living fossils" that, along with dinosaurs, were common in Mesozoic landscapes. Artificial flowers could also be purchased; many immigrants made them in their homes as piecework, using crafting techniques and tools that were originally perfected in Europe and introduced for American use.

In *The American Woman's Home* (1869), Catharine Beecher and Harriet Beecher Stowe provided instructions for assembling rustic decorations such as twig frames, fern planters, and a wooden flower stand; they also suggested training ivy to frame windows and twine decoratively about a room and encouraged women to cultivate plants in conservatories and Wardian cases. Floral designs were used frequently in the needlework, painted china, and botanical art produced by women with leisure time, as well as in many household decorations that were mass-produced during the nineteenth century. Needlework included mottoes that were worked in wool or silk on perforated cardboard; some of these contained botanical references, such as "I Am the Vine," "Consider the Lilies," "Nothing But the Leaves," and "The Old Oaken Bucket," which was a reminder of temperance. They were ornamented with the appropriate stitched vegetation. Decorative carvings on furniture depicted grapes and other flowers and fruits (see chapters 4 and 8). Antimacassars were crocheted, knotted, or tatted in various designs and used to protect upholstered parlor furniture from the stains of macassar oil hairdressings. These were fragrant commercial mixtures of non-drying oils produced by trees native to tropical Asia, made from combinations of pure macassar oil (pressed from the seeds of *Schleichera trijuga*) and ylang ylang (distilled from the flowers of *Cananga odorata*). Henry Ward Beecher (1859) complained that women spent too much time engaged in fashionable pursuits such as "lace-work, painting rice paper, casting wax flowers so ingeniously that no mortal can tell what is meant," which he believed led to nervous disorders, feebleness, and morbid tastes. He preferred that women spend more time outdoors and suggested friendly horticultural rivalries that would encourage industry and piety. He also would have approved of the popular nineteenth century hobby of collecting red, brown, and green feathery, filamentous algae from tide pools and beaches for mounting and parlor display. These were floated in water onto heavy paper and dried, where the filaments adhered as a result of the cell wall hydrocolloids similar to those in Irish moss (see chapter 6).

Of course, some of the earliest American household decorations were wallpapers illustrating plant diversity, exotic habitats, and landscape design. These were first imported from Europe and China, but after 1760 American patterns were block-printed on paper made of cloth fibers and then hand-colored. The nationwide colonial shortage of cotton and linen rags was at first a limiting factor (see chapter 8), but an increasing population resulted in more cloth that was available to be recycled into paper of various types. Popular eighteenth century patterns depicted urns with stylized shrubbery and columnar trees, and floral and vine patterns favored stylized, radially symmetrical flowers such as roses, poppies, pinks, and peonies. Many patterns resembled the patterns printed on chintz fabric with birds, insects, and fanciful or exotic plants such as bamboo and palms; others were damask-type patterns that were printed with symmetrical floral vases. Wallpapers with repeating patterns were also used to line trunks and cover the various sorts of boxes used for storage in early nineteenth century homes. Some oriental wallpapers were hand-painted, nonrepeating instructional scenes that illustrated stages in tea, rice, and silk production, while others were block-printed with the color added by hand. Mansion houses were sometimes decorated with printed landscape murals depicting trees, shrubbery, and perhaps ancient ruins.

The notion of a flower language originated in early nineteenth century France, but many books were published in nineteenth century America that listed the connotations of various flowers, usually in a romantic context. Parlor bookshelves included books that were decorative and sentimental, which typically combined botany with religious and moral lessons, flower poetry, and the language of flowers. Even botany texts such as Phelps's *Botany for Beginners* devoted a few pages to "examples of attaching sentiments to flowers," a list of several familiar plants and flowers with the meanings that they convey; it is the last subject covered in the volume and perhaps was added as an afterthought or by suggestion. Some were predictable, such as lilies suggesting purity, while others were less obvious, such as "a message for you" signified by iris. Boxwood signified constancy, cockscomb suggested foppery, jasmine meant gentleness, and the list continued. Other volumes were devoted primarily to floral meanings along with poetry and illustrations, such as *The Rural Wreath: or, Life Among the Flowers* (1854) by Laura Greenwood, described as "a Table Book for the Parlor, of a sentimental character, to diversify the monotony of a long winter evening . . . containing the Language of Flowers and Sentiments of the Heart." Sarah Josepha Hale's *Flora's Interpreter: or, the American Book of Flowers and Sentiments* (1832) also included a few pages of botanical instruction, and the Latin name and classification was included for each genus, so the book did have some improving merit beyond romantic amusement.

In 1995, author Beverly Seaton published a detailed social analysis (*The Language of Flowers*) in which she noted that floral meanings often varied depending on the source and that many found the "language" confusing. She concluded that probably very few Americans presented bouquets to each other that were assembled according to floral meanings; flower books were displayed on the center parlor table, but whether they were ever used in American homes beyond casual reading is unlikely. Volumes such as *Wild Flowers and Where They Grow* (1882) by Amanda Harris presented native and introduced wildflowers arranged by season, and the illustrations included some precise pen-and-ink drawings that would have been useful in identification. The language was flowery and judgmental, as in her description of the pitcher plant (*Sarracenia purpurea*) and sundew (*Drosera rotundifolia*): "our superb sarracenia entraps insects in that fatal cup, and destroys, devours, absorbs, or in some way annihilates them. And a curious little sundew . . . does the same unnatural thing." The most useful and accurate of the nineteenth century popular books on wildflowers was *How To Know the Wild Flowers*, first published in 1893 by Mrs. William Starr Dana. The first printing sold out in five days, and to satisfy demand, the book was revised twice and reprinted several times over the next four decades. Dana provided accurate plant descriptions, along with useful lore and notes about medicinal uses; she included a few references to poetry and sentiment but avoided classification, preferring instead to group plants artificially by flower color for ease in identification. Her various editions had decorative covers (the 1898 volume was imprinted with a banner reading "New leaf new life the days of frost are o'er"), but *How To Know the Wild Flowers* was a working field guide rather than a book for parlor display.

The nineteenth century affinity for plants resulted in conservation concerns. For years, many followed the advice of nineteenth century household manuals and gathered wildflowers, ferns, and greenery for gardens, ferneries, Wardian cases, and other sorts of decorations. Before greenhouses supplied the commercial need for flowers and greenery, the florist trade used immigrant labor to gather wild plants; ferns, trailing arbutus, and mountain laurel were all used in large quantities by urban florists, and by the 1890s botanists and naturalists noted that many New England native plants were becoming increasingly rare. Newspaper articles illuminated this early conservation concern; the Boston *Transcript* (16 August 1900) noted that "many of our wild flowers, formerly met with in such profusion even in the vicinity of large cities, are now rarely seen . . . many kinds of wild flowers are decreasing noticeably in quantity." Conservation groups organized, such as the Society for the Protection of Native Plants (now the New England Wild Flower Society) founded by Amy Folsom and a group of conservation-minded Boston women. Jane Loring Gray, widow of botanist Asa Gray, became their honorary president in 1901, and Merritt Lyndon Fernald was a trustee. As part of their effort to protect the native wild flora, they printed weather-resistant muslin signs to discourage the digging or uprooting of rare plants and the disappearance of populations from lack of seed:

> Spare the flowers.
> Thoughtless people are
> destroying the flowers by pulling them up
> or by picking too many of them.
> *Cut* what flowers you take
> and leave plenty to go to seed.

Christmas Botany

New England Puritans did not celebrate Christmas as a secular holiday, and during the seventeenth century those who displayed Christmas greenery were subject to a fine; the hanging of conifers, holly, and mistletoe predated Christianity in Europe, and Puritans arriving in North America considered their use to be an expression of heathenism and pagan sympathies. Puritans recognized December 25 only as a date that marked a sacred event in the church year, and this influence persisted in New England. Even after the Civil War, Boston public school pupils who were absent from school on Christmas day were disciplined, but by the 1880s immigration from Europe helped to dissolve anti-Christmas sentiment. Seasonal greenery was collected in rural areas and sold in American cities, and the American popular press took up the notion of Christmas.

The most obvious botanical manifestation of Christmas are evergreen conifers adorned with decorations, a tradition introduced to most American homes only in the relatively recent past. Christmas trees are attributed to the German custom that was introduced to England by Prince Albert; in 1846, the *Illustrated London News* published a sketch of Queen Victoria, Prince Albert, and their children gathered around a Christmas tree, and *Godey's Magazine and Lady's Book* in 1850 published a similar illustration. The royal connection resulted in the widespread adoption of the custom in America, which some rationalized with the religious justification that a Christmas tree symbolized the tree of knowledge in the Garden of Eden. For others, a Christmas tree was merely a decoration and perhaps the center of a larger seasonal display. The press had directed Victorian American attention to the Christmas tree custom, but in fact German immigrants in Pennsylvania had Christmas trees in their homes since the end of the eighteenth century.

Ornaments were made from paper or glass, and they were used with tinsel and spun glass to transform a forest conifer into a Christmas tree. Glass ornaments were German imports, as were the first artificial trees that appeared in the 1890s; concern in Germany about loss of native fir trees motivated the manufacture of artificial trees from goose feathers dyed green and bound to wooden dowels. The candles used for illumination were a commonplace hazard, and by the 1890s, consumers were able to purchase the first electrical lights, which were advertised as a safe alternative to the flames and dripping wax that easily ignited a dry, combustible conifer. Christmas trees were cut, hauled from the forests, and shipped by train to cities, where they were sold in marketplaces along with wreaths and other greenery. In New England, the conifers often came from woodlots in Vermont and the Berkshire Mountains of western Massachusetts, and by the end of the century conifer conservation was also an American concern, even though a minority of homes (perhaps one in four) erected a Christmas tree annually.

The most common Christmas trees were pines (*Pinus* spp.), spruces (*Picea* spp.), and firs (*Abies* spp.); the preferred trees had pyramidal shapes and resinous, fragrant needles that remained on the branches even when dry. Balsam fir (*Abies balsamea*), native to Canada and eastern North America, was particularly prized as a fragrant Christmas tree. Conifers produce separate pollen and seed cones, typically on the same tree; the pollen cones are ephemeral and last only long enough to release the pollen in the spring, while the longer lived woody seed cones became Christmas decorations in their own right. The appeal of conifers was in their evergreen needles, foliage adaptations that prevent damage and desiccation during the winter. Needles have minimum surface area, which decreases the loss of water vapor from the epidermis. They are also structurally tough, with a waxy cuticle and stomata that occur in slightly depressed grooves, which also slows desiccation. Resin ducts in the needles contain the aromatic sap or pitch, which made conifers desirable trees in domestic landscapes because of their presumed healthful qualities (see chapter 9). Wreaths were also assembled from conifer branches, and according to Eric Sloane's *A Reverence for Wood* (1965), the old New England custom was to hang the wreath on the barn door because of the association with the manger.

Clubmosses (*Lycopodium* spp.) native to northeastern forests were also widely known as groundpines, small evergreen plants that were frequently

European holly (*Ilex aquifolium*) and mistletoe (*Viscum album*) were illustrated in "Not a Creature Was Stirring, Not Even a Mouse," by Thomas Nast in *Harper's Weekly* on 25 December 1886.

mistaken for conifer seedlings. They were known to Linnaeus as cryptogams, spore-producing plants with "hidden" (microscopic) sexual parts. Now they are interpreted botanically as "living fossils," the modern descendants of extinct "coal plants," the arboreal lycopods that inhabited Carboniferous swamps. Their upright shoots grow from long runners, which suited club-mosses for use in evergreen wreaths and as "festooning material" for American homes. A farmer's wife in Monmouth County, New Jersey, had been the first person to gather the plant and ship it along to New York markets, but by 1888 an editorial in *Garden and Forest* magazine noted that both clubmosses and native holly had become scarce from over-collection for Christmas greenery.

Holly and mistletoe were used in wreaths, vases, and garlands on mantles or tucked behind picture frames. The shiny leaves of English holly (*Ilex aquifolium*) are more handsome than the leaves of American holly (*I. opaca*), but the English species was not hardy during northern winters. English holly was cultivated in the South, and holly was exported from America to England, but which species was supplied is unclear; American holly was used for greenery because it was available, especially in the Northeast, but the holly often shown by Nast and other nineteenth century illustrators more closely resembles English holly. Holly was prized because it is an evergreen tree with broad leaves, and its red "berries" (the fruit are actually stone fruits or drupes) added to its appeal. Perhaps to reform its pagan history, some tried to interpret holly with religious symbolism; the sharp points on the leaf margins were likened to the crown of thorns, and the red fruits were compared to drops of Christ's blood. The trees are dioecious, with separate staminate and pistillate trees, and the two sexes have to grow in close proximity for effective insect pollination to occur. Staminate hollies produce only pollen. The red drupes are produced only by trees that produce flowers with fertile pistils; the pistillate hollies often suffered from over-collection to supply markets with greenery, which became a cottage industry. In New Jersey and other states with native holly, women and children made wreaths and garlands that were sent to market along with holly branches, and sprigs were also used to cover wire forms in the shapes of crosses, anchors, and stars. Holly was the plant most frequently illustrated in nineteenth century periodicals, early Christmas cards, and on the printed pasteboard boxes used for presents.

Mistletoe (*Viscum album*) was another commonly used evergreen plant, which first arrived in American markets packed in wooden crates shipped by steamer from Europe. It had subtle morphology compared to holly, with pale green leaves, brittle stems, and nearly translucent white berries, but its use as winter greenery dated from ancient times. The native North American mistletoe (*Phoradendron flavescens*) was harvested from Appalachian hardwood trees by the 1890s, and both genera share a peculiar botany as partial parasites on trees. Mistletoes extend their roots (haustoria) into the branches of their host trees, where they absorb water and minerals and may cause wood deformations and the abnormal shoot growth known as a witch's broom. Their widespread distribution occurs because their berries are eaten by birds, and the seeds germinate on tree branches where they are deposited. Mistletoe on the branches of host trees becomes noticeable in the fall, when the deciduous oaks drop their leaves while mistletoe retains its pale foliage. Parasitic growth probably explains the mythology and claims of magical powers associated with European mistletoe; in contrast, Christian legends considered mistletoe as the source of the wood used to make the cross and attribute its present reduced size to punishment and shame. Mistletoe was used like holly in vases and on mantles, and mistletoe branches also appeared frequently in nineteenth century illustrations.

Conifers, holly, and mistletoe originated as temperate species that did not require intensive cultivation or horticultural management to have a ready supply in December. In contrast, poinsettias (*Euphorbia pulcherrima*) were introduced from Mexico in 1825 by Joel Poinsett, the first United States ambassador to Mexico, and these required glasshouse culture. The small flowers of poinsettias are surrounded by red pigmented bracts (modified leaves), which is their primary attraction as a Christmas plant. Physiologically speaking, they are short-day plants that will not flower and produce bracts unless they receive at least twelve hours of darkness each day. In nature, poinsettias grow into shrubs, and Poinsett collected cuttings from wild plants, which he cultivated in a glasshouse at his South Carolina plantation. He propagated and sent poinsettias to botanical gardens, as well as nurseryman John Bartram of Philadelphia, who in turn shared them with Robert Buist, a Pennsylvania florist and nurseryman. Buist recognized their horticultural value and was the first to sell poinsettias, but they are conspicuously absent from the Christmas illustrations that appeared in nineteenth century periodicals such as *Harper's Weekly*. The article on Poinsett in *Johnson's Universal Cyclopaedia* (1885–86) makes no mention of poinsettia, which apparently did not become widely familiar as a horticultural Christmas plant until glasshouses were available for its cultivation.

Journey's End

Death is inevitable, and the customary connections between the end of life and various aspects of botany were considerable. Seventeenth century colonists followed English traditions, often burying family members in shrouds and adorned with silk ribbons tied at the wrists. The archeological examination of the 1681 grave of Anne Wolseley Calvert of Saint Mary's City, Maryland, revealed remnants of rosemary, the herb traditionally associated with remembrance. Her grave was in an early chapel, and her remains were discovered among its ruins. Some rural families had their own burial grounds in a remote area of their property, while village cemeteries in early New England towns were situated on land with a pleasant aspect and often with a good view. The collective memory of crowded, unsanitary European and colonial graveyards motivated Americans to want better burial places, and nineteenth century reformers produced the rural cemetery movement. By the 1840s, burials under churches and in crowded graveyards were considered unsanitary. Years earlier, concerned Bostonians had encouraged Dr. Jacob Bigelow to propose an "ornamental cemetery" that would combine a horticultural collection, parklike landscaping, and burial plots. Mount Auburn cemetery was consecrated in 1831 with the help of the newly incorporated Massachusetts Horticultural Society and funded in part with money from subscribers who agreed to purchase lots for sixty dollars each. Bigelow's plans included an elaborate front gate of Egyptian design, avenues and paths, ponds, lawns, dells, a Gothic chapel, statuary, and a garden for horticultural experimentation. A grid design was rejected in lieu of landscaping with the contours of the site, and trees and shrubbery were planted in abundance to mature into a virtual arboretum. The avenues were named for trees (magnolia, cedar, beech, rosebay, and so on), and the smaller paths were named for shrubs and herbaceous plants (mimosa, camellia, yarrow, heliotrope, and so on). As planned, within a few years, Mount Auburn became a popular spot; in 1839, "The Picturesque Pocket Companion" was printed for visitors, the first of many pamphlets, maps, and tree lists to assist visitors with botanizing while walking the avenues and paths.

The rural cemetery movement began in Boston, but this new type of functional landscape garden was promoted nationwide as essential to community well-being. Reformers suggested dignified but cheerful surroundings with landscaping in conformity with the character of the site, and the garden cemeteries evolved as places where families visited the graves of departed relatives, picnicked, and walked. A garden environment contrasted with the old notion of a decrepit burial ground and required a considerable investment in ornamental plantings; cemeteries were often the first place that new cultivars and introductions were planted, and as botanical gardens they evolved into repositories of horticultural diversity. Vegetation was also valued because it was thought to counteract the escape of dangerous miasmata from graves, a particular concern during infectious epidemics. Rural cemeteries were planned in most major cities during the 1840s and 1850s, including Green-Wood in Brooklyn and Laurel Hill in Philadelphia. As botanical gardens that happened to include graves, American cemeteries served as the model for urban public parks in American cities. In *The Horticulturist and Journal of Rural Art and Rural Taste* (July 1849), Andrew Jackson Downing encouraged the planting of large public parks based on the rural cemetery model:

> One of the most remarkable illustrations of the popular taste, in this country, is to be found in the rise and progress of our rural cemeteries. Twenty years ago, nothing better than a common grave-yard, filled with high grass, and a chance sprinkling of weeds and thistles, was to be found in the Union. . . . Eighteen years ago, Mount Auburn, about six miles from Boston, was made a rural cemetery. It was then a charming natural site, finely varied in surface, containing about 80 acres of land, and admirably clothed by groups and masses of native forest trees. It was tastefully laid out, monuments were built, and whole highly embellished. No sooner was attention generally roused to the charms of this place than the idea of rural cemeteries took the public mind by storm. . . . Not twenty years have passed since that time; and, at the present moment, there is scarcely a city of note in the whole country that has not its rural cemetery. . . . If the road to Mount Auburn is now lined with coaches, continually carrying the inhabitants of Boston by thousands and tens of thousands, is it not likely that such a garden, full of the most varied instruction, amusement, and recreation, would be ten times more visited?

Headstones and monuments often included symbolic plants that were carved into stone to commemorate the dead and comfort the survivors. Weeping willows were a familiar early carving on headstones, but other botanical images were more subtle in their meaning. Gourds symbolized the mortal body, while pomegranates suggested immortality; some gourds were also carved to resemble breasts, which, according to Laurel Thatcher Ulrich in her book *Good Wives, Images and Reality of the Lives of Women in Northern New England 1650–1750* (1982), signified the milk of the gospel in the early eighteenth century. During the nineteenth century, poppies suggested prolonged sleep, roses and lilies signified purity, and ivy implied fidelity. A sheaf of wheat represented the harvest of a life well-lived, while a plucked olive branch or flower bud symbolized the death of a child. After 1880, stones were carved to resemble entire rustic tree trunks; trees represented wisdom and long life, but they suggested a premature death if they were carved from stone to resemble a tree stump or broken limb. Other plants held personal significance—for instance, the Mount Auburn memorial to novelist Sarah Payson Willis (died 1872), which consists of a stone cascade of ferns; her well-known pseudonym was Fanny Fern, and she wrote *Ruth Hall, Fern Leaves from Fanny's Port-Folio,* and other satirical domestic novels (see chapter 2). She was educated at Catharine Beecher's seminary in Hartford, where she was probably exposed to the domestic science promoted in the Beecher cookbooks and household manuals, all fodder for her novels.

Mourning also had its own botany. During the eighteenth and nineteenth centuries, funeral biscuits were baked and often stamped with molds carved with stylized designs of roses and other fragrant funeral flowers; these were a custom introduced from Yorkshire, where they were baked with tansy seeds and caraway. The biscuits were offered as repast for mourners, along with wine and funeral breads baked with caraway or dried fruits; this was customary particularly in

The pendulous growth habit of weeping willow trees (*Salix babylonica*) suggested grief and was incorporated into headstone carvings and memorial needlework.

Pennsylvania and Virginia, where some bakeries specialized in funeral provisions as a reliable source of business. Nineteenth century undertakers hung a black floral wreath to indicate the parlor wake that followed a death, and fragrant fresh flowers helped to dispel odors and perhaps seemed reassuringly hygienic before disease transmission by microorganisms was understood. Eighteenth and nineteenth century memorial needlework often featured willows; the pendulous growth form of a weeping willow beside an urn suggested grief and was introduced from England in the late 1700s. Needlework mourning pictures were stitched for both family members and public figures, and following the death of George Washington in 1799, most American homes displayed a memorial to the first president. A typical piece of silk-embroidered needlework featured Washington's portrait and tomb surrounded by weeping willows, with laurel wreaths and roping for ornamentation.

American customs were influenced by Queen Victoria, who entered a period of sustained mourning following the death of Prince Albert in 1861. Dressing in black for a long period was part of the protocol, and the black mourning jewelry that became popular was made from jet, the mineral remains of extinct Jurassic conifers. Tree trunks became waterlogged and compressed in swamps; hard jet formed in marine habitats, while mineralization in freshwater produced soft jet. Large jet deposits were mined for centuries at Whitby along the north Yorkshire coast. Romans had known of the mineral and used it for jewelry, and during the Middle Ages it was used for crosses and other religious articles. Jet can be cut into facets, carved, and polished to a high gloss, which led to its use in mourning jewelry. The Victorian demand for large mourning brooches and other types of mourning jewelry made jet a highly desirable lightweight material, and specimens were exhibited at London's Great Exhibition of 1851. The English jet industry supplied both the worldwide market until jet supplies dwindled and black glass and other materials served as substitutes.

Mourners visited graves, particularly the graves of children who did not survive childhood. In *The Mother's Book* (1831), Lydia Maria Child wrote

> So important do I consider cheerful associations with death, that I wish to see our grave-yards laid out with walks and trees, and beautiful shrubs, as places of public promenade. We ought not to draw such a line of separation between those who are living in this world, and those who are alive in another.

Flowers were the norm, along with the notion of death as a prolonged sleep; the lyrics of parlor songs like "Willie's Grave" (1857) by Henry DeLafayette Webster described the scene:

> The grass is growing on the turf,
> Where Willie sleeps,
> Among the flow'rs we've planted there,
> The soft wind creeps.

"The Ash Grove" was another parlor favorite; the lyrics by Welsh poet William Talhaiarn were translated by Thomas Oliphant and described a woody thicket, typical of the landscapes envisioned by Downing and familiar in the rural cemeteries that were consecrated and planted during the nineteenth century:

> The ash grove, how graceful, how plainly 'tis speaking,
> The wind through it playing has language for me.
> Whenever the light through its branches is breaking
> A host of kind faces is gazing on me.
> The friends of my childhood again are before me,
> Each step wakes a memory as freely I roam.
> With soft whispers laden its leaves rustle o'er me,
> The ash grove, the ash grove alone is my home.
>
> My laughter is over, my step loses lightness,
> Old countryside measures steal soft on my ears;
> I only remember the past and its brightness,
> The dear ones I mourn for again gather here.
> From out of the shadows their loving looks greet me,
> And wistfully searching the leafy green dome,
> I find other faces fond bending to greet me,
> The ash grove, the ash grove alone is my home.

In European folklore, ash trees (*Fraxinus* spp.) were considered to be protective trees, which Talhaiarn likely knew. Infants were fed a spoonful of ash sap to assure a long life, and mothers often hung cradles from an ash limb, confident that the tree would protect infants from snakes and other dangers. The folklore practice of passing a child through a cleft in an ash tree to cure weaknesses or disease tied the well-being of an individual to an ash tree. Victorian Americans singing these lyrics had likely long forgotten the traditional association of *Fraxinus* with life, but over time botanical histories are often lost. Few recount the stories of New World survival in primitive colonies, which depended on the botanical foods, medicines, fuel, timber, and fibers that the earliest colonists carried with them or discovered in their new land. Yet when we eat corn, apples, or maple syrup; flavor foods with tropical spices; use a herbal remedy; drink imported tea; build a fireplace fire; or plant heirloom vegetables or flowers, we revisit many of these botanical traditions. They are part of our shared heritage, which over generations and centuries melded European and Native American plants into a unique tradition of American household botany.

Bibliography

Abell, L. G. 1852. *Skilful Housewife's Book.* New York: O. Judd.

Allen, Lewis F. 1852. *Rural Architecture: Being a Complete Description of Farm Houses, Cottages, and Out Buildings.* New York: C. M. Saxton.

Ames, Kenneth L. 1992. *Death in the Dining Room and Other Tales of Victorian Culture.* Philadelphia: Temple University Press.

Bailey, L. H. 1949. *Manual of Cultivated Plants.* New York: MacMillan.

Ballard, Martha. 1785–1812. Martha Ballard's Diary Online. From DoHistory, Harvard University, Cambridge, Massachusetts. http://www.dohistory.com/diary/index.html. Accessed 2000.

Barnard, Frederick A. P. and Arnold Guyot. 1885–86. *Johnson's Universal Cyclopaedia.* vols. I–VIII. New York: A. J. Johnson.

Bartram, William. 1791. *Travels Through North and South Carolina, Georgia, East and West Florida, the Cherokee Country, the Extensive Territories of the Muscogulees, or Creek Confederacy, and the Country of the Chactaws.* Reprint. Mineola, New York: Dover, 1955.

Beecher, Catharine E. 1858. *Miss Beecher's Domestic Receipt-Book.* New York: Harper and Brothers.

Beecher, Catharine E., and Harriet Beecher Stowe. 1869. *The American Woman's Home.* New York: J. B. Ford.

Beecher, Henry Ward. 1859. *Plain and Pleasant Talks about Fruits, Flowers, and Farming.* New York: Derby and Jackson.

Benes, Peter, ed. 1984. *Foodways in the Northeast.* In The Dublin Seminar for New England Folklife Annual Proceedings, 1982. Boston: Boston University.

———. 1987. *Families and Children.* In The Dublin Seminar for New England Folklife Annual Proceedings, 1985. Boston: Boston University.

———. 1988. *The Farm.* In The Dublin Seminar for New England Folklife Annual Proceedings, 1986. Boston: Boston University.

———. 1992. *Medicine and Healing.* In The Dublin Seminar for New England Folklife Annual Proceedings, 1990. Boston: Boston University.

———. 1996. *Plants and People.* In The Dublin Seminar for New England Folklife Annual Proceedings, 1995. Boston: Boston University.

Bermejo, J. E. Hernandez, and J. Leon. 1994. *Neglected Crops, 1492 from a Different Perspective.* FAO Plant Production and Protection Series No. 26. Rome: Food and Agricultural Organization of the United Nations.

Betts, Edwin M. 1944. *Thomas Jefferson's Garden Book.* Philadelphia: The American Philosophical Society.

Betts, Edwin M., and Hazleurst Bolton Perkins, with Peter Hatch. 1986. *Thomas Jefferson's Flower Garden at Monticello.* Charlottesville: University Press of Virginia.

Beverly, Robert. 1705. *The History and Present State of Virginia.* Reprint. Chapel Hill: University of North Carolina, 1947.

Billing, Jennifer, and Paul W. Sherman. 1998. Antimicrobial functions of spices: Why some like it hot. *Quarterly Review of Biology* 73: 3–49.

Breck, Joseph. 1833. *The Young Florist.* Reprint. Bedford, Massachusetts: Applewood, 1989.

———. 1851. *The Flower-Garden: or, Breck's Book of Flowers.* Boston: J.P. Jewett and Company.

———. 1866. *New Book of Flowers.* New York: Orange, Judd.

Brown, O. Phelps. 1897. *The Complete Herbalist: or the People Their Own Physicians.* Jersey City, New Jersey: O. P. Brown.

Buist, Robert. 1839. *The American Flower Garden Directory.* Philadelphia: Carey and Hart.

Byrd, William. 1737. *Natural History of Virginia, or The Newly Discovered Eden.* Reprint. Edited and translated from German by Richmond Croom Beatty and William J. Mulloy. Richmond, Virginia: Dietz Press, 1940.

Campenella, Thomas J. 2001. Henry David Thoreau and the Yankee Elm. *Arnoldia* 61 (2): 26–31.

Child, Lydia Maria. 1831. *The Mother's Book.* Reprint. Bedford, Massachusetts: Applewood, 1989.

———. 1837. *The Family Nurse.* Reprint. Bedford, Massachusetts: Applewood, 1997.

———. 1844. *The American Frugal Housewife.* Reprint. Mineola, New York: Dover, 1999.

Coats, Alice M. 1968. *Flowers and Their Histories.* New York: McGraw-Hill.

Colonial Dames of America. 1995. *Herbs and Herb Lore of Colonial America.* Reprint. Mineola, New York: Dover. First published in 1970 under the title *Simples, Superstitions, and Solace: Plant Material Used in Colonial Living.*

Confederate Receipt Book: A Compilation of Over One Hundred Receipts Adapted to the Times. 1863. Richmond, Virginia: West & Johnson. From

Documenting the American South, University of North Carolina, Chapel Hill. http://docsouth.unc.edu/receipt/receipt.html. Accessed 2000.
Dana, Mrs. William Starr. 1893. *How to Know the Wild Flowers.* Reprint. Boston: Houghton Mifflin, 1989.
Downing, Andrew Jackson. 1845. *The Fruits and Fruit Trees of America.* New York: John Wiley.
———. 1849. *A Treatise on the Theory and Practice of Landscape Gardening.* New York: G. P. Putnam.
———. 1854. *Rural Essays.* New York: Leavitt and Allen.
Druggists' Hand-book of American and Foreign Drugs, Preparations, etc. 1870. Boston: Charles Roberts.
Duke, James A. 1985. *Handbook of Medicinal Herbs.* Boca Raton, Florida: CRC Press.
Dupree, A. Hunter. 1968. *Asa Gray.* New York: Atheneum.
Earle, Alice Morse. 1898. *Child Life in Colonial Days.* Reprint. Lee, Massachusetts: Berkshire House, 1993.
———. 1898. *Home Life in Colonial Days.* New York: Macmillan.
———. 1901. *Old-Time Gardens.* New York: Macmillan.
Eastlake, Charles L. 1868. *Hints on Household Taste in Furniture, Upholstery and Other Details.* London: Lonmans, Green, and Company.
Erichsen-Brown, Charlotte. 1989. *Medicinal and Other Uses of North American Plants: A Historical Survey with Special Reference to the Eastern Indian Tribes.* Reprint. New York: Dover Publications. First published in 1979 under the title *Use of Plants for the Past 500 Years.*
Evelyn, John. 1699. *Acetaria: A Discourse of Sallets.* Reprint. Brooklyn, New York: Brooklyn Botanic Garden, 1937.
Ewan, Joseph, ed. 1969. *A Short History of Botany in the United States.* New York: Hafner.
Farrar, Eliza Ware. 1836. *The Young Lady's Friend, by a Lady.* Boston: John B. Russell.
Favretti, Rudy, and Joy Favretti. 1990. *For Every House a Garden.* Hanover, New Hampshire: University Press of New England.
Fernald, Merritt Lyndon. 1950. *Gray's Manual of Botany.* 8th ed. New York: American Book Company.
Fischer, David Hackett. 1989. *Albion's Seed: Four British Folkways in America.* New York: Oxford University Press.
Fitch, George W. 1859. *Colton and Fitch's Modern School Geography.* New York: Ivison and Phinney.
Francatelli, John Elme. 1877. *The Modern Cook: A Practical Guide to the Culinary Art in All Its Branches.* Philadelphia: T. B. Peterson and Brothers.

Gerard, John. 1633. *The Herball, or Generall Historie of Plantes.* Reprint. Revised and enlarged by Thomas Johnson, based on the original 1597 edition. Mineola, New York: Dover Publications, 1975.

Glasse, Hannah. 1805. *The Art of Cookery Made Plain and Simple.* Reprint. Bedford, Massachusetts: Applewood, 1997.

Goodrich, Samuel G. 1840. *Peter Parley's Illustrations of the Vegetable Kingdom: Trees, Plants and Shrubs.* Boston: B. B. Mussey.

Graham, Ada, and Frank Graham Jr. 1995. *Kate Furbish and the Flora of Maine.* Gardiner, Maine: Tilbury House.

Gray, Asa. 1857. *First Lessons in Botany and Vegetable Physiology.* New York: Ivison, Phinney, Blakeman and Company.

———. 1858. *How Plants Grow: A Simple Introduction to Structural Botany.* New York: American Book Company.

Green, Harvey. 1983. *The Light of the Home: An Intimate View of the Lives of Women in Victorian America.* New York: Pantheon Books.

Greenwood, Laura. 1854. *The Rural Wreath: or, Life Among the Flowers.* Boston: Dayton and Wentworth.

Griggs, Barbara. 1981. *Green Pharmacy: The History and Evolution of Western Herbal Medicine.* Rochester Vermont: Healing Arts Press.

Hale, Sarah Josepha. 1832. *Flora's Interpreter: or, the American Book of Flowers and Sentiments.* Boston: Marsh, Capen, and Lyon.

———. 1841. *The Good Housekeeper: or the Way to Live Well, and to Be Well While We Live.* 6th ed. Boston: Otis, Broaders.

———. 1853. *The New Household Receipt-Book.* New York: H. Long and Brother.

Hariot, Thomas. 1590. *A Brief and True Report of the New Found Land of Virginia.* Reprint. Mineola, New York: Dover, 1972.

Harland, Marion. 1875. *Breakfast, Lunch, and Tea.* New York: Charles Scribner's.

———. 1883. *The Cottage Kitchen: A Collection of Practical and Inexpensive Receipts.* New York: Charles Scribner's.

Harris, Amanda B. 1882. *Wild Flowers and Where They Grow.* Boston: D. Lothrop and Company.

Heath, Dwight B. 1622. *Mourt's Relation: A Journal of the Pilgrims at Plymouth.* Reprint. Bedford, Massachusetts: Applewild, 1986. First published in 1622 under the name *Mourt's Relation: A Relation or Journal of the English Plantation settled at Plymouth.*

Henderson, Mary F. 1878. *Practical Cooking, and Dinner Giving.* New York: Harper and Brothers.

Hill, Albert F. 1952. *Economic Botany.* 2d ed. New York: McGraw-Hill.

Husmann, George C. 1866. *The Cultivation of the Native Grape.* New York: G. E. and F. W. Woodward.

Jewett, Sarah Orne. 1881. *Country By-ways.* Boston: Houghton Mifflin and Company.

Josselyn, John. 1672. *New-Englands Rarities.* Reprint. Boston, Massachusetts: William Veazie, 1865.

———. 1675. *Account of Two Voyages to New England.* Reprint. Boston, Massachusetts: William Veazie, 1865.

Kalm, Pehr. 1770. *Peter Kalm's Travels in North America.* Reprint. Mineola, New York: Dover, 1987.

Kiple, Kenneth F., and Kriemhild Conee Ornelas. 2000. *The Cambridge World History of Food.* Cambridge: Cambridge University Press.

Langstroth, Lorenzo L. 1853. *The Hive and the Honey-Bee: a Bee Keeper's Manual.* Northampton, Massachusetts: Hopkins, Bridgman & Co.

Lawson, John. 1709. *A New Voyage to Carolina.* Reprint. Chapel Hill: University of North Carolina Press, 1967.

Leighton, Ann. 1970. *Early American Gardens.* Amherst: University of Massachusetts Press.

———. 1976. *American Gardens in the Eighteenth Century.* Amherst: University of Massachusetts Press.

———. 1987. *American Gardens of the Nineteenth Century.* Amherst: University of Massachusetts Press.

Leslie, Eliza. 1843. *The House Book: A Manual of Domestic Economy.* Philadelphia: Carey & Hart.

———. 1847. *The Indian Meal Book.* Philadelphia: E. L. Carey and A. Hart.

———. 1851. *Miss Leslie's Complete Cookery: Directions for Cookery.* Philadelphia: E. L. Carey and A. Hart.

———. 1857. *Miss Leslie's New Cookery Book.* Philadelphia: T. B. Peterson and Brothers.

Lewis, Walter H., and Memory P. F. Elvin-Lewis. 1977. *Medical Botany: Plants Affecting Man's Health.* New York: John Wiley and Sons.

Lynn, Catherine. 1980. *Wallpaper in America.* New York: W. W. Norton.

Livingston, A. W. 1893. *Livingston and the Tomato.* Reprint. Columbus: Ohio State University Press, 1998.

Markham, Gervase. 1615. *The English House-wife.* Reprint. Edited by Michael R. Best. Montreal: McGill-Queen's University Press, 1994.

Matossian, Mary Kilbourne. 1991. *Poisons of the Past: Molds, Epidemics, and History.* New Haven: Yale University Press.

McMahon, Bernard. 1806. *The American Gardener's Calendar.* Philadelphia: B. Graves.

Miller, Amy Bess. 1976. *Shaker Herbs: A History and a Compendium.* New York: Clarkson Potter.

———. 1998. *Shaker Medicinal Herbs: A Compendium of History, Lore, and Uses.* Pownal, Vermont: Storey Books.

Millspaugh, Charles. F. 1892.*Medicinal Plants.* Philadelphia: John. C. Yorston.

Moerman, Daniel. 1998. *Native American Ethnobotany.* Portland, Oregon: Timber Press.

Newhall, Charles Stedman. 1890. *The Trees of Northeastern America.* New York: G. P. Putnam's Sons.

Nylander, Jane C. 1994. *Our Own Snug Fireside: Images of the New England Home 1760–1860.* New York: Alfred A. Knopf.

Peattie, Donald Culross. 1948. *A Natural History of Trees of Eastern and Central North America.* Boston: Houghton Mifflin.

Phelps, Mrs. Lincoln. 1833. *Botany for Beginners: An Introduction to Mrs. Lincoln's Lectures on Botany.* New York: F. J. Huntington and Company.

———. 1852. *Familiar Lectures on Botany, with a New and Full Description of the Plants of the United States.* New York: Huntington and Savage.

Porcher, Francis Peyre. 1863. *Resources of the Southern Fields and Forests.* Charleston: Evans and Cogswell. From Documenting the American South, University of North Carolina, Chapel Hill. http://docsouth.unc.edu/imis/porcher/porcher.html. Accessed 2003.

Putnam, Mrs. E. 1860. *Mrs. Putnam's Receipt Book and Young Housekeeper's Assistant.* New York: Phinney, Blakeman, and Mason.

Rafinesque, C. S. 1828–1830. *Medical Flora or Manual of Medical Botany of the United States.* Philadelphia: Samuel C. Atkinson.

Randolph, Mary. 1824. *The Virginia House-Wife.* Reprint. Annotated and edited by Karen Hess. Columbia: University of South Carolina Press, 1984.

Root, Waverly. 1980. *Food: An Authoritative, Visual History and Dictionary of the Foods of the World.* New York: Simon and Schuster.

Rorer, Sarah Tyson. 1886. *Mrs. Rorer's Philadelphia Cookbook.* Philadelphia: Arnold and Company.

Ruth, John A. 1883. *Decorum: A Practical Treatise on Etiquette and Dress of the Best American Society.* New York: Union Publishing House.

Scott, Frank J. 1870. *The Art of Beautifying Suburban Home Grounds of Small Extent.* New York: D. Appleton and Company.

Scott, Sarah E. 1868. *Every-Day Cookery for Every Family, Containing 1,000 Recipes Adapted to Moderate Incomes.* Philadelphia: Davis and Brother.

Seaton, Beverly. 1995. *The Language of Flowers, A History.* Charlottesville: University Press of Virginia.

Shaw, Robert. 2000. *American Baskets.* New York: Clarkson Potter.

Shepherd, Sue. 2000. *Pickled, Potted, and Canned: How the Art and Science of Food Preservation Changed the World.* New York: Simon and Schuster.

Sherman, Paul W., and Jennifer Billings. 1999. Darwinian gastronomy: Why we use spices. *BioScience* 49: 453–463.

Simmons, Amelia. 1796. *The First American Cookbook.* Reprint of *American Cookery.* Mineola, New York: Dover, 1984.

Sloane, Eric. 1965. *A Reverence for Wood.* New York: Wilfred Funk.

Smallzried, Kathleen Ann. 1956. *The Everlasting Pleasure: Influences on America's Kitchens, Cooks, and Cookery.* New York: Appleton-Century-Crofts.

Stage, Sarah. 1979. *Female Complaints, Lydia Pinkham and the Business of Women's Medicine.* New York: W. W. Norton and Company.

Strasser, Susan. 1982. *Never Done: A History of American Housework.* New York: Pantheon.

Sturtevant, Edward Lewis. 1919. *Sturtevant's Edible Plants of the World.* Reprint. Edited and compiled by U. P. Hedrick. Mineola, New York: Dover, 1972. First published in 1919 under the name *Sturtevant's Notes on Edible Plants.*

Sumner, Judith. 2000. *The Natural History of Medicinal Plants.* Portland, Oregon: Timber Press.

Swan, Susan Burrows. 1977. *Plain and Fancy: American Women and Their Needlework, 1700–1850.* New York: Holt, Rinehart, and Winston.

Taylor, Raymond L. 1996. *Plants of Colonial Days.* Reprint. Mineola, New York: Dover. First published in 1952 under the name *A Guide to One Hundred and Sixty Flowers, Shrubs, and Trees in the Gardens of Colonial Williamsburg.*

Thoreau, Henry David. 1996. *Faith in a Seed: The Dispersion of Seeds and Other Late Natural History Writings.* Washington, D.C.: Island Press.

———. 1999. *Wild Fruits: Thoreau's Rediscovered Last Manuscript.* New York: W. W. Norton.

Ulrich, Laurel Thatcher. 1982. *Good Wives: Images and Reality of the Lives of Women in Northern New England 1650–1750.* New York: Random House.

Usher, George. 1974. *Dictionary of Plants Used by Man.* New York: Hafner Press.

Vaughan, J. G., and C. Geissler. 1997. *The New Oxford Book of Food Plants.* Oxford: Oxford University Press.

Wakefield, Priscilla. 1811. *An Introduction to Botany in a Series of Familiar Letters.* 6th ed. Philadelphia: Kimber and Conrad.

Walden, J. H. 1858. *Soil Culture, Containing a Comprehensive View of Agriculture, Pomology, Domestic Animals, Rural Economy, and Agricultural Literature.* New York: Robert Sears.

Warder, J. A. 1867. *American pomology. Apples.* New York: Orange Judd and Company.

Waring, George E. 1877. *Village Improvements and Farm Villages.* Boston: J. R. Osgood and Company.

Washington, Martha. 1996. *Martha Washington's Booke of Cookery and Booke of Sweetmeets.* Annotated and edited by Karen Hess. New York: Columbia University Press.

Weaver, William Woys. 1989. *American Eats: Forms of Edible Folk Art.* New York: Harper and Row.

Webster, Mrs. A. L. 1853. *The Improved Housewife, or Book of Receipts.* Boston: Phillips, Sampson and Company.

Wilbur, C. Keith. 1992. *Homebuilding and Woodworking in Colonial America.* Guilford, Connecticut: Globe Pequot Press.

Williams, Roger. 1643. *A Key into the Language of America.* Reprint. Bedford, Massachusetts: Applewood, 1997.

Williams, Susan. 1985. *Savory Suppers and Fashionable Feasts.* New York: Pantheon.

Winslow, Edward. 1622. *Good Newes from New England.* Reprint. Bedford, Massachusetts: Applewood, 1996.

Index